LB 2846
A23
2011

John A. Graziano Memorial Library
Samuel Merritt University
400 Hawthorne Avenue
Oakland, CA 94609
510-869-8900

DATE DUE

GAYLORD	PRINTED IN U.S.A.

D1617158

/20/2011 AS

UNDERSTANDING EDUCATIONAL STATISTICS USING MICROSOFT EXCEL® AND SPSS®

UNDERSTANDING EDUCATIONAL STATISTICS USING MICROSOFT EXCEL® AND SPSS®

MARTIN LEE ABBOTT
Department of Sociology
Seattle Pacific University
Seattle, Washington

WILEY

A JOHN WILEY & SONS, INC., PUBLICATION

Copyright © 2011 by John Wiley & Sons, Inc. All rights reserved

Published by John Wiley & Sons, Inc., Hoboken, New Jersey
Published simultaneously in Canada

No part of this publication may be reproduced, stored in a retrieval system, or transmitted in any form or by any means, electronic, mechanical, photocopying, recording, scanning, or otherwise, except as permitted under Section 107 or 108 of the 1976 United States Copyright Act, without either the prior written permission of the Publisher, or authorization through payment of the appropriate per-copy fee to the Copyright Clearance Center, Inc., 222 Rosewood Drive, Danvers, MA 01923, (978) 750-8400, fax (978) 750-4470, or on the web at www.copyright.com. Requests to the Publisher for permission should be addressed to the Permissions Department, John Wiley & Sons, Inc., 111 River Street, Hoboken, NJ 07030, (201) 748-6011, fax (201) 748-6008, or online at http://www.wiley.com/go/permission.

Limit of Liability/Disclaimer of Warranty: While the publisher and author have used their best efforts in preparing this book, they make no representations or warranties with respect to the accuracy or completeness of the contents of this book and specifically disclaim any implied warranties of merchantability or fitness for a particular purpose. No warranty may be created or extended by sales representatives or written sales materials. The advice and strategies contained herein may not be suitable for your situation. You should consult with a professional where appropriate. Neither the publisher nor author shall be liable for any loss of profit or any other commercial damages, including but not limited to special, incidental, consequential, or other damages.

For general information on our other products and services or for technical support, please contact our Customer Care Department within the United States at (800) 762-2974, outside the United States at (317) 572-3993 or fax (317) 572-4002.

Wiley also publishes its books in a variety of electronic formats. Some content that appears in print may not be available in electronic formats. For more information about Wiley products, visit our web site at www.wiley.com.

Library of Congress Cataloging-in-Publication Data:

ISBN: 978-0-470-88945-9

Printed in Singapore

10 9 8 7 6 5 4 3 2 1

To those who seek a deeper understanding of the world as it appears and of what lies beyond.

CONTENTS

Preface xix

Acknowledgments xxi

1 Introduction 1

    Approach of the Book, 1
    Project Labs, 2
    Real-World Data, 3
    Research Design, 3
    "Practical Significance"—Implications of Findings, 4
    Coverage of Statistical Procedures, 5

2 Getting Acquainted with Microsoft Excel® 7

    Data Management, 7
        Rows and Columns, 8
        Data Sheets, 8
    The Excel® Menus, 9
        Home, 9
        Insert Tab, 9
        Page Layout, 9
        Formulas, 10
        Data, 11
        Review and View Menus, 16

3 Using Statistics in Excel®   17

Using Statistical Functions, 17
Entering Formulas Directly, 17
Data Analysis Procedures, 20
Missing Values and "0" Values in Excel® Analyses, 20
Using Excel® with Real Data, 20
    School-Level Achievement Database, 20
    TAGLIT Data, 21
    The STAR Classroom Observation Protocol™ Data, 22

4 SPSS® Basics   23

Using SPSS®, 23
General Features, 24
Management Functions, 26
    Reading and Importing Data, 26
    Sort, 26
Additional Management Functions, 30
    Split File, 30
    Transform/Compute (Creating Indices), 32
    Merge, 34
Analysis Functions, 39

5 Descriptive Statistics—Central Tendency   41

Research Applications—Spuriousness, 41
Descriptive and Inferential Statistics, 44
The Nature of Data—Scales of Measurement, 44
    Nominal Data, 45
    Ordinal Data, 46
    Interval Data, 48
    Ratio Data, 50
    Choosing the Correct Statistical Procedure for the Nature of Research Data, 50
Descriptive Statistics—Central Tendency, 51
    Mean, 52
    Median, 52
    Mode, 54
Using Excel® and SPSS® to Understand Central Tendency, 56
    Excel®, 56
    SPSS®, 58
Distributions, 61
Describing the Normal Distribution, 62
    Central Tendency, 63
    Skewness, 63

Kurtosis, 65
Descriptive Statistics—Using Graphical Methods, 66
  Frequency Distributions, 66
  Histograms, 67
Terms and Concepts, 71
Real-World Lab I: Central Tendency, 74
Real-World Lab I: Solutions, 75
  Results, 75
  Results, 76

6 Descriptive Statistics—Variablity 81

Range, 82
Percentile, 82
Scores Based on Percentiles, 83
Using Excel® and SPSS® to Identify Percentiles, 84
Note, 86
Standard Deviation and Variance, 87
Calculating the Variance and Standard Deviation, 88
  The Deviation Method, 88
  The Average Deviation, 89
The Computation Method, 91
  The Sum of Squares, 91
  Sample SD and Population SD, 92
  Obtaining SD from Excel® and SPSS®, 94
Terms and Concepts, 96
Real-World Lab II: Variability, 97
Real-World Lab II: Solutions, 97
  Results, 97

7 The Normal Distribution 101

The Nature of the Normal Curve, 101
The Standard Normal Score: z Score, 103
The z-Score Table of Values, 104
Navigating the z-Score Distribution, 105
  Calculating Percentiles, 108
  Creating Rules for Locating z Scores, 108
Calculating z Scores, 111
Working with Raw Score Distributions, 114
Using Excel® to Create z Scores and Cumulative Proportions, 115
  STANDARDIZE Function, 115
  NORMSDIST Function, 117
  NORMDIST Function, 118
Using SPSS® to Create z Scores, 119
Terms and Concepts, 121

Real-World Lab III: The Normal Curve and z Scores, 121
Real-World Lab III: Solutions, 122

8 The Z Distribution and Probability — 127

Transforming a z Score to a Raw Score, 128
Transforming Cumulative Proportions to z Scores, 128
Deriving Sample Scores from Cumulative Percentages, 130
Additional Transformations Using the Standard Normal Distribution, 131
   Normal Curve Equivalent, 131
   Stanine, 131
   T Score, 132
   Grade Equivalent Score, 132
Using Excel® and SPSS® to Transform Scores, 132
Probability, 134
   Determinism Versus Probability, 135
   Elements of Probability, 136
   Probability and the Normal Curve, 136
   Relationship of z Score and Probability, 137
   "Inside" and "Outside" Areas of the Standard Normal Distribution, 139
   Outside Area Example, 140
"Exact" Probability, 141
From Sample Values to Sample Distributions, 143
Terms and Concepts, 144
Real-World Lab IV, 144
Real-World Lab IV: Solutions, 145

9 The Nature of Research Design and Inferential Statistics — 147

Research Design, 148
   Theory, 149
   Hypothesis, 149
Types of Research Designs, 150
   Experiment, 150
   Post Facto Research Designs, 153
The Nature of Research Design, 154
   Research Design Varieties, 154
   Sampling, 155
Inferential Statistics, 156
   One Sample from Many Possible Samples, 156
   Central Limit Theorem and Sampling Distributions, 157
   The Sampling Distribution and Research, 160
   Populations and Samples, 162
   The Standard Error of the Mean, 162
   "Transforming" the Sample Mean to the Sampling Distribution, 163
   Example, 163

Z Test, 166
   The Hypothesis Test, 167
   Statistical Significance, 168
   Practical Significance: Effect Size, 168
   Z-Test Elements, 169
Terms and Concepts, 169
Real-World Lab V, 171
Real-World Lab V: Solutions, 172

10 The *T* Test for Single Samples   175

Z Versus *T*: Making Accommodations, 175
Research Design, 176
   Experiment, 177
   Post Facto Comparative Design, 178
Parameter Estimation, 178
   Estimating the Population Standard Deviation, 178
   A New Symbol: s_x, 180
   Biased Versus Unbiased Estimates, 181
   A Research Example, 181
T Test for a Single Mean, 182
   Example Calculations, 184
   Degrees of Freedom, 185
   The *T* Distribution, 187
   The Hypothesis Test, 188
Type I and Type II Errors, 189
   Type I (Alpha) Errors (α), 189
   Type II (Beta) Errors (β), 190
Effect Size, 191
   Another Measurement of the (Cohen's *d*) Effect Size, 192
Power, Effect Size, and Beta, 193
One- and Two-Tailed Tests, 193
   Two-Tailed Tests, 194
   One-Tailed Tests, 194
   Choosing a One- or Two-Tailed Test, 196
A Note About Power, 196
Point and Interval Estimates, 197
   Calculating the Interval Estimate of the Population Mean, 197
The Value of Confidence Intervals, 199
Using Excel® and SPSS® with the Single-Sample *T* Test, 200
   SPSS® and the Single-Sample *T* Test, 200
   Excel® and the Single Sample *T* Test, 203
Terms and Concepts, 204
Real-World Lab VI: Single-Sample *T* Test, 205
Real-World Lab VI: Solutions, 206

11 Independent-Samples T Test 209

A Lot of "T's", 209
Research Design, 210
   Experimental Designs, 210
   Pretest or No Pretest, 213
   Post Facto Designs, 214
Independent T Test: The Procedure, 215
   Creating the Sampling Distribution of Differences, 216
   The Nature of the Sampling Distribution of Differences, 217
   Calculating the Estimated Standard Error of Difference, 218
   Using Unequal Sample Sizes, 220
   The Independent T Ratio, 221
Independent T-Test Example, 222
   The Null Hypothesis, 222
   The Alternative Hypothesis, 223
   The Critical Value of Comparison, 223
   The Calculated T Ratio, 224
   Statistical Decision, 225
   Interpretation, 226
Before–After Convention with the Independent T Test, 226
Confidence Intervals for the Independent T Test, 227
Effect Size, 228
Equal and Unequal Sample Sizes, 229
The Assumptions for the Independent-Samples T Test, 229
   The Excel® "F-Test Two Sample for Variances" Test, 230
   The SPSS® "Explore" Procedure for Testing the Equality of Variances, 233
   The Homogeneity of Variances Assumption for the Independent T Test, 235
   A Rule of Thumb, 236
Using Excel® and SPSS® with the Independent-Samples T Test, 236
   Using Excel® with the Independent T Test, 236
   Using SPSS® with the Independent T Test, 239
Parting Comments, 242
Nonparametric Statistics, 243
Terms and Concepts, 246
Real-World Lab VII: Independent T Test, 247
   Procedures, 247
Real-World Lab VII: Solutions, 248

12 Analysis of Variance 257

A Hypothetical Example of ANOVA, 258
The Nature of ANOVA, 259

The Components of Variance, 260
The Process of ANOVA, 261
Calculating ANOVA, 262
   Calculating the Variance: Using the Sum of Squares (SS), 262
   Using Mean Squares (MS), 265
   Degrees of Freedom in ANOVA, 266
   Calculating Mean Squares (MS), 266
   The F Ratio, 267
   The F Distribution, 269
Effect Size, 269
Post Hoc Analyses, 271
   "Varieties" of *Post Hoc* Analyses, 272
   The *Post Hoc* Analysis Process, 273
   Tukey's HSD (Range) Test Calculation, 273
   Means Comparison Table, 275
   Compare Mean Difference Values from HSD, 276
   Post Hoc Summary, 276
Assumptions of ANOVA, 276
Additional Considerations with ANOVA, 277
A Real-World Example of ANOVA, 277
Are the Assumptions Met?, 278
Hand Calculations, 281
   Calculating SS_T, 283
   Calculating SS_B, 283
   Calculating SS_W, 283
   The Hypothesis Test, 283
   Effect Size, 284
   Post Hoc Analysis, 284
Using Excel® and SPSS® with One-Way ANOVA, 285
   Excel® Procedures with One-Way ANOVA, 285
   SPSS® Procedures with One-Way ANOVA, 287
The Need for Diagnostics, 292
Nonparametric ANOVA Tests, 293
Terms and Concepts, 296
Real-World Lab VIII: ANOVA, 296
Real-World Lab VIII: Solutions, 297

13 Factorial Anova   307

Extensions of ANOVA, 307
   Within-Subjects ANOVA, 307
   Two-Way Within-Subjects ANOVA, 308
   ANCOVA, 308
Multivariate ANOVA Procedures, 309
   MANOVA, 309

MANCOVA, 309
Factorial ANOVA, 309
    Interaction Effects, 309
    An Example of $2 \times$ ANOVA, 310
    Charting Interactions, 311
    Simple Effects, 312
The Example DataSet, 312
Calculating Factorial ANOVA, 312
    Calculating the Interaction, 315
    The $2 \times$ ANOVA Summary Table, 315
    Creating the MS Values, 316
    The Hypotheses Tests, 317
    The Omnibus F Ratio, 317
    Effect Size for $2 \times$ ANOVA: Partial η^2, 318
    Discussing the Results, 319
Using SPSS® to Analyze $2 \times$ ANOVA, 321
    The "Plots" Specification, 323
    Omnibus Results, 325
    Simple Effects Analyses, 325
Summary Chart for $2 \times$ ANOVA Procedures, 327
Terms and Concepts, 327
Real-World Lab IX: $2 \times$ ANOVA, 329
Real-World Lab IX: $2 \times$ ANOVA Solutions, 330

14 Correlation   337

The Nature of Correlation, 338
    Explore and Predict, 338
    Different Measurement Values, 338
    Different Data Levels, 338
    Correlation Measures, 338
The Correlation Design, 339
Pearson's Correlation Coefficient, 340
    Interpreting the Pearson's Correlation, 340
    The Fictitious Data, 341
    Assumptions for Correlation, 342
Plotting the Correlation: The Scattergram, 342
    Patterns of Correlations, 343
    Strength of Correlations in Scattergrams, 344
Creating the Scattergram, 345
    Using Excel® to Create Scattergrams, 345
    Using SPSS® to Create Scattergrams, 347
Calculating Pearson's r, 348
The Z-Score Method, 349
The Computation Method, 351

Evaluating Pearson's r, 353
   The Hypothesis Test for Pearson's r, 353
   The Comparison Table of Values, 354
   Effect Size: The Coefficient of Determination, 354
Correlation Problems, 356
   Correlations and Sample Size, 356
   Correlation is Not Causation, 357
   Restricted Range, 357
   Extreme Scores, 358
   Heteroscedasticity, 358
   Curvilinear Relations, 358
The Example Database, 359
   Assumptions for Correlation, 360
   Computation of Pearson's r for the Example Data, 363
   Evaluating Pearson's r: Hypothesis Test, 365
   Evaluating Pearson's r: Effect Size, 365
Correlation Using Excel® and SPSS®, 366
   Correlation Using Excel®, 366
   Correlation Using SPSS®, 367
Nonparametric Statistics: Spearman's Rank-Order Correlation (r_s), 369
   Variations of Spearman's Rho Formula: Tied Ranks, 371
   A Spearman's Rho Example, 373
Terms and Concepts, 374
Real-World Lab X: Correlation, 376
Real-World Lab X: Solutions, 377

15 Bivariate Regression   383

The Nature of Regression, 384
The Regression Line, 385
Calculating Regression, 388
   The Slope Value b, 389
   The Regression Equation in "Pieces", 389
   A Fictitious Example, 389
   Interpreting and Using the Regression Equation, 390
Effect Size of Regression, 391
The Z-Score Formula for Regression, 392
   Using the Z-Score Formula for Regression, 392
   Unstandardized and Standardized Regression Coefficients, 394
Testing the Regression Hypotheses, 394
The Standard Error of Estimate, 394
   Calculating s_{est}, 395
Confidence Interval, 396
Explaining Variance through Regression, 397
Using Scattergrams to Understand the Partitioning of Variance, 399

A Numerical Example of Partitioning the Variation, 400
Using Excel® and SPSS® with Bivariate Regression, 401
   The Excel® Regression Output, 402
   The SPSS® Regression Output, 404
Assumptions of Bivariate Linear Regression, 408
Curvilinear Relationships, 409
Detecting Problems in Bivariate Linear Regression, 412
A Real-World Example of Bivariate Linear Regression, 413
   Normal Distribution and Equal Variances Assumptions, 413
   The Omnibus Test Results, 414
   Effect Size, 414
   The Model Summary, 415
   The Regression Equation and Individual Predictor Test of Significance, 415
   The Scattergram, 416
Advanced Regression Procedures, 417
   Multiple Correlation, 417
   Partial Correlation, 418
   Multiple Regression, 419
   Additional Considerations, 419
Terms and Concepts, 419
Real-World Lab XI: Bivariate Linear Regression, 420
Real-World Lab XI: Solutions, 422

16 Introduction to Multiple Linear Regression   429

The Elements of MLR, 429
   Same Process as Bivariate Regression, 430
   Similar Assumptions, 430
   Statistical Significance, 430
   Effect Size, 430
   Coefficients, 430
   Scatterdiagrams, 431
Some Differences Between Bivariate Regression and MLR, 431
   Multiple Coefficients, 431
   Multicollinearity, 431
   Explanation of R^2, 431
   Entry Schemes, 432
Stuff Not Covered, 432
   Using MLR with Categorical Data, 432
   Curvilinear Regression, 433
   Multilevel Analysis, 433
MLR Extended Example, 433
Are the Assumptions Met?, 434
The Findings, 437

The SPSS® Findings, 438
   The Unstandardized Coefficients, 442
   The Standardized Coefficients, 442
   Collinearity Statistics, 443
   The Squared Part Correlation, 443
   Conclusion, 444
Terms and Concepts, 445
Real-World Lab XII: Multiple Linear Regression, 445
Real-World Lab XII: MLR Solutions, 445

17 Chi Square and Contingency Table Analysis   453

Contingency Tables, 453
The Chi Square Procedure and Research Design, 454
   Post Facto Designs, 455
   Experimental Designs, 455
Chi Square Designs, 455
Goodness of Fit, 455
   Expected Frequencies—Equal Probability, 456
   Expected Frequencies—*A Priori* Assumptions, 456
The Chi Square Test of Independence, 456
A Fictitious Example—Goodness of Fit, 457
   Frequencies Versus Proportions, 460
Effect Size—Goodness of Fit, 460
Chi Square Test of Independence, 461
   Two-Way Chi Square, 461
   Assumptions, 462
A Fictitious Example—Test of Independence, 462
   Creating Expected Frequencies, 462
   Degrees of Freedom for the Test of Independence, 464
Special 2×2 Chi Square, 466
   The Alternate 2×2 Formula, 467
   Effect Size in 2×2 Tables: Phi, 467
   Correction for 2×2 Tables, 468
Cramer's V: Effect Size for the Chi Square Test of Independence, 469
Repeated Measures Chi Square, 470
   Repeated Measures Chi Square Table, 472
Using Excel® and SPSS® with Chi Square, 472
Using Excel® for Chi Square Analyses, 475
   Sort the Database, 475
   The Excel® Count Function, 476
   The Excel® CHITEST Function, 476
   The Excel® CHIDIST Function, 477
Using SPSS® for the Chi Square Test of Independence, 478
   The Crosstabs Procedure, 478

Analyzing the Contingency Table Data Directly, 481
Interpreting the Contingency Table, 483
Terms and Concepts, 483
Real-World Lab XIII: Chi Square, 484
Real-World Lab XIII: Solutions, 484
Hand Calculations, 484
Using Excel® for Chi Square Analyses, 485
Using SPSS® for Chi Square Solutions, 486

18 Repeated Measures Procedures: T_{dep} and ANOVA$_{ws}$ — 489

Independent and Dependent Samples in Research Designs, 490
Using Different T Tests, 491
The Dependent T-Test Calculation: The Long Formula, 491
Example, 492
Results, 494
Effect Size, 494
The Dependent T-Test Calculation: The Difference Formula, 495
The T_{dep} Ratio from the Difference Method, 496
T_{dep} and Power, 496
Using Excel® and SPSS® to Conduct the T_{dep} Analysis, 496
T_{dep} with Excel®, 497
T_{dep} with SPSS®, 498
Within-Subjects ANOVA (ANOVA$_{ws}$), 499
Experimental Designs, 499
Post Facto Designs, 501
Within-Subjects Example, 501
Using SPSS® for Within-Subjects Data, 501
Sphericity, 501
The SPSS® Procedure, 502
The SPSS® Output, 504
The Omnibus Test, 506
Effect Size, 507
Post Hoc Analyses, 507
The Interpretation, 507
Nonparametric Statistics, 508
Terms and Concepts, 509

References — 511

Appendix: Statistical Tables — 513

Index — 523

PREFACE

I have written this book many times in my head over the years! As I conducted research and taught statistics (graduate and undergraduate) in many fields, I developed an approach to helping students understand the difficult concepts in a new way. I find that the great majority of students are visual learners, so I developed diagrams and figures over the years that help create a conceptual picture of the statistical procedures that are often problematic to students (like sampling distributions!).

The other reason I wanted to write this book was to give students a way to understand statistical computing without having to rely on comprehensive and expensive statistical software programs. Because most students have access to Microsoft Excel®,[1] I developed a step-by-step approach to using the powerful statistical procedures in Excel® to analyze data and conduct research in each of the statistical topics I cover in the book.

I also wanted to make those comprehensive statistical programs more approachable to statistics students, so I have also included a hands-on guide to SPSS® in parallel with the Excel® examples. In some cases, SPSS® has the only means to perform some statistical procedures; but in most cases, both Excel® and SPSS® can be used.

Last, like my other work dealing with applied statistical topics (Abbott, 2010), I included real-world data in this book as examples for the procedures I discuss. I introduce extended examples in each chapter that use these real-world datasets, and I conclude the chapters with a Real-World Lab in which I present data for students

---

[1] Excel® references and screen shots in this book are used with permission from Microsoft.

to use with Excel® and SPSS®. Each Lab is followed by the Real World Lab: Solutions section so that students can examine their work in greater depth.

One limitation to teaching statistics through Excel® is that the data analysis features are different, depending on whether the user is a Mac user or a PC user. I am using the PC version, which features a Data Analysis suite of statistical tools. This feature may no longer be included in the Mac version of Excel® you are using.

I am posting the datasets for the real-world labs at the Wiley Publisher ftp site. You can access these datasets there to complete the labs instead of entering the data from the tables in the chapters. You may note some slight discrepancies in the results if you enter the data by hand rather than downloading the data due to rounding of values. The data in the chapters are typically reported to two decimal places, whereas the analyses reported in the Labs are based on the actual data that both Excel® and SPSS® carry to many decimal places even though you may only see a value with two decimal places. Despite any slight differences resulting from rounding, the primary findings should not change. You may encounter these types of discrepancies in your research with real data as you move data from program to program to page.

The John Wiley & Sons Publisher ftp address is as follows:

ftp://ftp.wiley.com/public/sci_tech_med/educational_statistics. You may also want to visit my personal website at the following address:

http://myhome.spu.edu/mabbott/.

MARTIN LEE ABBOTT

Seattle, Washington

ACKNOWLEDGMENTS

I would like to thank everyone who reviewed this manuscript. In particular, Nyaradzo Mvududu's thorough critique was invaluable throughout the process. Adrianna Bagnall reviewed the manuscript and provided help in a great many other ways, especially with the tables. Dominic Williamson's outstanding work on the figures and graphic design was a critical feature of my approach to conceptual understanding of complex processes. I am especially grateful for his design of the image on the book cover. Kristin Hovaguimian again provided outstanding support for the Index—not an easy task with a book of this nature. My graduate students in Industrial/Organizational Psychology were kind to review the Factorial ANOVA chapter (Chapter 13).

I also want to thank Duane Baker (The BERC Group, Inc.) and Liz Cunningham (T.E.S.T., Inc.) for approval to use their data in this book as they did for my former work (Abbott, 2010). Using real-world data of this nature will be very helpful to readers in their efforts to understand statistical processes.

I especially want to recognize Jacqueline Palmieri and Stephen Quigley at John Wiley & Sons, Inc. for their continuing encouragement. They have been steadfast in their support of this approach to statistical analysis from the beginning of our work together.

MARTIN LEE ABBOTT

1

INTRODUCTION

Many students and researchers are intimidated by statistical procedures. This may in part be due to a fear of math, problematic math teachers in earlier education, or the lack of exposure to a "discovery" method for understanding difficult procedures. Readers of this book should realize that they have the ability to succeed in understanding statistical processes.

APPROACH OF THE BOOK

This is an introduction to statistics using EXCEL® and SPSS® to make it more understandable. Ordinarily, the first course leads the student through the worlds of descriptive and inferential statistics by highlighting the formulas and sequential procedures that lead to statistical decision making. We will do all this in this book, but I place a good deal more attention on conceptual understanding. Thus, rather than memorizing a specific formula and using it in a specific way to solve a problem, I want to make sure the student first understands the nature of the problem, why a specific formula is needed, and how it will result in the appropriate information for decision making.

By using statistical software, we can place more attention on understanding how to *interpret findings*. Statistics courses taught in mathematics departments, and in some social science departments, often place primary emphases on the formulas/ processes themselves. In the extreme, this can limit the usefulness of the analyses to the practitioner. My approach encourages students to focus more on how to understand and make applications of the results of statistical analyses. EXCEL®

Understanding Educational Statistics Using Microsoft Excel® and SPSS®. By Martin Lee Abbott.
© 2011 John Wiley & Sons, Inc. Published 2011 by John Wiley & Sons, Inc.

and other statistical programs are much more efficient at performing the analyses; the key issue in my approach is how to interpret the results in the context of the research question.

Beginning with my first undergraduate course through teaching statistics with conventional textbooks, I have spent countless hours demonstrating how to conduct statistical tests by hand and teaching students to do likewise. This is not always a bad strategy; performing the analysis by hand can lead the student to understand how formulas treat data and yield valuable information. However, it is often the case that the student gravitates to memorizing the formula or the steps in an analysis. Again, there is nothing wrong with this approach as long as the student does not stop there. The outcome of the analysis is more important than memorizing the steps to the outcome. Examining the appropriate output derived from statistical software shifts the attention from the nuances of a formula to the wealth of information obtained by using it.

It is important to understand that I do indeed teach the student the nuances of formulas, understanding why, when, how, and under what conditions they are used. But in my experience, forcing the student to scrutinize statistical output files accomplishes this and teaches them the appropriate use and limitations of the information derived.

Students in my classes are always surprised (ecstatic) to realize they can use their textbooks, notes, and so on, on my exams. But they quickly find that, unless they really understand the principles and how they are applied and interpreted, an open book is not going to help them. Over time, they come to realize that the analyses and the outcomes of statistical procedures are simply the ingredients for what comes next: building solutions to research problems. Therefore, their role is more detective and constructor than number juggler.

This approach mirrors the recent national and international debate about math pedagogy. In my recent book, *Winning the Math Wars* (Abbott et al., 2010), my colleagues and I addressed these issues in great detail, suggesting that, while traditional ways of teaching math are useful and important, the emphases of reform approaches are not to be dismissed. Understanding and memorizing detail are crucial, but problem solving requires a different approach to learning.

PROJECT LABS

Labs are a very important part of this course since they allow students to take charge of their learning. This is the "discovery learning" element I mentioned above. Understanding a statistical procedure in the confines of a classroom is necessary and helpful. However, learning that lasts is best accomplished by students directly engaging the processes with actual data and observing what patterns emerge in the findings that can be applied to real research problems.

In this course, we will have several occasions to complete Project Labs that pose research problems on actual data. Students take what they learn from the book material and conduct a statistical investigation using EXCEL® and SPSS®. Then, they have the opportunity to examine the results, write research summaries, and compare findings with the solutions presented at the end of the book.

These are labs not using data created for classroom use but instead using real-world data from actual research databases. Not only does this engage students in the learning process with specific statistical processes, but it presents real-world information in all its "grittiness." Researchers know that they will discover knotty problems and unusual, sometimes idiosyncratic, information in their data. If students are not exposed to this real-world aspect of research, it will be confusing when they engage in actual research beyond the confines of the classroom.

The project labs also introduce students to two software approaches for solving statistical problems. These are quite different in many regards, as we will see in the following chapters. EXCEL® is widely accessible and provides a wealth of information to researchers about many statistical processes they encounter in actual research. SPSS® provides additional, advanced procedures that educational researchers utilize for more complex and extensive research questions. The project labs provide solutions in both formats so the student can learn the capabilities and approaches of each.

REAL-WORLD DATA

As I mentioned, I focus on using real-world data for many reasons. One reason is that students need to be grounded in approaches they can use with "gritty" data. I want to make sure that students leave the classroom prepared for encountering the little nuances that characterize every research project.

Another reason I use real-world data is to familiarize students with contemporary research questions in education. Classroom data often are contrived to make a certain point or show a specific procedure, which are both helpful. But I believe that it is important to draw the focus away from the procedure per se and understand how the procedure will help the researcher resolve a research question. The research questions are important. Policy reflects the available information on a research topic, to some extent, so it is important for students to be able to generate that information as well as to understand it. This is an "active" rather than "passive" learning approach to understanding statistics.

RESEARCH DESIGN

People who write statistics books have a dilemma with respect to research design. Typically, statistics and research design are taught separately in order for students to understand each in greater depth. The difficulty with this approach is that the student is left on their own to synthesize the information; this is often not done successfully.

Colleges and universities attempt to manage this problem differently. Some require statistics as a prerequisite for a research design course, or vice versa. Others attempt to synthesize the information into one course, which is difficult to do given the eventual complexity of both sets of information. Adding somewhat to the problem is the approach of multiple courses in both domains.

I do not offer a perfect solution to this dilemma. My approach focuses on an in-depth understanding of statistical procedures for actual research problems. What this means is that I cannot devote a great deal of attention in this book to research design apart from the statistical procedures that are an integral part of it. However, I try to address the problem in two ways.

First, wherever possible, I connect statistics with specific research designs. This provides an additional context in which students can focus on using statistics to answer research questions. The research question drives the decision about which statistical procedures to use; it also calls for discussion of appropriate design in which to use the statistical procedures. We will cover essential information about research design in order to show how these might be used.

Second, I am making available an online course in research design as part of this book. In addition to databases and other research resources, you can follow the web address in the Preface to gain access to the online course that you can take in tandem with reading this book or separately.

"PRACTICAL SIGNIFICANCE"—IMPLICATIONS OF FINDINGS

I emphasize "practical significance" (effect size) in this book as well as statistical significance. In many ways, this is a more comprehensive approach to uncertainty, since effect size is a measure of "impact" in the research evaluation. It is important to measure the likelihood of chance findings (statistical significance), but the extent of influence represented in the analyses affords the researcher another vantage point to determine the relationship among the research variables.

I call attention to problem solving as the important part of statistical analysis. It is tempting for students to focus so much on using statistical procedures to create meaningful results (a critical matter!) that they do not take the next steps in research. They stop after they use a formula and decide whether or not a finding is statistically significant. I strongly encourage students to think about the findings in the context and words of the research question. This is not an easy thing to do because the meaning of the results is not always cut and dried. It requires students to think beyond the formula.

Statisticians and practitioners have devised rules to help researchers with this dilemma by creating criteria for decision making. For example, squaring a correlation yields the "coefficient of determination," which represents the amount of variance in one variable that is accounted for by the other variable. But the next question is, How much of the "accounted for variance" is meaningful?

Statisticians have suggested different ways of helping with this question. One such set of criteria determines that 0.01 (or 1% of the variance accounted for) is considered "small" while 0.05 (5% of variance) is "medium," and so forth. (And, much to the dismay of many students, there are more than one set of these criteria.)

But the material point is that these criteria do not apply equally to every research question.

If a research question is, "Does class size affect math achievement," for example, and the results suggest that class size accounts for 1% of the variance in math achievement, many researchers might agree it is a small and perhaps even inconsequential impact. However, if a research question is, "Does drug X account for 1% of the variance in AIDS survival rates," researchers might consider this to be much more consequential than "small"!

This is not to say that math achievement is any less important than AIDS survival rates (although that is another of those debatable questions researchers face), but the researcher must consider a range of factors in determining meaningfulness: the intractability of the research problem, the discovery of new dimensions of the research focus, whether or not the findings represent life and death, and so on.

I have found that students have the most difficult time with these matters. Using a formula to create numerical results is often much preferable to understanding what the results mean in the context of the research question. Students have been conditioned to stop after they get the right numerical answer. They typically do not get to the difficult work of what the right answer *means* because it isn't always apparent.

COVERAGE OF STATISTICAL PROCEDURES

The statistical applications we will discuss in this book are "workhorses." This is an introductory treatment, so we need to spend time discussing the nature of statistics and basic procedures that allow you to use more sophisticated procedures. We will not be able to examine advanced procedures in much detail. I will provide some references for students who wish to continue their learning in these areas. It is hoped that, as you learn the capability of EXCEL® and SPSS®, you can explore more advanced procedures on your own, beyond the end of our discussions.

Some readers may have taken statistics coursework previously. If so, my hope is that they are able to enrich what they previously learned and develop a more nuanced understanding of how to address problems in educational research through the use of EXCEL® and SPSS®. But whether readers are new to the study or experienced practitioners, my hope is that statistics becomes meaningful as a way of examining problems and debunking prevailing assumptions in the field of education.

Often, well-intentioned people can, through ignorance of appropriate processes promote ideas in education that may not be true. Furthermore, policies might be offered that would have a negative impact even though the policy was not based on sound statistical analyses. Statistics are tools that can be misused and influenced by the value perspective of the wielder. However, policies are often generated in the absence of compelling research. Students need to become "research literate" in order to recognize when statistical processes should be used and when they are being used incorrectly.

2

GETTING ACQUAINTED WITH MICROSOFT EXCEL®

Microsoft Excel® is a powerful application for education researchers and students studying educational statistics. Excel® worksheets can hold data for a variety of uses and therefore serve as a database. We will focus primarily on its use as a spreadsheet, however. This book discusses how students of statistics can use Excel® menus to create specific data management and statistical analysis functions.

I will use Microsoft® Office Excel® 2007 for all examples and illustrations in this book.[1] Like other software, Excel® changes occasionally to improve performance and adapt to new standards. As I write, other versions are projected, however, most all of my examples use the common features of the application that are not likely to undergo radical changes in the near future.

I cannot hope to acquaint the reader with all the features of Excel® in this book. Our focus is therefore confined to the statistical analysis and related functions called into play when using the data analysis features. I will introduce some of the general features in this chapter and cover the statistical applications in more depth in the following chapters.

DATA MANAGEMENT

The opening spreadsheet presents the reader with a range of menu choices for entering and managing data. Like other spreadsheets, Excel® consists of rows and

---

[1] Used with permission from Microsoft, as per "Use of Microsoft Copyrighted Content" approvals.

Understanding Educational Statistics Using Microsoft Excel® and SPSS®. By Martin Lee Abbott.
© 2011 John Wiley & Sons, Inc. Published 2011 by John Wiley & Sons, Inc.

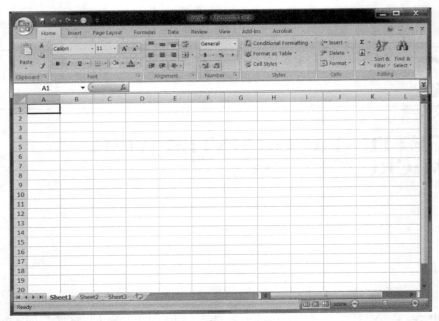

FIGURE 2.1 The initial Excel® spreadsheet.

columns for entering and storing data of various kinds. Figure 2.1 shows the spreadsheet with its menus and navigation bars. I will cover much of the available spreadsheet capacity over the course of discussing our statistical topics in later chapters. Here are some basic features:

Rows and Columns

Typically, rows represent cases in statistical analyses, and columns represent variables. According to the Microsoft Office® website, the spreadsheet can contain over one million rows and over 16,000 columns. We will not approach either of these limits; however, you should be aware of the capacity in the event you are downloading a large database from which you wish to select a portion of data. One practical feature to remember is that researchers typically use the first row of data to record variable names in each of the columns of data. Therefore, the total dataset contains (rows −1) cases, which takes this into account.

Data Sheets

Figure 2.1 shows several "Sheet" tabs on the bottom of the spreadsheet. These are separate worksheets contained in the overall workbook spreadsheet. They can be used independently to store data, but typically the statistical user puts a dataset on one Sheet and then uses additional Sheets for related analyses. For example, as we

THE EXCEL® MENUS

will discuss in later chapters, each statistical procedure will generate a separate "output" Sheet. Thus, the original Sheet of data will not be modified or changed. The user can locate the separate statistical findings in separate Sheets. Each Sheet tab can be named by "right-clicking" on the Sheet. Additional Sheets can be created by clicking on the small icon to the right of "Sheet3" shown in Figure 2.1.

THE EXCEL® MENUS

The main Excel® menus are located in a ribbon at the top of the spreadsheet beginning with "Home" and extending several choices to the right. I will comment on each of these briefly before we look more comprehensively at the statistical features.

Home

The "Home" menu includes many options for formatting and structuring the entered data, including a font group, alignment group, cells group (for such features as insert/delete options), and other such features.

One set of sub-menus is particularly useful for the statistical user. These are listed in the "Number" category located in the ribbon at the bottom of the main set of menus. The default format of Number is typically "General" shown in the highlighted box (see Figure 2.1). If you select this drop-down menu, you will be presented with a series of possible formats for your data among which is one entitled "Number"—the second choice in the sub-menu. If you click this option, Excel® returns the data in the cell as a number with two decimal points.

When you double-click on the "Number" option, however, you can select from a larger sub-menu that allows you many choices for your data, as shown in Figure 2.2. (The additional choices for data formats are located in the "Category:" box located on the left side of this sub-menu.) We will primarily use this "Number" format since we are analyzing numerical data, but we may have occasion to use additional formats. You can use this sub-menu to create any number of decimal places by using the "Decimal places:" box. You can also specify different ways of handling negative numbers by selecting among the choices in the "Negative numbers:" box.

Insert Tab

I will return to this menu many times over the course of our discussion. Primarily, we will use this menu to create the visual descriptions of our analyses (graphs and charts).

Page Layout

This menu is helpful for formatting functions and creating the desired "look and feel" of the spreadsheet.

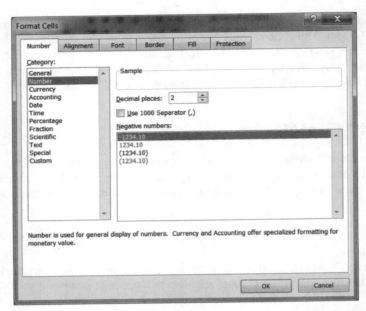

FIGURE 2.2 The variety of cell formats available in the Number sub-menu.

Formulas

The Formulas menu is a very important part of the statistical arsenal of Excel®. We will discuss specific functions as we get to them in the course of our study; for now, I will point out that the first section of this menu is the "Function Library" that contains a great many categories of functions (i.e., "Financial," "Logical," "Text," etc.). Selecting any of these results in a sub-menu of choices for formulas specific to that category of use. There are at least two ways to create statistical formulas, which we will focus on in this book.

1. *The "More Functions" Tab.* This tab presents the user with additional categories of formulas, one of which is "Statistical." As you can see when you select it, there are a great many choices for handling data. Essentially, these are embedded formulas for creating specific statistical output. For example, "AVERAGE" is one of the first formulas listed when you choose "More Functions" and then select "Statistical." This formula returns the mean value of a set of selected data from the spreadsheet.

2. *"Insert Functions" Tab.* A second way to access statistical (and other) functions from the Function Library is using the "Insert Function" sub-menu that, when selected, presents the user with the screen shown in Figure 2.3.

Choosing this feature is the way to "import" the function to the spreadsheet. The screen in Figure 2.3 shows the "Insert Function" box I obtained from my computer. As you can see, there are a variety of ways to choose a desired function. The "Search for a function:" box allows the user to describe what they want to do with

THE EXCEL® MENUS

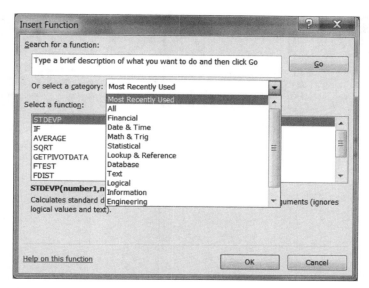

FIGURE 2.3 The "Insert Function" sub-menu of the "Function Library."

their data. When selected, the program will present several choices in the "Select a function:" box immediately below it, depending on which function you queried.

The "Or select a category:" box lists the range of function categories available. The statistical category of functions will be shown if double-clicked (as shown in Figure 2.3). Accessing the list of statistical functions through this button will result in the same list of functions obtainable through the "More Functions" tab.

When you use the categories repeatedly, as we will use the "Statistical" category repeatedly, Excel® will show the functions last used in the "Select a function" box as shown in Figure 2.3.

Data

This is the main menu for our discussion in this book. Through the sub-menu choices, the statistical student can access the data analysis procedures, sort and filter data in the spreadsheet, and provide a number of data management functions important for statistical analysis. Figure 2.4 shows the sub-menus of the Data menu.

The following are some of the more important sub-menus that I will explain in detail in subsequent chapters.

Sort and Filter. The Sort sub-menu allows the user to rearrange the data in the spreadsheet according to a specific interest or statistical procedure. For example, if you had a spreadsheet with three variables—Gender, Reading achievement, and Math achievement—you could use the "sort" key to arrange the values of the variables according to gender. Doing this would result in Excel® arranging the gender

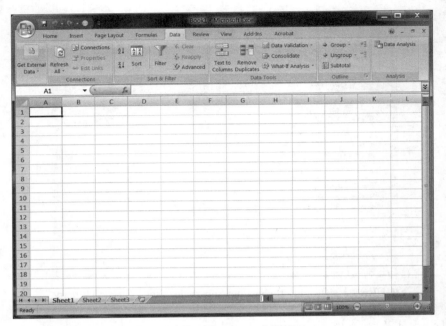

FIGURE 2.4 The sub-menus of the Data menu.

categories, "M" and "F," in ascending or descending order (alphabetically, depending on whether you proceed from "A to Z" or from "Z to A") with the values of the other variables linked to this new arrangement. Thus, a visual scan of the data would allow you to see how the achievement variables change as you proceed from male to female students. The following two figures show the results of this example. Figure 2.5 shows the unsorted variables.

As you can see from Figure 2.5, you cannot easily discern a pattern to the data, depending on whether males or females have better math and reading scores in this sample.[2] Sorting the data according to the Gender variable may help to indicate relationships or patterns in the data that are not immediately apparent. Figure 2.6 shows the same three variables sorted according to gender (sorted "A to Z" resulting in the Female scores listed first).

Figure 2.6 shows the data arranged according to the categories of the Gender variable. Viewed in this way, you can detect some general patterns. It appears, generally, that female students performed much better on math and just a bit higher on reading than the male students. Of course, this small sample is not a good indicator of the overall relationship between gender and achievement. For example, the math scores for the last male in the dataset ("10") and for the third female student ("24") exert a great deal of influence in this small dataset; a much larger sample would not register as great an influence.

---

[2] The example data in these procedures are taken from the school database we will use throughout the book. The small number of cases is used to explain the procedures, not to make research conclusions.

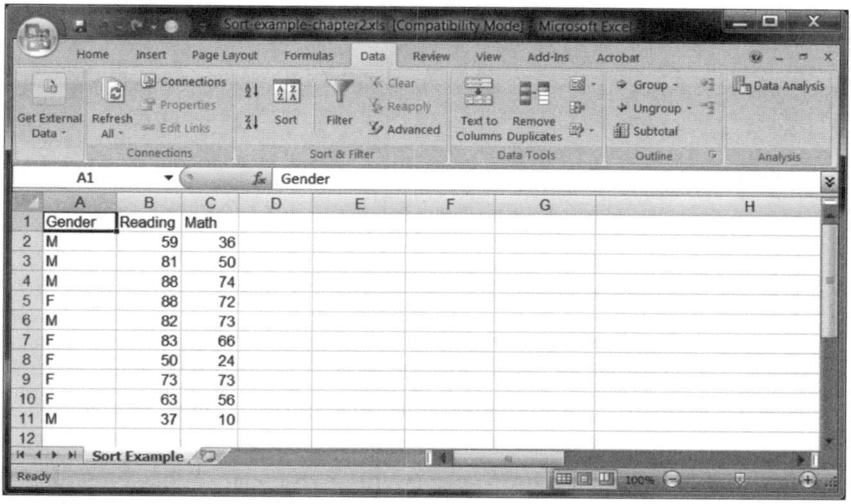

THE EXCEL® MENUS

FIGURE 2.5 Unsorted data for the three-variable database.

An important operational note for sorting is to first "select" the entire database before you sort any of the data fields. If you do not sort the entire database, you can inadvertently only sort one variable, which may result in the values of this variable disengaging from its associated values on adjacent variables. In these cases, the values for each case may become mixed. Selecting the entire database before any sort ensures that the values of a given variable remain fixed to the values of all the variables for each of the cases. The "Filter" sub-menu is useful in this regard. Excel® adds drop-down menus next

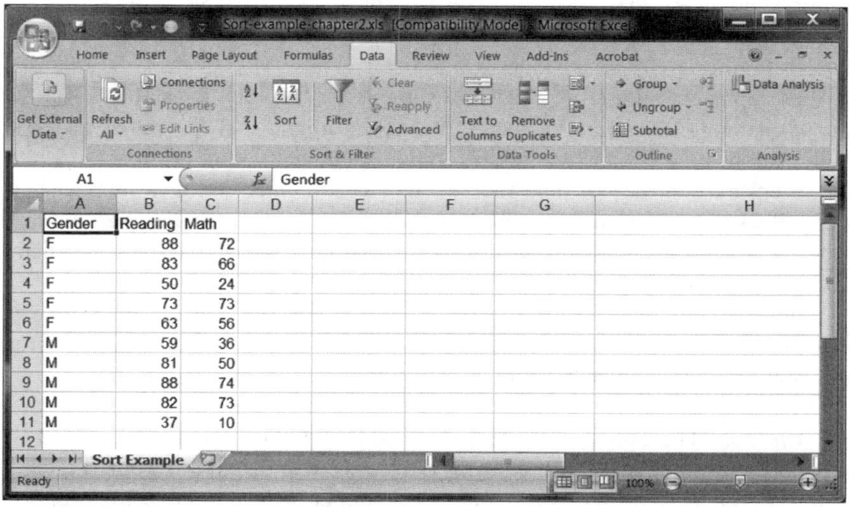

FIGURE 2.6 Using the "Sort" function to arrange values of the variables.

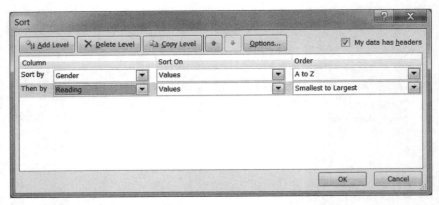

FIGURE 2.7 The Excel® sub-menu showing a sort by multiple variables.

to each variable when the user selects this sub-menu. When you use the menus, you can specify a series of ways to sort the variables in the database without "disengaging" the values on the variables.

You can also perform a "multiple" sort in Excel® using the Sort menu. Figure 2.7 shows the sub-menu presented when you choose Sort. As you can see from the screen, choosing the "Add Level" button in the upper left corner of the screen results in a second sort line ("Then by") allowing you to specify a second sort variable. This would result in a sort of the data first by Gender, and then the values of Reading would be presented low to high within both categories of gender.

Excel® also records the nature of the variables. Under the "Order" column on the far right of Figure 2.7, the variables chosen for sorting are listed as either "A to Z," indicating that they are "alphanumeric" or "text" variables, or "Smallest to Largest," indicating they are numerical variables. Text variables are composed of values (either letters or numbers) that are treated as letters and not used in calculations. In Figure 2.6, gender values are either "F" or "M," so there is little doubt that they represent letters. If I had coded these as "1" for "F" and "2" for "M" without changing the format of the cells, Excel® might treat the values differently in calculations (since letters cannot be added, subtracted, etc.). In this case I would want to ensure that the "1" and the "2" would be treated not as a number but as letters. Be sure to format the cells properly (from the "Number" group in the Home menu) so that you can be sure the values are treated as you intend them to be treated in your analyses.

Figure 2.8 shows the resulting sort. Here you can see that the data were first sorted by Gender (with "F" presented before "M") and then the values of "Reading" were presented low to high in value within both gender categories.

Data Analysis. This sub-menu choice (located in the "Data" tab in the "Analysis" group) is the primary statistical analysis device we will use in this book. Figure 2.4 shows the "Data Analysis" sub-menu in the upper right corner of the menu bar. Choosing this option results in the box shown in Figure 2.9.

THE EXCEL® MENUS

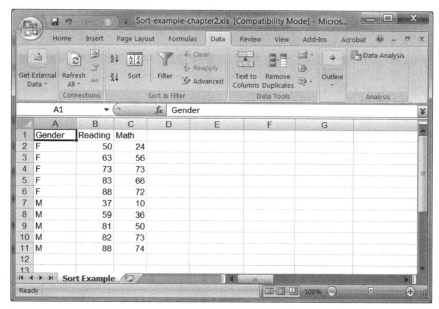

FIGURE 2.8 The Excel® screen showing the results of a multiple sort.

Figure 2.9 shows the statistical procedures available in Excel®. The scroll bar to the right of the screen allows the user to access several additional procedures. We will explore many of these procedures in later chapters.

You may not see the Data Analysis sub-menu displayed when you choose the Data menu on the main Excel® screen. That is because it is often an "add-in" program. Not everyone uses these features so Excel® makes them available as an "adjunct".[3]

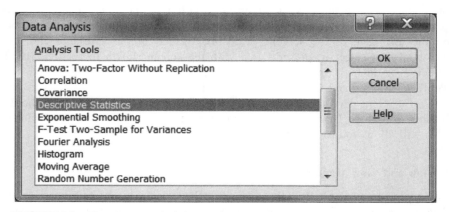

FIGURE 2.9 The "Data Analysis" sub-menu containing statistical analysis procedures.

---

[3] Mac users may not have access to the Data Analysis features since they were removed in previous versions.

16 GETTING ACQUAINTED WITH MICROSOFT EXCEL®

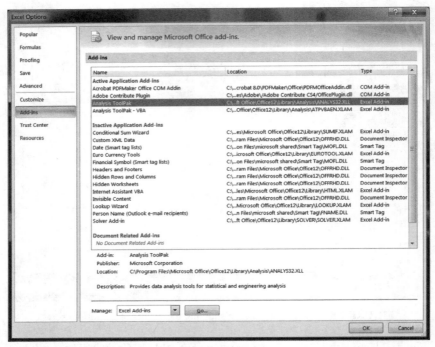

FIGURE 2.10 The Add-In options for Excel®.

If your Excel® screen does not show the Data Analysis sub-menu in the right edge of the menu bar when you select the Data menu, you can add it to the menu. Select the "Office Button" in the upper left corner of the screen and then you will see an "Excel® Options" button in the lower center of the screen. Choose this and you will be presented with several options in a column on the left edge of the screen. "Add-Ins" is one of the available choices, which, if you select it, presents you with the screen shown in Figure 2.10. I selected "Add-Ins" and the screen in Figure 2.10 appeared with "Analysis ToolPak" highlighted in the upper group of choices. When you select this option (you might need to restart Excel® to give it a chance to add), you should be able to find the Data Analysis sub-menu on the right side of the Data Menu. This will allow you to use the statistical functions we discuss in the book.

Review and View Menus

These two tabs available from the main screen have useful menus and functions for data management and appearance. I will make reference to them as we encounter them in later chapters.

3

USING STATISTICS IN EXCEL®

The heart of the statistical uses of Excel® is in the Data Analysis sub-menu that I described in Chapter 2. I will introduce many of these statistical tools in later chapters as we encounter different statistical topics. However, before we delve into those specific topics, I want to point out other ways that we can build statistical formulas directly into the spreadsheet.

USING STATISTICAL FUNCTIONS

In Chapter 2 I described several ways in which users can enter statistical formulas directly from the available sub-menus (see especially the "Formulas" section). As I mentioned, there are several statistical formulas available that we will use extensively in this book. Most are single-procedure formulas like calculating "AVERAGE" or "STDEV" (Standard Deviation), for example. Other procedures are more complex like the "FTEST" that calculates the equivalence in variance in two sets of data.

ENTERING FORMULAS DIRECTLY

Another very important use of Excel® is to "embed" formulas directly into the worksheet so that you can devise whatever calculation you need. The functions we discussed above are simply common calculations that have been arranged so that if

---

Understanding Educational Statistics Using Microsoft Excel® and SPSS®. By Martin Lee Abbott.
© 2011 John Wiley & Sons, Inc. Published 2011 by John Wiley & Sons, Inc.

you have repeated need for a certain calculation, you can use them more quickly than entering the formulas by hand.

Choosing the "=" key notifies Excel® that what follows is a user-created formula. Thereafter, you can enter the calculation you want as a string of characters. For example, using the sample of Gender, Reading, and Math scores shown in Figure 2.6, the following commands (user-created formulas) would yield the average value for "Math" scores: =Sum(C2:C11)/10

In this example, there are three main components of the formula:

- "=" informs Excel® that the user is entering a formula.
- "Sum(C2:C11)" calls for adding the values together from cell C2 to C11.
- "/10" divides the summed Math scores by 10 (the total number of scores), yielding the average Math score (53.46).

Figure 3.1 shows how this looks.

The results of entering the formula are shown in cell E2 (or whatever cell you used to enter the formula) in Figure 3.1. The formula you entered is shown in the formula bar directly above the spreadsheet. As you can see, it appears exactly as I described above. The "answer" of the formula appears in the cell, but Excel® remembers the formula and attaches it to the cell you chose to enter it. If any of the scores change, the average calculation will automatically adjust to reflect the change in values.

There are several ways to get the same result for most formulas you might want to enter. For example, you could use the menu system I described above

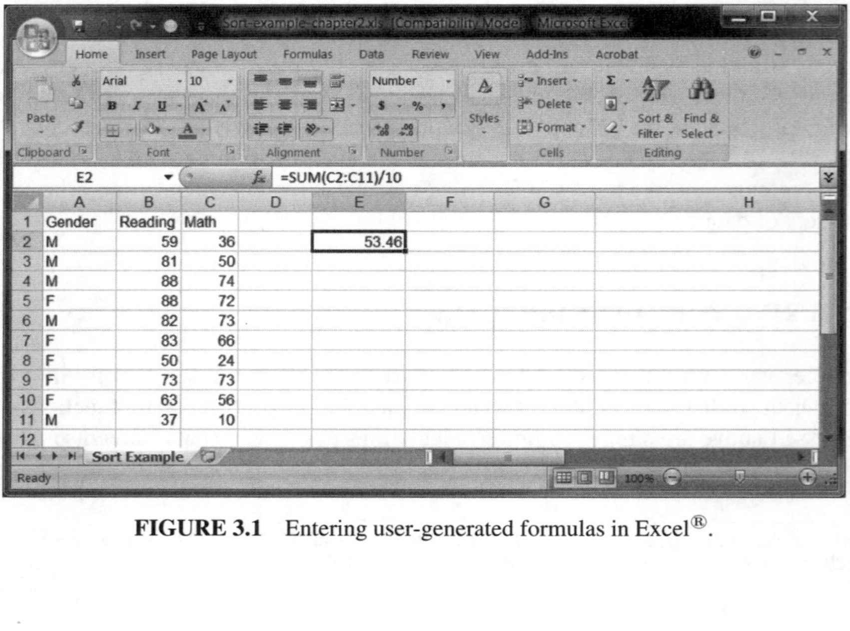

FIGURE 3.1 Entering user-generated formulas in Excel®.

ENTERING FORMULAS DIRECTLY

to enter a function to create the "AVERAGE," which is what we did using our own formula. Look at Figure 2.3 again, and you will see that "AVERAGE" (listed in the column on the left side of the screen) is one of the choices from the functions menu.

Another way to help create your own formulas is to use the "Σ" button shown on the "Home" tab. Look at Figure 2.1 and you will see this symbol in the third to last column (at the top) from the right side of the figure. The symbol means "sum of" and we will use it extensively in our discussion in later chapters since it is such an important function for statistical analyses. Figure 3.2 shows the result of clicking this symbol when the cursor is in cell E4. As you can see, when you select the symbol, it creates a formula calling for summing a series of cells you choose. In the example below, I selected the string of Math values (cells C2 to C11) with the cursor, which Excel® then added to the formula. You can see the selected cells enclosed in a dashed box surrounding the Math values.

Figure 3.2 also shows a "Screen Tip" box that appears when you choose the "Σ" button. Directly below the selected cell where the formula is entered, you will see the "help" bar explanation of the function: "=Sum(**number1**,[number2, . . .)." This shows that the sum symbol enters the Sum function wherein the numbers from the selected cells are added.

I used the Σ button in this example to demonstrate that it is helpful if you are building your own formula. Had we wanted to complete the formula for the average value of the math values, we would simply place the "/" figure at the end of the SUM function listed in the formula window. This would create the same formula we created directly, shown in Figure 3.1.

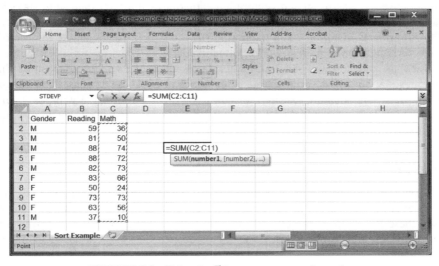

FIGURE 3.2 Using the Σ button to create a formula.

DATA ANALYSIS PROCEDURES

The Data Analysis sub-menu is a more comprehensive and extensive list of statistical procedures available in Excel®. Typically, this involves several related and linked functions and specialized formulas that statisticians and researchers use repeatedly. These are more complex than each separate function (e.g., Average, Standard Deviation, etc.), and in fact they may use several functions in the computation of the formulas. We will start with Descriptive Statistics in a later chapter (a Data Analysis sub-menu choice) and then move to several inferential procedures also represented in the sub-menu (e.g., t-Test, Correlation, ANOVA, Regression, etc.).

MISSING VALUES AND "0" VALUES IN EXCEL® ANALYSES

Some Excel® procedures you use may encounter difficulty if you are using large data sets or have several missing cases. In particular, you need to be careful about how to handle *missing cases and zeros*. Some procedures do not work well with missing values in the dataset. Also, be careful about how '0' values are handled. Remember, missing cases are not "0" values, and vice versa.

USING EXCEL® WITH REAL DATA

Over the next several chapters, I will introduce you to several databases that we will use to understand the different statistical procedures. I find it is always better to use real-world data when I teach statistics since students and researchers must, at some point, leave the classroom and venture into situations calling for the use of statistical procedures on actual research problems. I take this same approach in my book *The Program Evaluation Prism* (Abbott, 2010), in which I demonstrate the use of multiple regression using real-world evaluation data.

I will use three primary databases in this book, although I will introduce others in special situations as I describe the statistical procedures. These databases are related to my work in evaluation research for educational reform efforts. I will post them on the website identified in the Preface so that you can practice what you know with this real-world data.

School-Level Achievement Database

The state of Washington has a comprehensive database detailing school- and district-level data that we will use to describe several statistical procedures. I have used these in several evaluation research projects and find them very informative for many research applications.[1] The state website is easy to use and contains a

---

[1] You can review several technical reports using this database in our Washington School Research Center website: http://www.spu.edu/orgs/research/

variety of data over several years that can be downloaded. We will learn to download large datasets using this website.

The website is found in The Office of the Superintendent of Public Instruction in Washington State.[2] There are several files containing a variety of variables relating to school performance and description. For example, school-level achievement files are available for a number of subjects (e.g., reading and math) across several levels of grades and years. Additional files provide demographic and descriptive information on those same schools so that, when merged, a database is created that will allow primary research analyses.

The limitation of these data is that they are "aggregate" data. That is, each variable represents the average score across all the students in a particular school. Thus, reading achievement is not listed by student, but rather as the "percent passing the reading assessment" at various grade levels. Privacy laws prevent student level information being posted, so the researcher must be content with the aggregated scores.

Aggregate scores can be very helpful in identifying patterns or trends not easily seen otherwise. But we must always use these data with the caveat that we cannot make conclusions at a student level, but rather at a school level. Therefore, if we discover a relationship between reading achievement and class size, for example, we cannot say that students are better at reading in smaller (or larger) classes, but rather that reading achievement is higher in schools with smaller (or larger) class sizes. There may be features of the classes other than size that affect individual reading achievement.

Nevertheless, aggregate scores are helpful in pointing out patterns that can lead to further studies at the individual level. Later courses in statistics help students and researchers work at both levels simultaneously for a much more accurate and reliable way of understanding individual behavior and the influence of "larger" or external conditions on individual behavior. Raudenbush and Bryk (2002) discussed hierarchical linear modeling, for example, as one way to appreciate both levels and their interaction. I will discuss this a bit further in a later chapter.

Another reason I like to use this database is that it affords the student and researcher the opportunity to learn to download data for use in evaluation. The downloadable databases on the website are in either Excel® or text formats. In the exercises ahead, I will use the Excel® format to build a sample database.

TAGLIT Data

This database consists of several related databases addressing the impact of technology on different aspects of the classroom. It is a national-level database gathering data at the individual student, teacher, and school administrator level. The data files

---

[2] The website address is http://www.k12.wa.us/. The data are used courtesy of the Office of the Superintendent of Public Instruction, Olympia, Washington.

are from a national study in 2003, and all the data are used by permission from T.E.S.T., Inc.[3]

Because the data files are so massive and extensive, I will use a TAGLIT database that contains an aggregated set of data from teachers and students, primarily at the high-school level. I will review and explain the variables as I introduce various statistical procedures in later chapters.

The STAR Classroom Observation Protocol™ Data[4]

This is a remarkable dataset compiled at the individual student and teacher levels based on individual classroom observations. The BERC Group, Inc. collected thousands of these observations in the attempt to understand the impact of teaching and learning in the classroom. To what extent does model teaching affect how well learning proceeds? The heart of the STAR Protocol™ consists of a standardized method of measuring "Powerful Teaching and Learning™" in various subjects among elementary and middle schools over several years. We can connect these individual-level observations with school-level variables such as achievement, income level, and other important variables to understand the connections between classroom learning and other "environmental" variables.

---

[3] The author acknowledges the kind approval of T.E.S.T., Inc., the owner and manager of TAGLIT data, for the use of TAGLIT databases in this book. (http://www.testkids.com/taglit/).
[4] This dataset is used by permission of The BERC Group, Inc.

4

SPSS® BASICS

This book explores the use of statistical procedures in both Excel® and SPSS®. Therefore, I included the following sections to provide some familiarity with the basic functions of SPSS® along with those in Excel®. I will introduce the specific menus in later chapters that correspond to the statistical procedures we discuss.[1]

USING SPSS®

In a book such as this, it is important to understand the nature and uses of a statistical program like SPSS®. There are several statistical software packages available for manipulation and analysis of data, however, in my experience, SPSS® is the most versatile and responsive program. Because it is designed for a great many statistical procedures, we cannot hope to cover the full range of tools within SPSS® in our treatment. I will cover, in as much depth as possible, the general procedures of SPSS®, especially those that provide analyses for the statistical procedures we discuss in this book. The wide range of SPSS® products is available for purchase online (http://www.SPSS.com/).

The calculations and examples in this book require a basic familiarity with SPSS®. Generations of social science students and evaluators have used this statistical software, making it somewhat a standard in the field of statistical analyses. In

---

[1] Portions of this chapter are adapted from my book (Abbott, 2010) dealing with SPSS® applications in multiple regression, by permission of the publisher.

Understanding Educational Statistics Using Microsoft Excel® and SPSS®. By Martin Lee Abbott.
© 2011 John Wiley & Sons, Inc. Published 2011 by John Wiley & Sons, Inc.

the following sections, I will make use of SPSS® output with actual data in order to explore the power of statistics for discovery. I will illustrate the SPSS® menus so it is easier for you to negotiate the program. The best preparation for the procedures we discuss, and for research in general, is to become acquainted with the SPSS® data managing functions and menus. Once you have a familiarity with these processes, you can use the analysis menus to help you with more complex methods.

Several texts use SPSS® exclusively as a teaching tool for important statistical procedures. If you wish to explore all the features of SPSS® in more detail, you might seek out references such as Green and Salkind (2008) and Field (2005).

GENERAL FEATURES

Generally, SPSS® is a large spreadsheet that allows the evaluator to enter, manipulate, and analyze data of various types through a series of drop-down menus. The screen in Figure 4.1 shows the opening page where data can be entered. The tab on the bottom left of the screen identifies this as the "Data View" so you can see the data as they are entered.

A second view is available when first opening the program as indicated by the "Variable View" also located in the bottom left of the screen. As shown in Figure 4.2, the Variable View allows you to see how variables are named, the width of the column, number of decimals, variable labels, any values assigned to data, missing number identifiers, and so on. The information can be edited within the cells or by the use of the drop-down menus, especially the "Data"

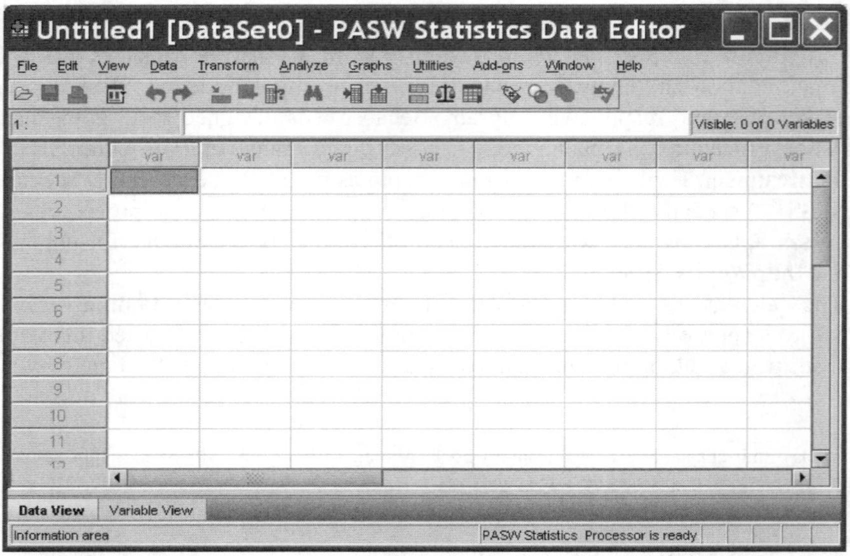

FIGURE 4.1 SPSS® screen showing data page and drop-down menus.

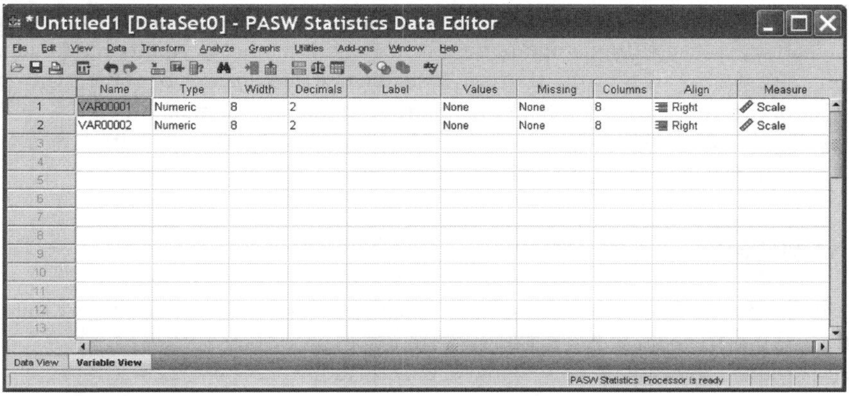

GENERAL FEATURES

FIGURE 4.2 SPSS® screen showing the variable view and variable attributes.

menu at the top of the screen. One of the important features on this page is the "Type" column, which allows the evaluator to specify whether the variable is "numeric" (i.e., a number), "String" (a letter, for example), or some other form (a date, currency, etc.).

Figure 4.3 shows the sub-menu available if you click on the right side of the "Type" column in the Variable View. This menu allows you to specify the nature of the data. For most analyses, having the data defined as numeric is required,

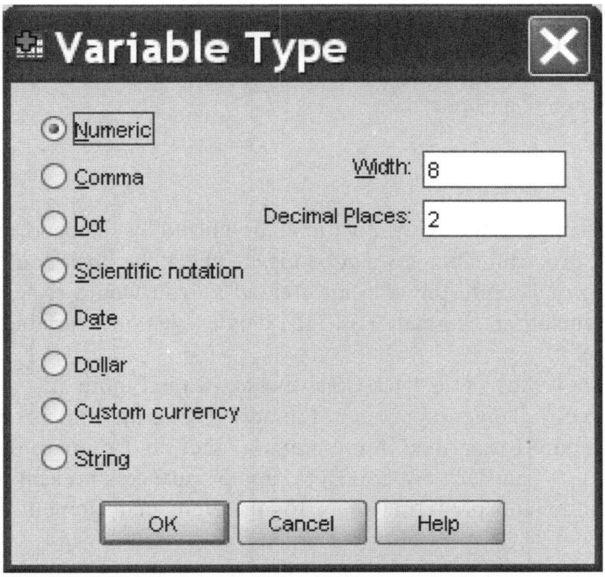

FIGURE 4.3 SPSS® screen showing sub-menu for specifying the type of variable used in the data field.

since most (parametric) statistical analyses require a number format. The "String" designation, shown below at the bottom of the choices, allows you to enter data as letters and words, such as quotes from research subjects, names of subject groups, and so on. If you use a statistical procedure that requires numbers, make sure the variable is entered as a "numeric" variable, or you will receive an error message and your requested procedure will not be executed.

MANAGEMENT FUNCTIONS

In this chapter, I will cover the essential functions that will allow you to get started right away with your analyses. Before a statistical procedure is created, however, it is important to understand how to manage the data file.

Reading and Importing Data

Data can be entered directly into the "spreadsheet" or it can be read by the SPSS® program from different file formats. The most common format for data to be imported to SPSS® is through such data programs as Microsoft Excel®, or simply an ASCII file where data are entered and separated by tabs. Using the drop-down menu command "File-Open-Data" will create a screen that enables the user to specify the type of data to be imported (e.g., Excel®, Text, etc.). The user will then be guided through an import wizard that will translate the data to the SPSS® spreadsheet format.

Figure 4.4 shows the screens from my computer that allow you to select among a number of "Files of Type" when you want to import data from SPSS®. These menus resulted from choosing "File" in the main menu and then choosing "Open Data." The small drop-down menu allows you to choose "Excel®" among other types of files when you import data.

Sort

It is often quite important to view a variable organized by size, and so on. You can run a statistical procedure, but it is a good idea to check the "position" of the data in the database to make sure the data are treated as you would expect. In order to create this organization, you can "sort" the data entries of a variable in SPSS® as you did in Excel®.

Figure 4.5 shows the same database as that shown in Figure 2.5, which was generated from Excel®. To get this screen, I imported the data from Excel® to SPSS® using the procedure I described in the previous section. Compare these figures to get a sense of how similar they are. Both are spreadsheets presenting the data in rows and columns and preparing them for different management and statistical analyses procedures.

Sorting the data in SPSS® is a bit easier than in Excel® because we do not need to select the entire database first. In SPSS®, the user can select "Data" from the main menu bar and then select "Sort." This results in the screens shown in Figure 4.6.

MANAGEMENT FUNCTIONS

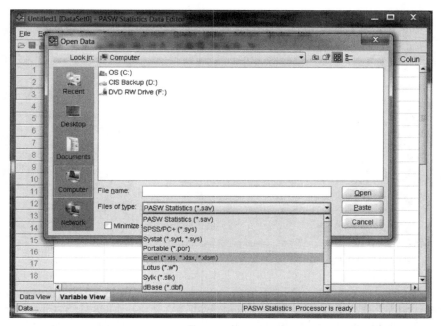

FIGURE 4.4 The SPSS® screens showing import data choices.

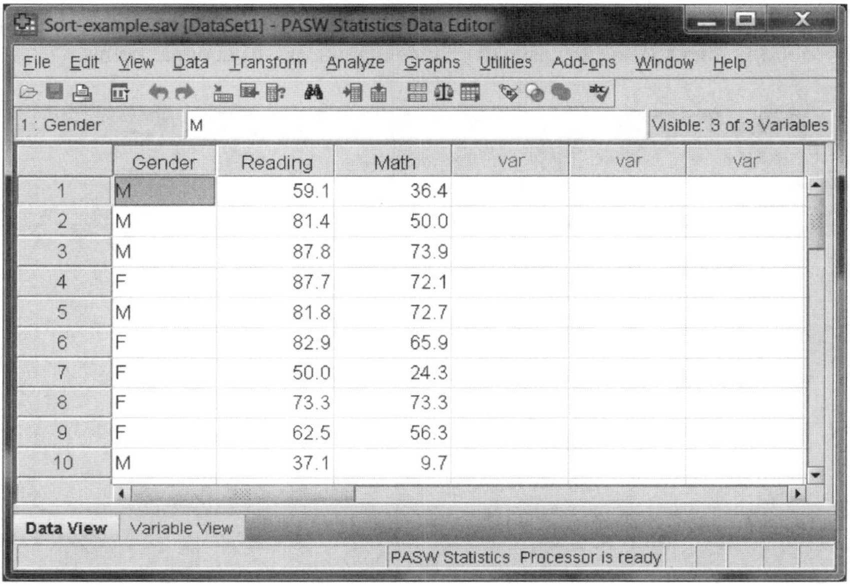

FIGURE 4.5 The SPSS® screen showing the unsorted "Sort-example" data.

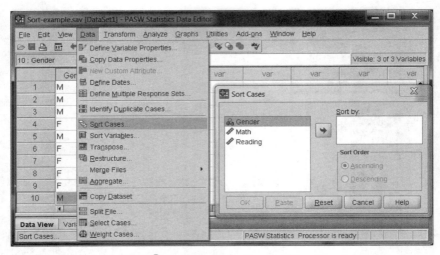

FIGURE 4.6 SPSS® data screens showing the "Sort Cases" function.

If we choose Gender, as shown in Figure 4.6, we can specify a sort that is either "Ascending" (alphabetical order beginning with "A" if the variable is a string variable, or starting with the lowest value if it is a numerical value) or "Descending." Selecting "Gender—Ascending" results in the screen shown in Figure 4.7. Compare this screen with the one shown in Figure 2.6 generated from Excel®. As you can see, they are very similar. The only difference between the data screens is that the values of the numerical variables in the SPSS® screen ("Reading" and "Math") are expressed with one decimal point, whereas the Excel® data are expressed in whole numbers. In either format it is easy to change the number of decimal points.

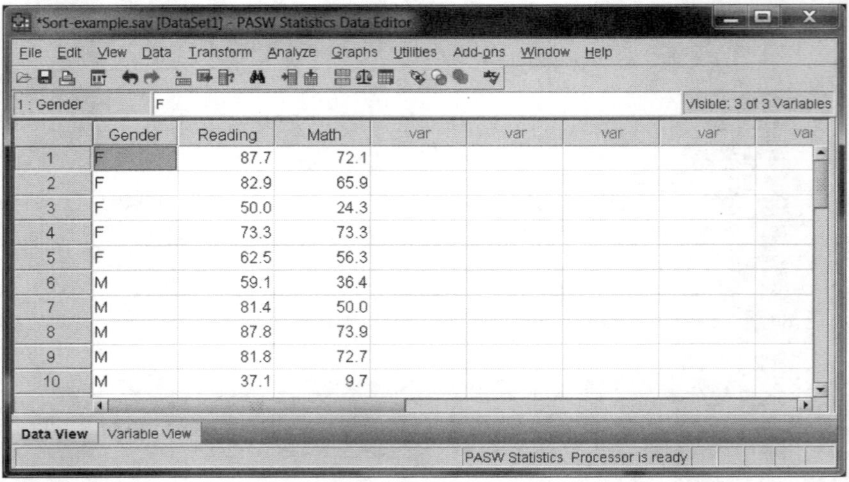

FIGURE 4.7 The SPSS® screen showing variables sorted by Gender.

MANAGEMENT FUNCTIONS

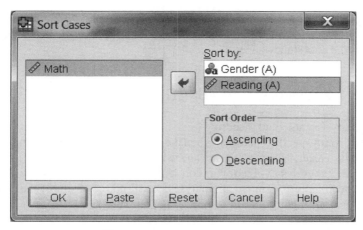

FIGURE 4.8 The SPSS® Sort Cases window showing a sort by multiple variables.

As with the Excel® data in Figure 2.6, the user can then visually inspect the data to observe any patterns that might exist among the variables. You can also sort the other variables easily in SPSS® using the same procedure since you do not need first to select the entire database as with Excel®.

SPSS® allows multiple sorts as does Excel®. Figure 2.8 shows the example database sorted first by Gender and then by Reading. We can generate a similar result in SPSS® by simply listing multiple variables in the "Sort Cases—Sort by" window. Figure 4.8 shows this specification in SPSS®.

Figure 4.8 also indicates the nature of the variable (string or numeric) by the small symbols next to each variable. Sorting by ascending order will therefore appropriately arrange the variables according to their type. Figure 4.9 shows the result

	Gender	Reading	Math
1	F	50.0	24.3
2	F	62.5	56.3
3	F	73.3	73.3
4	F	82.9	65.9
5	F	87.7	72.1
6	M	37.1	9.7
7	M	59.1	36.4
8	M	81.4	50.0
9	M	81.8	72.7
10	M	87.8	73.9

FIGURE 4.9 The SPSS® screen showing the results of sorting by multiple variables.

of this multiple sort. As you can see, the Reading values are "nested" within each category of Gender, as they are in Excel® (see Figure 2.8). This makes visual inspection of the data somewhat easier.

ADDITIONAL MANAGEMENT FUNCTIONS

SPSS® is very versatile with handling large datasets. There are several useful functions that perform specific operations to make the analyses and subsequent interpretation of data easier. Many of these can be accommodated in Excel®; however, these specific functions evolved with constant use by researchers since SPSS® is designed specifically for statistical analyses. I will not cover all of these, but the following sections highlight some important operations.

Split File

A useful command for students and researchers that we will use in subsequent chapters is "split file," which allows the user to arrange output specifically for the different values of a variable. Using our "sort" example from Figure 4.5, we could use the "split file" command to create two separate files according to the Gender variable and then call for separate statistical analyses on each variable of interest, for example the Math scores in Figure 4.5.

By choosing the "Data" drop-down menu, I can select "Split File" from a range of choices that enable me to perform operations on my existing data. Figure 4.10 shows the sub-menu for "Data" with "Split File" near the bottom.

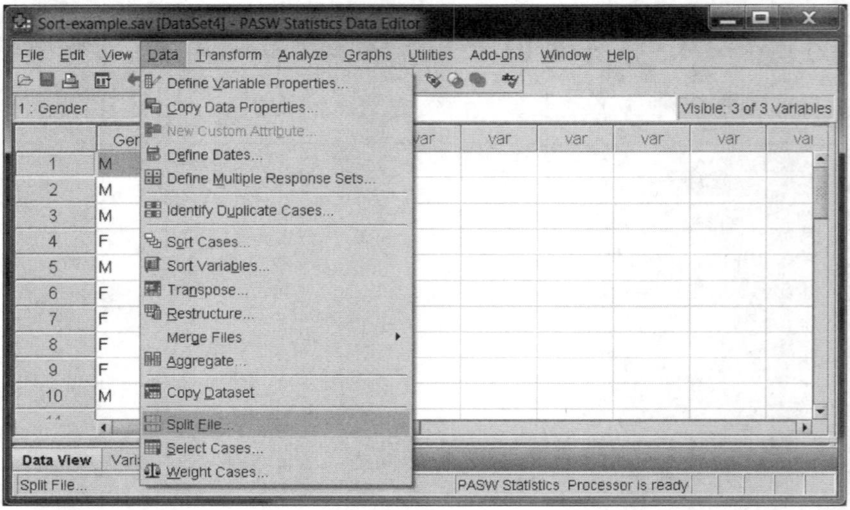

FIGURE 4.10 The "Split-File" option in SPSS®.

ADDITIONAL MANAGEMENT FUNCTIONS

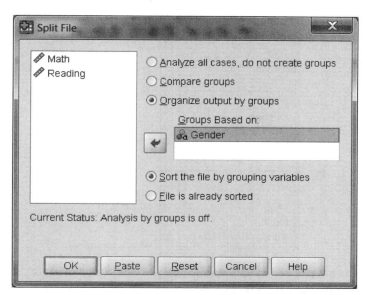

FIGURE 4.11 Steps for creating separate output using "Split File."

When I choose "Split File," I can then select which variable to use to create the separate data files. This is the "Organize output by groups" button shown in Figure 4.11. As you can see, if you choose this button, you can specify "Gender" by clicking on it in the left column and moving it to the "Groups Based on:" box by clicking the arrow button.

As Figure 4.11 shows, I selected the option "Organize output by groups" and then clicked on the variable "Gender" in the database. By these choices, I am issuing the command to create (in this case) two separate analyses for whatever statistical procedure I call for next since there are two values for the Group variable ("M" and "F"). When I perform a split file procedure in SPSS®, it does not change the database; rather, it simply creates separate output according to whatever statistical procedure you want to examine (e.g., descriptive statistics, correlation, etc.). I will discuss each of these procedures further in the chapters ahead. For now, it is important to understand that SPSS® has this useful function.

For example, I could next call for SPSS® to create means for each of the categories of gender. When I do so, SPSS® generates an "output" file showing separate results of the analysis by gender groups. Table 4.1 shows these results. As you can see, the output contains separate listings for Female and Male students. The top portion of Table 4.1 shows the Math and Reading scores for Females (58.38 and 71.28, respectively), and the bottom portion shows the scores for Males (48.54 for Math and 69.44 for reading).

Please note that when you use this procedure, it is necessary to "reverse" the steps you used after you have created the desired output. Otherwise, you will continue to get "split results" with every subsequent statistical analysis you call for.

TABLE 4.1 Split File Results for Gender and Achievement in SPSS®

Gender = F

	N	Mean
Math	5	58.380
Reading	5	71.280
Valid N (listwise)	5	

Gender = M

	N	Mean
Math	5	48.540
Reading	5	69.440
Valid N (listwise)	5	

SPSS® will continue to provide split file analyses until you "turn it off" by selecting the first option, "Analyze all cases, do not create groups" at the top of the option list in the Split File sub-menu. You can see this option near the top of the sub-menu in Figure 4.11.

Transform/Compute (Creating Indices)

One of the more useful management operations is the Compute function, which allows the user to create new variables. For this example, I am using a very small number of cases (15 of 3968 cases) from the TAGLIT database of middle and high schools from across the United States. The example case consists of three variables: "NumStudents," the number of students at the school; "NumTeachers," the number of teachers at the school; and "NumComputers" the number of computers at the school.

If I want to report a ratio of the number of computers available for the students at the schools, I can compute a new variable using the menus in SPSS®. At the main menu (see Figure 4.1), I can access this function by selecting the "Transform" and then "Compute Variable" option. This will result in a dialog box like the one shown in Figure 4.12.

In this example, I am creating a new variable ("CompStudent") by dividing the current "NumComputers" variable by the "NumStudents" variable. The first step is to name the new variable by entering it into the "Target Variable:" window at the upper left of the screen. Then, you can create a formula in the "Numeric Expression:" window. As Figure 4.12 shows, I clicked on NumComputers and placed it in the window by clicking on the arrow button. Then, I entered a "/" mark using the keypad below the window. Last, I placed NumStudents in the window to complete the formula: NumComputers/NumStudents.

This procedure will result in a new variable with values that represent the ratio of computers to students by school. Figure 4.13 shows the resulting database that now

ADDITIONAL MANAGEMENT FUNCTIONS 33

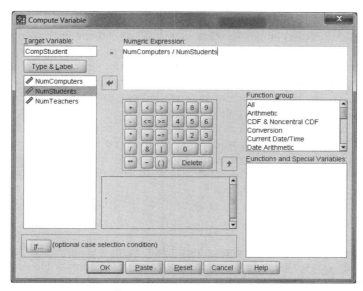

FIGURE 4.12 SPSS® screen showing "Transform Compute" functions.

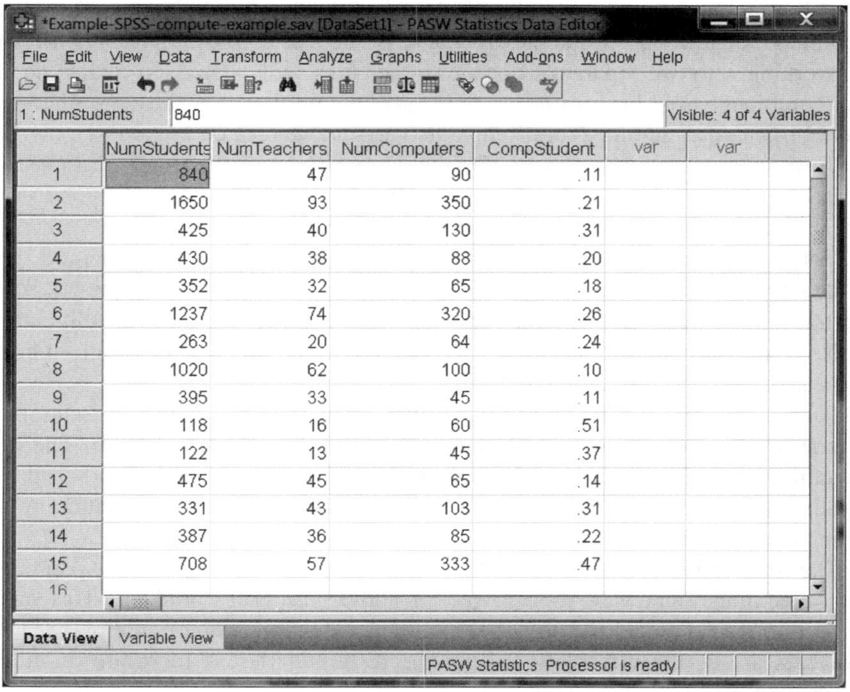

FIGURE 4.13 Data file showing a new variable, "CompStudent."

includes the new variable. As you can see, the computer-to-student ratio ranges from .10 to .51 (or one computer per ten students to one computer per two students).

As you can see from the screen in Figure 4.12, you can use the keypad in the center of the dialog box for entering arithmetic operators, or you can simply type in the information in the "Numeric Expression:" window at the top. You will also note that there are several "Function group:" options at the right in a separate window. These are operations grouped according to type. Scrolling to the bottom allows the user to specify "statistical functions" such as means, standard deviations, and so on. You can select whichever operation you need and then enter it into the Numeric Expression window by clicking on the up arrow next to the Function Group window. This feature of SPSS® is similar to the functions available in Excel® that we discussed in Chapter 3.

Merge

The merge function is one of the most useful, but the most misunderstood, functions in SPSS®. I have yet to see any accurate treatment of the appropriate steps for this procedure in any resource book. I will attempt to provide a brief introduction to the procedure here because it is so important, but experience is the best way to master the technique. I recommend that you create two sample files and experiment with how to use it.

The merge function allows you to add information to one file from another using a common identifier on which the procedure is "keyed." For example, suppose you are working with two separate school-level data files and you need to create one file that combines variables located on the separate data files. Perhaps one file has a school ID number and the NumComputers variable we created in the last example, while a second file has a school ID (the same values as the other file) and the NumStudents and NumTeachers variables. The merge function allows you to add variables from one of the files to the other using the common ID number.

You can approach the merge in several ways, but my preferred method is to choose one file as the "master" to which the separate information is brought. After the transfer, you can save this file separately as the master file since it will contain both sets of information. SPSS® allows you to specify which information to bring to the separate file in a dialog box.

In this example, I will merge two separate files with a common School ID number ("IDnum"). Figures 4.14 and 4.15 show the separate data files. The first contains the variables NumStudents and NumTeachers whereas the second file has the NumComputers variable. We want to create a master file containing all three variables.

The first step is to make sure both "IDnum" variables are sorted (Ascending) and saved in the same way. (See the section above on sorting variables.) This variable is the one on which the sort is keyed, and the merge cannot take place if the variables are sorted differently within the different files, if there are duplicate numbers, missing numbers, and so on.

Second, in the file you identify as the master file (usually one of the files to be merged), choose "Data-Merge-Add Variables" as shown in the dialog box in

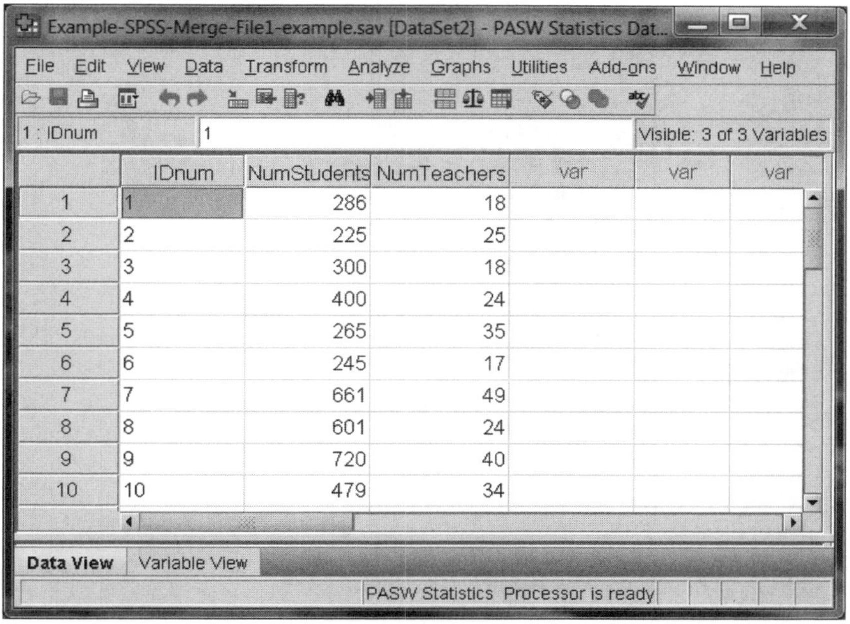

ADDITIONAL MANAGEMENT FUNCTIONS

FIGURE 4.14 Datafile #1 for merge example.

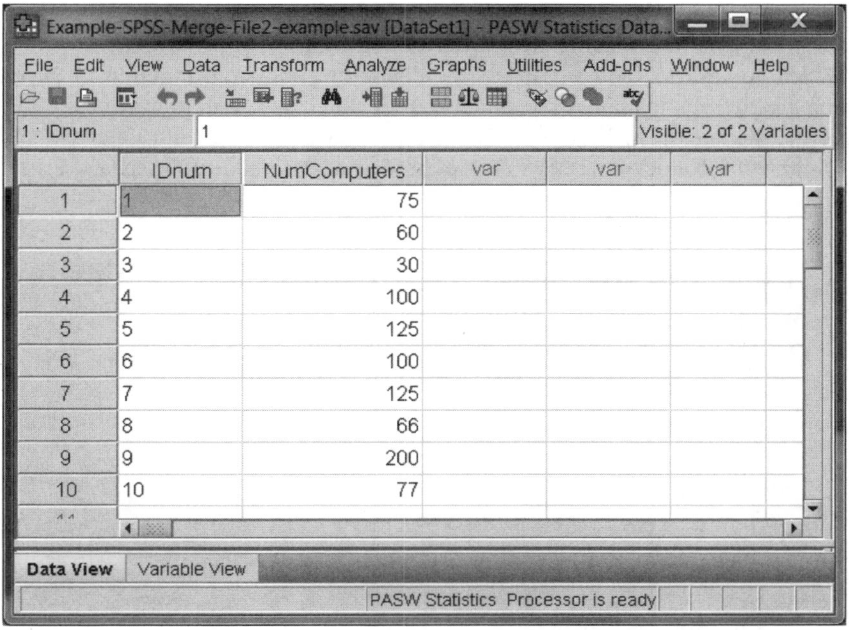

FIGURE 4.15 Datafile #2 for merge example.

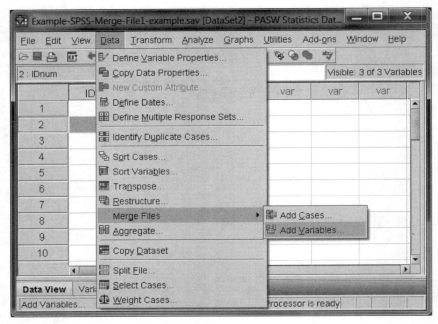

FIGURE 4.16 SPSS® screen showing the "Merge" options.

Figure 4.16. This will allow you to move entire variables from one file to another. The other option "Add Cases" allows you to append all the cases in one file to the cases in the other file, a completely different function.

When you ask to "add variables" the dialog box shown in Figure 4.17 appears which enables you to choose the data file from which you wish to move the desired variable. As shown in Figure 4.17, we are choosing the second database since it contains the NumComputers variable that we wish to add to the database in the first file.

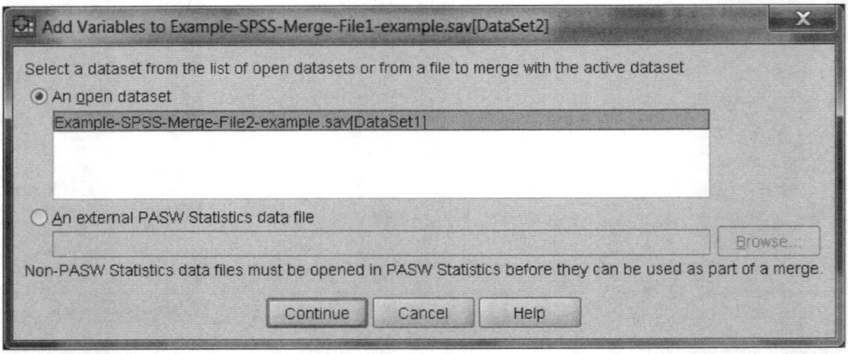

FIGURE 4.17 SPSS® screen listing the available data files to merge.

ADDITIONAL MANAGEMENT FUNCTIONS

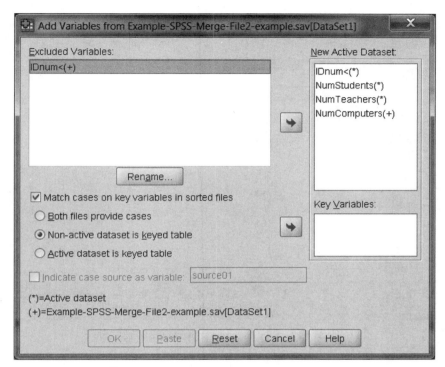

FIGURE 4.18 SPSS® screen used to identify the "key variable" on which to base the merge.

The next dialog box that appears is shown in Figure 4.18, in which you can specify which variable is the "key variable" on which to base the merge. In the current example, I have used the first file containing IDnum, NumStudents, and NumTeachers as the master file and called for a merge from the second file containing IDnum and NumComputers. The IDnum variable can be chosen from the top left dialog box. It is the only variable chosen because it is found in both files. After selecting the SchoolID variable, you can select the "Match cases on key variables in sorted files" box, along with the middle choice "Non-active dataset is keyed table." This tells SPSS® that you want the second file to be the one from which the new variable is to be chosen and placed in the master file. You can see that the new master file will consist of the variables listed in the top right dialog window when the merge is complete.

In this example, the "New Active Dataset" window, you can see that the new master file will consist of the IDnum, NumStudents, nad NumTeachers variables from the first file (indicated by an "(∗)" after the variables) and the Numcomputers variable from the second file (indicated by a "(+)" after the variable).

The next step is very important. When you click on the bottom arrow, you inform SPSS® that the IDnum variable is the keyed variable. When you place IDnum in this window, it removes it from the list of variables in the "New Active Dataset"

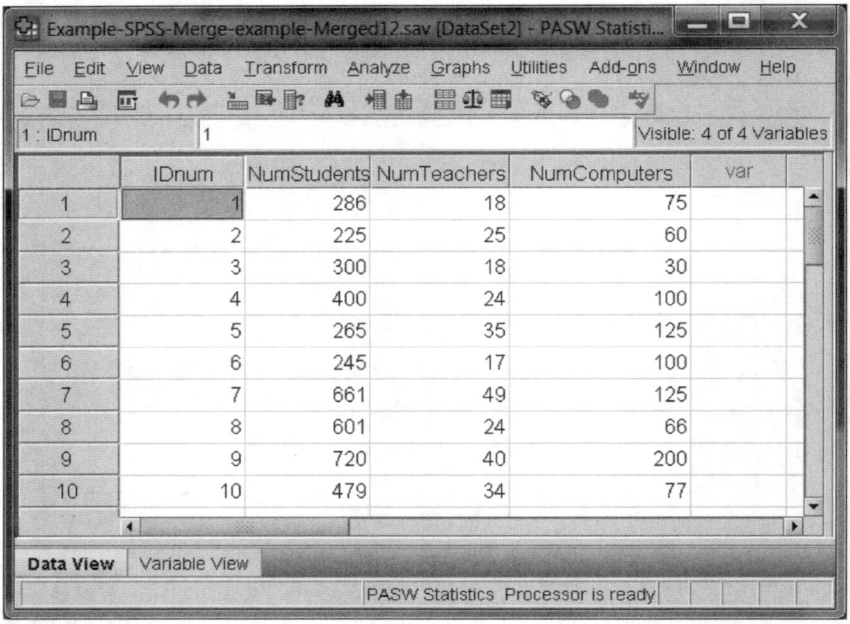

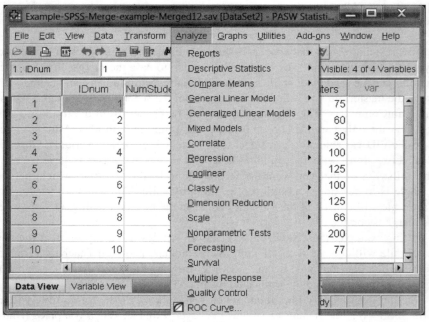

FIGURE 4.19 SPSS® screen showing the merged file.

FIGURE 4.20 The SPSS® Analyze menu options.

ANALYSIS FUNCTIONS 39

window since it is contained in both files. In this example, there will be no other variables listed in the "Excluded Variables" window since all will be included in the new master file.

Once you move the key variable to the window using the arrow key, you can choose "OK" and the desired variables will be added to the master file, which you can save under a different name. The panel in Figure 4.19 shows the new master file with the complete list of variables from both files keyed to the same school ID number.

This example used two simple files, but the same process can be used for more complex files. Once you have merged the files, it is often helpful to "eyeball" the data to make sure the variables merged with the variable values appropriately listed under the keyed variable.

ANALYSIS FUNCTIONS

Over the course of our study in this book, we will have extensive practice at conducting statistical procedures with SPSS®. All of these are accessible through the opening "Analyze" drop-down menu as shown in Figure 4.1. The screen in Figure 4.20 shows the contents of the Analyze menu. We will not be able to cover all of these in this book, but you will have the opportunity to explore several of the sub-menu choices. Many of these statistical functions are represented in the Excel® formulas that we described in Chapter 3. See Figure 2.9 for a partial list of those functions.

5

DESCRIPTIVE STATISTICS—CENTRAL TENDENCY

When I teach statistics to any group of students, I start by offering a series of questions that emphasize the importance of statistics for solving real research problems. Statistical formulas and procedures are logical and can be interesting, but the primary function for statistical analyses (at least, in my mind) is to bring clarity to a research question. As I discussed in a recent book dealing with statistics for program evaluation (Abbott, 2010), statistical procedures are best used to discover patterns in the data that are not directly observable. Bringing light to these patterns allows the student and the researcher to understand and engage in problem solving.

RESEARCH APPLICATIONS—SPURIOUSNESS

Do storks cause babies? This opening question usually results in a good laugh in introductory statistics classes. Of course they do not! But a time-worn example in sociology classes (see Bohrnstedt and Knoke, 1982) is that, in Holland several years ago, communities where more storks nested had higher birth rates than did communities with fewer nesting storks. This strong association has always been used to illustrate the fact that just because two things are strongly *related* to one another (correlation), one does not necessarily *cause* the other. There may be a third variable, not included in the analysis, that is related to both variables resulting in the appearance of a relationship.

Understanding Educational Statistics Using Microsoft Excel® and SPSS®. By Martin Lee Abbott.
© 2011 John Wiley & Sons, Inc. Published 2011 by John Wiley & Sons, Inc.

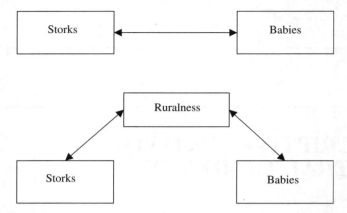

FIGURE 5.1 The spurious relationship between storks and babies.

What might that third variable be? Several possibilities exist. One might be called something like "ruralness" or "urbanness." That is, the more rural the area, the more feeding and nesting possibilities exist for storks. It also happens that there are more children outside of cities, since single individuals more than families often live in cities. Thus, the apparent relationship between storks and babies is really a function of this third variable that, when introduced into the equation, reduces the original relationship to zero. This is illustrated in Figure 5.1. The top panel shows the apparent relationship between babies and storks, with a two-way line connecting the variables indicating that the two are highly related to one another. The bottom panel shows that, when the third variable (Ruralness) is introduced, the relationship between storks and babies disappears.

This situation is known as "spuriousness" in research. "Spurious" carries the definition of "false," which fits a situation in which something apparent is not accurate or true. Spuriousness is the reason why "correlation does not equal causation" in research. Just because something is related does not mean that one causes the other. You can find spurious relationships in any field of research because it is often difficult to identify whether an apparent, calculated relationship between two variables is the result of another variable, or variables, not taken into account in the research. This is one of the reasons why statistics courses are important; we can learn statistical procedures that will allow us to study multiple relationships at the same time so as to identify potentially spurious situations.

Identifying potentially spurious relationships is often quite difficult, and it comes only after extended research with a database. The researcher must know their data intimately in order to make the discovery. An example of this is a study I reported in my program evaluation book (Abbott, 2010), this one in the field of industrial democracy. Initial data analyses on worker participation in an electronics firm suggested that if workers were given the ability to participate in decision making, they would have higher job satisfaction. This was a reasonable assumption, given similar assumptions in the research literature. However, the more I examined the data from individual workers, the more I questioned this assumption and decided to test it further.

I noticed from interviews that many workers did not want to participate in decision making, even though they had the opportunity to do so. They were content to do their work and leave at the end of the day without a role in deciding what the work at their plant should look like or how it should change. I therefore analyzed the original "participation—job satisfaction" but this time added a third variable, "desire for participation" to the analysis. I found that the apparent relationship was much more complex than previously assumed. One of my findings was that "attitudes toward management" was not related directly to worker satisfaction, but that "desire for management" affected both of the study variables and led to a different conclusion than the simple assumption that "participation causes satisfaction."

The popular press often presents research findings that are somewhat bombastic but which might possibly be spurious. The value of statistics is that it equips the student and researcher with the skills necessary to debunk simplistic findings. I have explored one such finding in education for many years and in a variety of study contexts. This is the assumption that low achievement in K–12 education is related to ethnicity, with some ethnic groups performing better than others. Research findings are reported that supposedly lend support to this assumption.

Over several years at the Washington School Research Center, a research center investigating educational reform, my colleagues and I have explored this assumption using available data from the state of Washington. What we found, in a variety of research studies, is that the ethnicity—achievement relationship is more complex than it appears. There may be other variables not in the analysis that might explain why it appears that one variable is related to the other.

Based on the research literature, we identified socioeconomic status as a potentially meaning variable that might help to illuminate this relationship. Gathering data on all the schools in the state, we analyzed the impact of ethnicity on achievement, but this time introduced "low income" into the analysis. The only variable we had to indicate low income was one that educational researchers use almost exclusively, namely, the percent of the students at the school qualified to receive free or reduced (F/R) lunch based on family income. Despite the potential problems with using this variable as an indicator of family income, it is nevertheless one available measure that allows us to see the effects of low income on the ethnicity—achievement relationship.

In virtually every study we conducted, we found that ethnicity, while a very important variable in the consideration of school-based achievement, was not the influence it was assumed to be. On closer inspection, we found that most of the relationship between ethnicity and achievement was really a function of low income. That is, the greater the percent of children in school qualified for F/R, the lower the school-based achievement. Ethnicity still influenced achievement, but in a very small way. Most of the influence of ethnicity was exerted through the avenue of low income. This relationship can be viewed in much the same way that the variables are viewed in Figure 5.1.

We will have the opportunity to explore this relationship further in our discussions in this book. I have made the original database available so that you can explore it on your own, and I will use many examples from the database to clarify the statistical procedures in subsequent chapters.

DESCRIPTIVE AND INFERENTIAL STATISTICS

Statistics, like other courses of study, is multifaceted. It includes "divisions" that are each important in understanding the whole. Two major divisions are Descriptive and Inferential statistics. Descriptive statistics are methods to summarize and "boil down" the essence of a set of information so that it can be understood more readily and from different vantage points. We live in a world that is bombarded with data; descriptive statistical techniques are ways of making sense of it. Using these straightforward methods allows the researcher to detect numerical and visual patterns in data that are not immediately apparent.

Inferential statistics are a different matter altogether. These methods allow you to make predictions about unknown values on the basis of small sets of "sample" values. In real life, we are presented with situations that cannot provide us with certainty: Would my method for teaching mathematics improve the scores of all students who take the course? Can we predict what a student might score on a standardized test? Inferential statistics allow us to infer or make an observation about an unknown value from values that are known. Obviously, we cannot do this with absolute certainty—we do not live in a totally predictable world. But we can do it within certain bounds of probability. It is hoped that statistical procedures will allow us to get closer to certainty than we could get without them.

THE NATURE OF DATA—SCALES OF MEASUREMENT

The first step in understanding complex relationships like the ones I described earlier is to be able to understand and describe the nature of what data are available to a researcher. We often jump into a research analysis without truly understanding the features of the data we are using. Understanding the data is a very important step because it can reveal hidden patterns and it can suggest custom-made statistical procedures that will result in the strongest findings.

One of the first realizations by researchers is that data come in a variety of sizes and shapes. That is, researchers have to work with available information to make statistical decisions, and that information takes many forms. The examples we discussed earlier illustrate several kinds of data:

1. Students are identified as either "qualified" or "not qualified" for free or reduced lunches.

2. Workers either "desire participation" or "do not desire participation."
3. Job satisfaction is measured by worker responses to several questionnaire items asking them to "Agree Strongly," "Agree," "Neither Agree nor Disagree," "Disagree," or "Disagree Strongly."
4. School leaders measure student achievement through reading and math tests in which students obtain percentages of right answers.

Nominal Data

The first two examples show that data can be "either–or" in the sense that they represent mutually exclusive categories. If you are qualified for F/R, for example, you do not fit the "not qualified" category; if you desire participation, you do not fit the category of others who do not desire participation. Other examples of this "categorical" data are sex (male and female) and experimental groups (treatment or control).

This kind of data, called "nominal," does not represent a continuum, with intermediate values. Each value is a separate category only related by the fact that they are categories of some larger value (e.g., male and female are both values of sex). These data are called nominal since the root of the word indicates "names" of categories. They are also appropriately called "categorical" data.

The examples of nominal data above can also be classified as "dichotomous" since they are nominal data that have only two categories. Nominal data also include variables with more than two categories such as schooling (e.g., public, private, home schooling). We will discuss later that dichotomous data can come in a variety of forms also, like (a) "true dichotomies" in which the categories naturally occur like sex and (b) "dichotomized variables" that have been created by the researcher from some different kind of data (like satisfied and not satisfied workers).

In all cases, nominal data represent mutually exclusive categories. Educators typically confront nominal data in classifying students by gender or race; or, if they are conducting research, they classify groups as "treatment" and "control." In order to quantify the variables, researchers assign numerical values to the categories, like treatment groups = 1 and control groups = 2. In these cases, the numbers are only categories; they do not represent actual values. Thus, a control group is not twice a treatment group. The numbers are only a convenient way of identifying the different categories.

Because nominal data are categorical, we cannot use the mathematical operations of addition, subtraction, multiplication, and division. It would make no sense to divide the treatment group by the control group to get one-half. Researchers must simply indicate the percentage of individuals who occupy the categories, for example, as a way of reporting what the data indicate. Thus, we might say that, in our data, 51% are male and 49% are female.

Ordinal Data

The third example listed above indicates another kind of data: ordinal data. These are data with a second characteristic of meaning, namely, position. There are categories, as in nominal data, but with these data the categories are related by "more than" and "less than." Some categories are placed above in value or below in value of some other category. Educational researchers typically find ordinal data in many places: student attitude surveys, teacher questionnaires, and in-class tests, for example. In these cases, one person's response can be more or less than another person's on the same measure. In example 3 above, job satisfaction can be measured by a question that workers answer about their work, such as the following: "I am happy with the work I do."

1. Agree Strongly (SA)
2. Agree (A)
3. Neither Agree nor Disagree (N)
4. Disagree (D)
5. Disagree Strongly (SD)

As you can see, one worker can be quite happy and can indicate "Agree Strongly" while another can report that they are less happy by indicating "Agree." Both workers are reporting different levels of happiness, with some being more or less happy than others. Teacher agreement or disagreement with certain school policies are similarly measured in educational studies. (These response scales are typically called 'Likert' scales.) In one of the WSRC studies, for example (Abbott et al., 2008), my colleagues and I reported on a teacher survey in which we asked respondents several questions about various changes in their schools. For example, respondents indicated their agreement (using the same scale noted above in the worker participation study) to the questionnaire item, "Teacher leaders have had a key role in improving our school."

The resulting data from questionnaire items such as these are ordinal data. They add greater than and less than to the nominal scale, and in this sense they represent a more complex set of data. Using such data provides a deeper understanding of people's attitudes than simply classifying their answers as nominal data (e.g., agree–disagree, true–false).

These kinds of questionnaire data are the stock-in-trade of social scientists because they provide such a convenient window into people's thinking. Educational researchers use them constantly in a variety of ways with students, teachers, parents, and school leaders.

There is a difficulty with these kinds of data for the researcher however. Typically, the researcher needs to provide a numerical referent for a person's response to different questionnaire items in order to describe how a group responded to the items. Therefore, they assign numbers to the response categories as shown in Table 5.1.

THE NATURE OF DATA—SCALES OF MEASUREMENT

TABLE 5.1   Typical Ordinal Response Scale

SA	A	N	D	SD
1	2	3	4	5

The difficulty arises when the researcher treats the numbers (1 through 5 in Table 5.1) as integers rather than ordinal indicators. If the researcher thinks of the numbers as integers, they typically create an average rating on a specific questionnaire item for a group of respondents. Thus, assume for example that four educators responded to the questionnaire item above ("Teacher leaders have had a key role in improving our school") with the following results: 2, 4, 3, 1. The danger is in averaging these by adding them together and dividing by four to get 2.5 as follows $(2+4+3+1)/4$. This result would mean that, on average, all four respondents indicated an agreement halfway between the 2 and the 3 (and, therefore, halfway between agree and neither). This assumes that each of the numbers has an equal distance between them—that is, that the distance between 4 and 3 is the same as the distance between 1 and 2. *This is what the scale in Table 5.1 looks like if you simply think of the numbers as integers.*

However, an ordinal scale makes no such assumptions. Ordinal data only assume that a 4 is greater than a 3, or a 3 is greater than a 2, *but not that the distances between the numbers are the same*. Table 5.2 shows a comparison between how an ordinal scale appears and how it might actually be represented in the minds of two different respondents.

According to Table 5.2, respondent 1 is the sort of person who is quite certain when they indicate SA. This same person, however, makes few distinctions between A and N and between D and SD (but they are certain that any disagreement is quite a distance from agreement or neutrality). Respondent 2, by contrast, doesn't make much of a distinction between SA, A, and N, but seems to make a finer distinction between areas of disagreement, indicating stronger feelings about how much further SD is from D.

It is hoped that this example helped you to see that the numbers on an ordinal scale do not represent an objective distance between the numbers, they are only indicators of ordinal categories and can differ between people on the same item. The upshot, for research, is that you cannot add the numbers and divide by the total to get an average because the distances between the numbers may be different for each respondent! Creating an average would then be based on different meanings of the numbers and would not accurately represent how all the respondents, as a group, responded to the item.

TABLE 5.2   Perceived Distances in Ordinal Response Items

Scale Categories	SA	A	N	D	SD
The way it appears	1	2	3	4	5
Respondent 1 *perception*	1		2 3		4 5
Respondent 2 *perception*		1 2 3		4	5

Interval Data

The majority of what we will study in this book uses interval data. These data are numbers that have the properties of nominal and ordinal data; but they add another characteristic, namely, equal distance between the numbers. Interval data are numbers that have equal distance between them, so that the difference between 90 and 91 is the same as the distance between 103 and 104; in both cases, the difference is one unit. The value of this assumption is that you can use mathematical operations (multiplication, addition, subtraction, and division) to analyze the numbers because they all have equal distances.

An example of an interval scale is a standardized test score such as an intelligence quotient (IQ). While psychologists and educational researchers disagree about what IQ really represents, the numbers share the equal distance property. With IQ, or other standardized tests, the respondent indicates their answers to a set of questions designed to measure the characteristic or trait studied.

Job satisfaction (JS) is another example. Here, respondents may indicate they strongly agree, agree, and so on, with a series of items measuring their attitudes toward their job. The Job Diagnostic Survey (Hackman and Oldham, 1980) includes the following item as part of the measurement of job satisfaction: "I am generally satisfied with the kind of work I do in this job" (response scale is Disagree Strongly, Disagree, Disagree Slightly, Neutral, Agree Slightly, Agree, and Agree Strongly).

The measurement of job satisfaction uses a series of these kinds of questions to measure a worker's attitude toward their job. What makes this kind of data different from ordinal data, which uses a similar response scale, is that the JDS uses a "standardized" approach to measurement. That is, the test was used with a number of different sets of workers, under the same directions, with the same materials, time, and general conditions. The JDS items measured job satisfaction among managerial, clerical, sales, machine trade, and other workers. With such wide application, the *set* of items comprising the job satisfaction index comes to represent a consistent score. In fact, there are specific statistical procedures that measure the extent to which the scores are consistent across usage.

The result of repeated, standardized use of these kinds of instruments is that the response scales come to be accepted as interval data. The distance between units comes to have meaning as equivalent distances. Thus, even though they may be based on the same kind of ordinal response scales, as we discussed above, the set of measures can be multiplied, divided, added, and subtracted with consistent results. This is the way interval measures are typically used by educational researchers. Whether or not we truly understand what one unit of IQ represents, we can proceed with measurement using recognized statistical procedures.

A difficulty in research is that all kinds of items might be thrown together by someone unaware of the nature of research to yield a "scale" that is then used in statistical procedures for problem solving. For example, a principal may wish to understand the attitudes of teachers regarding some educational policy like whether or not to change a math curriculum. She might ask other principals if they have done the same thing, or simply sit down and compose a few questions that she believes measures teacher opinion accurately. Typically, this involves compiling a series of Likert items (i.e., involving the response scale illustrated in Table 5.1). Leaving aside the issue of whether the items are written correctly, she might then distribute the questions to her teachers and compile the results. As you might imagine, she compiles the results by *assigning numbers to the response categories*, as in Table 5.1, and creating averages to each item across all the teacher respondents.

The difficulty with this procedure is, I hope, now obvious. Rather than averaging the response scores, the teacher should simply report the frequencies of teachers who report each category. Consider the example in Table 5.3 in which five teachers respond to a principal-created item like, "We should change the math curriculum." As you can see, the teachers indicated their attitudes by their choices using an "x" under the appropriate Likert scale category. If the principal were to average the responses (thereby treating the data as interval), the five teachers would indicate a 3.2 average, or slightly above Neutral. However, if the principal treated the data as ordinal, she would report the frequencies in the bottom row of Table 5.3. According to this report, 60% of the teachers who responded to the item were in agreement while 20% were strongly unfavorable and 20% were neutral. Using the data differently indicates different views of the teachers' responses.

This example illustrates several characteristics of numbers that we will discuss in subsequent chapters. However, I point out here that statistics students and researchers need to be careful to understand what kind of data they have available and how to treat it for answering a research question.

In the course of actual research, evaluators often treat ordinal data as if it were interval data. While from a purist standpoint this is not strictly accurate, researchers use this assumption, especially with a survey instrument they, or other researchers, have used repeatedly. Standardized instruments like IQ or some Job Satisfaction measures are widely accepted and used as interval data. Regardless, students and researchers need to carefully consider the nature of the items of any instrument used to measure attitudes and decide how most accurately to represent the results.

Confidence with this assumption increases with well-written items. The interested student should seek research publications that discuss the criteria for creating survey items. Babbie (2010) is one such resource that provides a thorough set of

TABLE 5.3   Comparison of Interval and Ordinal Scales

	1	2	3	4	5	
	SD	D	N	A	SA	Avg
Teacher 1	x					1
Teacher 2				x		4
Teacher 3			x			3
Teacher 4				x		4
Teacher 5				x		4
	20%	0	20%	60%		3.2

guidelines for creating appropriate questions. Creating items that conform to rules such as these provides a stronger foundation for treating these kinds of ordinal data as interval. Students and researchers should still exercise caution, especially with self-generated instruments.

Ratio Data

It is hard to imagine a worker with *no* satisfaction! Even if they are not completely enthralled with their work, or totally hate it, they have some attitude. Even "neutral" attitudes indicate a sort of ambivalence in which there are some positive and some negative aspects of the job. The fact is that it is difficult to imagine the *absolute absence* of some concepts, attitudes, and behaviors, even IQ. Someone low on the IQ scale still *has* an IQ, even if their IQ "score" is zero.

There are other variables that can be said to have the possibility of absolute zero; the amount of money in your pocket, the distance between two lines, number of credits of math, number of friends, age, and so on. In such cases, a "0" value means *none*. Percentages can have a meaningful zero, as in the case of what percentage of students go on to college once they graduate from high school, or what percentages of students pass the state math test. Often, interval scale measures have zeros, but they are not "true" zeros in the sense of absolute nothing!

The statistical value of ratio scales, those with absolute zeros, is that the researcher can make ratios, hence the derivation of the name. If there is a fixed and absolute zero, then two things can be referenced to one another since they have a common benchmark. Student A with 4 math credits can have *twice* the number of math credits as Student B who has 2 math credits, for example. If you express this relationship in a number, you can divide the 4 credits (student A) by 2 credits (Student B) to get 2, or twice the number of credits. Of course you can also express the ratio the other way by saying that Student B has only half the math credits as Student A (or $1/2$). Or, we can speak of School A evidencing twice the math completion rate of School B (assuming both schools use credits in the same way).

Choosing the Correct Statistical Procedure for the Nature of Research Data

An important rule to remember about statistics is to use appropriate statistical tools with the different kinds of data. In the following chapters we will learn about different methods for solving a particular problem using the approach that fits the available information. In this sense, statistics is like a collection of tools that we can use. We typically do not remove a screw with a sledge hammer (although I have been tempted at times!); we assess what particular tool works best with the screw that has to be removed. In the same way, we have to assess what statistical tool works best with the data we have available. Scales of measurement help us to classify the data in order to determine what the next steps might be to analyze it properly.

In real life, research data come in many forms: nominal, ordinal, interval, and ratio scales. It is hoped that you will gain familiarity with these as I discuss them in

DESCRIPTIVE STATISTICS—CENTRAL TENDENCY

this book. This primary step in statistics is often the one that confounds even experienced researchers. You cannot use certain methods with certain kinds of data—if you do, you will not get accurate or meaningful results. As I mentioned earlier, you cannot calculate a mean (you could, but it would be "meaningless") on nominal or ordinal data. Also, we shouldn't trust a mean calculated on the appropriate level of data (interval) if the data are of a certain kind (e.g., income, housing values, etc.). It takes practice to recognize the scale of measurement, but it is a step that cannot be missed.

DESCRIPTIVE STATISTICS—CENTRAL TENDENCY

Descriptive statistics include numerical and graphical procedures to assist the researcher to understand and see patterns in data. Typically, a researcher gathers data, which, unexamined, exists as a series of numbers with no discernable relationship. By using descriptive statistical techniques, the researcher can present the data in such a way that whatever patterns exist can be assessed numerically and visually.

The best way to understand these procedures is to begin with a real example. The following table of data is taken from the school database in Washington. The scores represent a sample of schools ($N = 40$) that report scores from fourth-grade students' state achievement test in 2009. Each case in Table 5.4 represents the aggregated scores for each school in which the given percentage of students passed the math achievement portion of the test.

Simply looking at the numbers is not the best way to understand the patterns that may exist. The numbers are in no particular order, so the researcher probably cannot discern any meaningful pattern. Are there procedures we can use to numerically understand these patterns?

Central tendency measures suggest that a group of scores, like those in Table 5.4, can be understood more comprehensively by using a series of numerical and graphical procedures. As these measures suggest, we can understand a lot about a set of

TABLE 5.4 Aggregated School Percentages of Students Passing the Math Standard

	Math Percent Met Standard		
38	59	35	50
37	74	73	79
46	50	69	89
63	62	66	50
51	25	24	42
30	53	34	73
36	63	40	56
50	72	58	10
40	50	49	56
41	77	28	27

data just by observing whether or not most of the scores cluster or build up around a typical score. That is, do the scores have a *tendency* to approach the middle from both ends? There will be scores spreading out around this central point (a topic explored in a later chapter), but it is helpful to describe the central point in different ways and for different purposes. The primary question the researcher asks here is, Can we identify a 'typical' score that represents most of the scores in the distribution? In the example above, the researcher needs to know what math achievement passing score is typical for this sample of schools.

Mean

Perhaps the most basic statistical analysis for numerically describing the central tendency is the mean, or arithmetic average of a set of scores. Remember from our discussion of the levels of data that the researcher needs at least interval data to create a mean score. This is because you need to be able to add, subtract, multiply, and divide numbers in order to calculate it. If you have less than interval data, it would not make sense to use these arithmetic operations since you could not assume the intervals between data points are equal. Thus, for example, you could not get an average gender (nominal) or an average opinion about the value of constructivist teaching (ordinal, unstandardized survey question).

Calculating the mean value uses one of the most basic formulas in statistics, the average: $\Sigma X/N$. This formula uses the "Σ" symbol, which means "sum of." Therefore, the average, or mean value, can be calculated by adding up the numbers, or summing them, and then dividing by how many numbers there are in the set: N. Using the values in Table 5.4, we can calculate the mean by summing the 40 numbers to get 2025. If we divide this number by 40, the amount of numbers in the set, we get 50.63.

$$\frac{\Sigma X}{N} = \frac{2025}{40} = 50.63$$

What does the mean of 50.63 indicate? If you inspect the data in Table 5.4, you will see that 10% of the students in one school passed the math assessment, while 89% of the students at another school passed. That is quite a difference! What is the *typical* percentage of students who passed the math assessment? That is, if you had to report one score that most typified all the scores, which would it be? This is the mean, or average value. It expresses a central value (towards the middle) that characterizes all the values.

Median

Another measure of central tendency is the median, or middle score among a set of scores. This isn't a calculation like the mean, but rather it identifies the score that lies directly in the middle of the set of scores when they are arranged large to small (or small to large). In our set of scores, the median is 50. If you were to rank order the set of scores by listing them small to large, you would find that the direct middle of the set of scores is between the twentieth (50) and twenty-first (50) numbers in the list. In order to identify the direct middle score, you would have to average these two numbers to get 50 [(50 + 50)/2]. An equal number of scores in the group of scores are above and below 50.

The median is important because sometimes the arithmetic average is not the most typical score in a set of scores. For example, if I am trying to find the typical housing value in a given neighborhood, I might end up with a lot of houses valued at a few hundred thousand dollars and five or six houses valued in the millions. If you added all these values up and divided by the number of houses, the resulting average would not really characterize the typical house because the influence of the million dollar homes would present an inordinately high value.

To take another example, the values in Table 5.5 are similar to those in Table 5.4, with the exception of seven values. In order to illustrate the effects of "extreme scores," I replaced each score over 70 with a score of 98. If you calculate an average on the adjusted values in Table 5.5, the resulting value is 54.35.

Changing seven values resulted in the mean changing from 50.63 to 54.35. But what happens to the median when we make this change? Nothing. The median remains 50, since it represents the middle of the group of scores, not their average value. In this case, which is the more *typical* score? The mean value registers the influence of these large scores, thereby "pulling" the average away from the center of the group. The median stays at the center.

This small example shows that only a few extreme scores can exert quite an influence on the mean value. It also shows that the median value in this circumstance might be the more typical score of all the scores since it stays nearer the center of the group. Researchers should be alert to the presence of extreme scores since they oftentimes strongly affect the measure of central tendency. This is especially true any time the values reflect money, like housing values, household income, and so on.

TABLE 5.5 Adjusted School Percentages

Math Percent Met Standard

38	59	35	50
37	98	98	98
46	50	69	98
63	62	66	50
51	25	24	42
30	53	34	98
36	63	40	56
50	98	58	10
40	50	49	56
41	98	28	27

Mode

The mode is the most frequently occurring score in a set of scores. This is the most basic of the measures of central tendency since it can be used with virtually any set of data. Referring to Table 5.4, you can arrange the scores and discover that 50 is the most frequently occurring score in the set. The mode is a typical score or category since data most often "mass up" around a central point, so it makes sense that the mode, at the greatest point of accumulation in the set, represents the most prevalent score.

Central Tendency and Levels of Data. Earlier, I stated that statistics are like tools and the researcher must use the most appropriate tool with the data available for their research. This is true with representing central tendency as well.

The mean is used with interval (or ratio) data since it is a mathematical calculation that requires equal intervals. The median and mode can be used with interval as well as "lower levels" of data (i.e., ordinal and nominal), whereas a mean cannot. Using either median or mode with interval data does not require a mathematical calculation; it simply involves rank ordering the values and finding the middle score or the most frequently occurring score, respectively. The mean cannot be used with ordinal or nominal data since we cannot use mathematical calculations involving addition, subtraction, multiplication, and division on these data, as I discussed above.

The median is a better indicator of central tendency than the mean with "skewed" or imbalanced distributions. We will have more to say shortly about skewed sets of scores, but for now we should recognize that a set of scores can contain extreme scores that might result in the mean being unfairly influenced and therefore not being the most representative measure of central tendency. Table 5.5 shows this situation as we discussed in the median section above. Even when the data are interval (as, for example, when the data are dealing with monetary value, or income), the mean is not always the best choice of central tendency despite the fact that it can use arithmetic calculations.

The mode, on the other hand, is helpful in describing when a set of scores fall into more than one distinct cluster ("bimodal distribution"). Consider Table 5.6, in which I "adjusted" just a few scores to illustrate the situation in which a set of scores has more than one most frequently occurring value. If you look closely, you will see that there are now two modes: 50 and 73.

In this situation, what is the most appropriate measure of central tendency? The data are interval, so we could calculate a mean. The mean for these adjusted scores is 50.80, just slightly higher than the mean (50.63) of the data in Table 5.4. However, would the resulting mean value truly be the most characteristic, or typical, score in the set of scores? No, because the scores in the set illustrated in Table 5.6 no longer cluster around a central point; they cluster around two central points.

Figure 5.2 shows how this looks graphically. As you can see, the math achievement values along the bottom axis show that there are two most common scores:

DESCRIPTIVE STATISTICS—CENTRAL TENDENCY 55

TABLE 5.6 Math Achievement Percentages for Bimodal Set

Math Percent Met Standard			
38	59	35	50
37	74	73	79
46	50	72	89
63	62	73	50
51	25	24	42
30	53	34	73
36	63	40	56
50	73	58	10
40	50	49	56
41	73	28	27

50 and 73. Clearly, this distribution does not look like a smooth normal curve. Rather, there are two distinct *humps* that make it problematic to define with just one characteristic value. In this case, if we reported that the mean value of 50.8, you can see that it is not the most characteristic score and would therefore be misleading.

These are real educational data. If we were using the data to describe the sample of schools' achievement, for example, we might report that the set of schools contain two separate groups. Several schools perform at around 50% met standard for achievement whereas another group performs around 74%. If this were the case, we might want to investigate the characteristics of the different groups of schools to see why they might perform so differently.

We will examine below the graphing procedures we used to create Figure 5.2. First, we will examine how to use Excel® and SPSS® to calculate central tendency values.

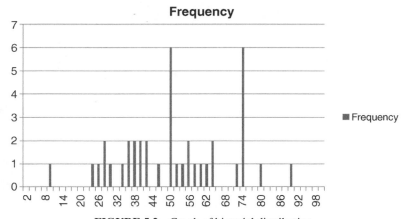

FIGURE 5.2 Graph of bimodal distribution.

USING EXCEL® AND SPSS® TO UNDERSTAND CENTRAL TENDENCY

Calculating measures of central tendency with Excel® and SPSS® is straightforward and simple.

Excel®

Starting with Excel®, you can see from Figure 5.3 that I created a simple spreadsheet listing the percentage values of school math achievement from Table 5.4. You cannot see all the values since there are 40 scores and they exceed one page on the spreadsheet.

Figure 5.3 shows the data listed with the heading, "Table 5.4 Percentages" in the "Data" menu screen. This screen contains the "Data Analysis" submenu that we discussed in Chapter 3. When you select Data Analysis, Excel® provides the list of tools that we can use to calculate central tendency values. This drop box, shown in Figure 5.3 overlaying the spreadsheet, lists the "Analysis Tools" in a separate menu. I selected "Descriptive Statistics" as shown in the highlighted section of Figure 5.3. Selecting Descriptive Statistics results in another drop box, as shown in Figure 5.4.

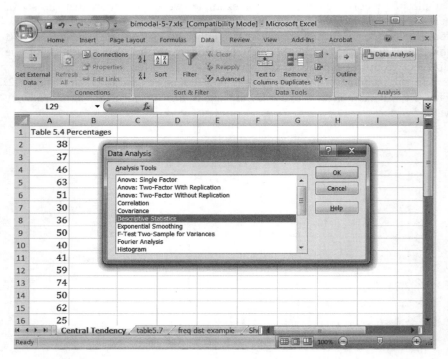

FIGURE 5.3 School achievement values for calculating central tendency.

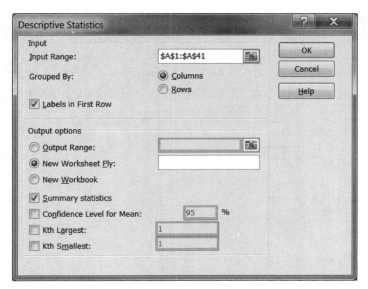

FIGURE 5.4 Descriptive statistics drop box for calculating central tendency.

There are several features to observe in Figure 5.4.

1. I selected the entire column of data, including the heading for the analyses, so I needed to "inform Excel®" not to include the heading in the analyses along with the percentages. You can see this selection in the "Labels in First Row" box near the top of the drop box.
2. The window to the right of the "Input Range:" row (second from the top) shows the cells that I selected in the spreadsheet. As you can see, the window shows "$A$1:$A$41"; this indicates that I chose the cells from A1 to A41 for my analysis. This includes the heading in A1 and then the percentage data from A2 to A41.
3. Near the middle of the drop box, the "New Worksheet Ply:" is selected. This instructs Excel® to present the results of the analyses on a separate worksheet within the spreadsheet. We could specify that the results be placed in the same worksheet by listing the appropriate location in the widow to the right of this selection.
4. I placed a check mark in the "Summary Statistics" option. This will create the central tendency calculations on a separate sheet within the spreadsheet.

When I select "OK," Excel® provides a separate worksheet with the results of the analysis I requested. This is shown in Figure 5.5, in which I renamed the active worksheet "Results." You can see a column of results, among which are the central tendency measures that were discussed above. The mean (highlighted in Figure 5.5) is 50.625, and the median and mode are both 50. There

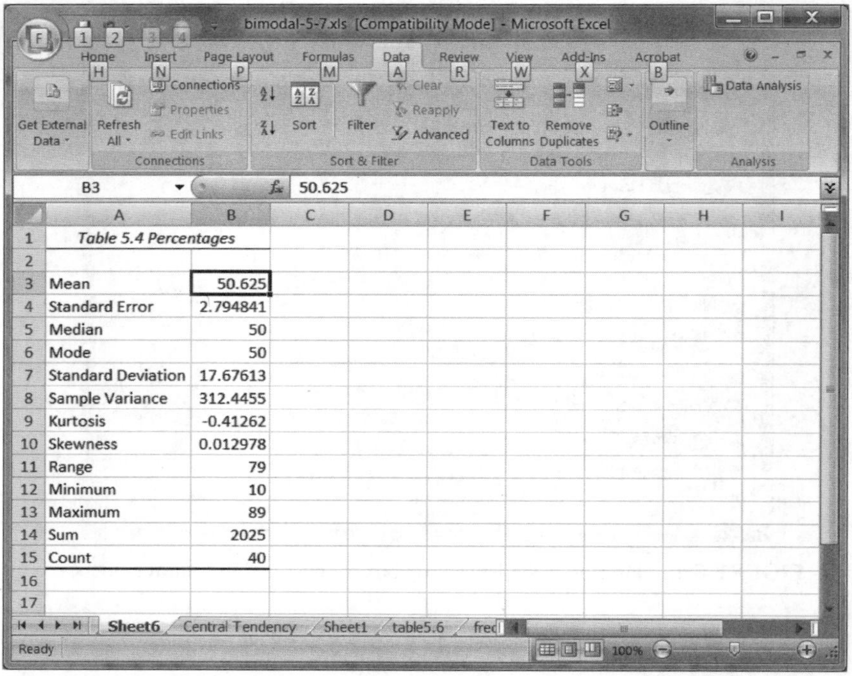

FIGURE 5.5 Descriptive statistics results worksheet.

are many other calculated values as well that we will discuss in subsequent chapters. You should note that Excel® does not make it clear if there is more than one mode. If you run the same analysis (descriptive statistics) using the data from Table 5.6, the results indicate that the mode is 50. No mention is made of the other mode of 73.

SPSS®

Providing the same results with SPSS® is just as easy as using Excel®. Figure 5.6 shows the screen in which I imported the data from Excel® to SPSS®. The variable label is now "Table54Percentages" since SPSS® does not use empty spaces in variable names (nor are most punctuation marks tolerated). In Figure 5.6 you can see that I selected the Analyze menu, and then I specified "Descriptive Statistics" and "Frequencies" in a secondary drop menu. You will also notice that there is another choice we could make if we wished to obtain descriptive statistics: "Descriptives," the choice immediately below Frequencies. Choosing Frequencies enables me to specify several more descriptive statistics, including the median and mode values, than the Descriptives submenu allows.

Figure 5.7 shows the SPSS® screens that result from choosing Frequencies. The first screen that appears allows you to specify which variable you wish to use in the analyses. Use the arrow in the middle of the screen to select and

USING EXCEL® AND SPSS® TO UNDERSTAND CENTRAL TENDENCY

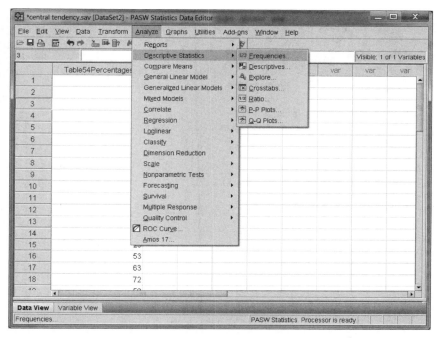

FIGURE 5.6 Descriptive frequencies menus in SPSS®.

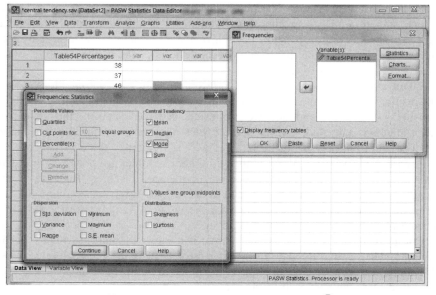

FIGURE 5.7 Frequencies submenus in SPSS®.

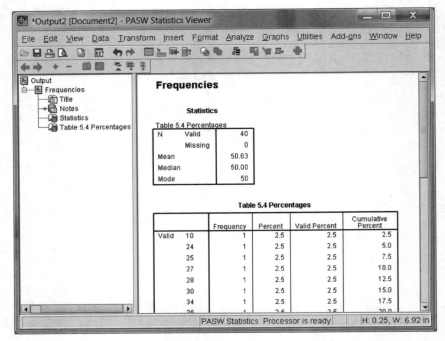

FIGURE 5.8 SPSS® frequency output.

move the variable into the "Variable(s)" window, as shown in Figure 5.7. The second screen comes from choosing "Statistics" in the Frequencies submenu. You can see this button just to the right of the "Variable(s)" screen. This second screen enables you to select Mean, Median, and Mode (among other analyses) as shown in Figure 5.7.

When you make these selections, SPSS® generates an output file showing the results of your request, as shown in Figure 5.8. The output file lists the measures of central tendency we requested. As you can see, the values of mean, median, and mode are the same values reported from the Excel® analysis. The additional output in SPSS® includes a frequency table in which each raw score value is listed along with the number of times it appears in the data ("Frequency") and the resulting percent of the set of scores. Because of the length of the SPSS® output, you cannot see all the raw score values, but the frequency table lists a frequency of 5 for the value 50, along with the resulting percent of the entire set of values (12.5%).

One further feature of the SPSS® output is a report on multiple modes. As an example, when I used the data from Table 5.6 (showing the bimodal values), the output report shows the table in Figure 5.9. Although the specific values of the multiple modes are not shown, the output includes a statement that there are multiple modes and the listed value is the lowest value.

Statistics

N valid	40
N missing	0
Mean	50.80
Median	50.00
Mode	50[a]

[a]Multiple modes exist. The smallest value is shown.

FIGURE 5.9 SPSS® output showing multiple mode report.

DISTRIBUTIONS

The purpose of descriptive statistics, like the measures of central tendency we have discussed thus far, is to explain the features of the set of values we have to work with as we seek to answer a research question. We used the sample of reading achievement values above to identify a typical score among a set of scores, for example. In the next chapter, we will explore visual ways of describing the same data.

Data in the real world present themselves to the researcher in patterned ways. We know from experience that people's attitudes and behaviors are fairly predictable. This doesn't mean that our actions are predetermined or fixed. Rather, it suggests that human beings approach life and experience in similar ways, with similar results. I discuss this assumption in more detail in my earlier book on the nature of patterns embedded in data (Abbott, 2010). In my classes, I stress the point that statistics cannot achieve certainty; it can only increase our understanding of uncertainty. Researchers focus their analyses at discovering the patterns of likelihood for certain actions and beliefs.

I will not belabor these points here, but I do want to point out that, because behavior and attitudes are somewhat predictable, we can study them scientifically. Part of this study is the recognition that data are typically "shaped" in recognizable patterns called *distributions*. Most people recognize the concept of "normal distribution," where data mass up around some middle value and taper off to the left and right of this value (or above and below it). A great many, if not most, human behaviors and attitudes are characterized by this distribution. (I will discuss the features of the normal distribution much more comprehensively in later chapters.)

Not all data are normally distributed, however. Although the values distributed are patterned and predictable, they often take many different shapes depending on

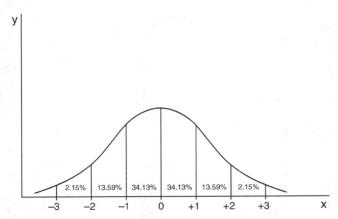

FIGURE 5.10 Example of the normal distribution of values.

what they measure. For example, a "Poisson" distribution, which measures the likelihood of rare events with large numbers of people (e.g., rare diseases across the United States), does not always have the appearance of the normal, "bell"-shaped distribution. There are many other distributions that you can study in advanced mathematical statistics texts if you are interested. In this course, we will base our statistical analyses on the normal distribution.

Think of the distribution as values being "dealt out" along some scale of values. The more values there are to be dealt out, the more the values pile up around a central value with fewer values falling above and below the central point. Figure 5.10 shows an example of a normal distribution. We will examine this in greater detail later, but for now you can see the "hump" in the middle with "tails" in the ends of the distribution of values. When we calculated the mean, median, and mode values earlier, we were calculating the value of the distributed values that characterized most of the values. If all of these values are the same, we can speak of a "perfect" normal distribution where there are known percentages of values distributed under the curve. If they are the same, then there would be only one mode, the median would be the exact middle of the set of scores, and the mean value would not reflect extreme scores that would imbalance the set of scores in one or the other direction.

DESCRIBING THE NORMAL DISTRIBUTION

As we will see, we can learn a great deal about the normal distribution by seeing it graphically with a set of actual data. I will show how to create a visual distribution by using the frequency information calculated in SPSS®. Before we enter that discussion, however, I want to point out several ways to describe the normal distribution using numerical calculations.

Central Tendency

We have already discussed these calculations (i.e., mean, mode, and median), but there are other ways to capture a characteristic central point among a set of values. In advanced statistics courses, you might learn some of these additional measures of central tendency. You may learn to create the "harmonic mean," the "geometric mean," or the "quadratic mean," depending on the nature of your data. In this course, we will describe how the data from a distribution of scores "clusters up" using the arithmetic mean.

Skewness

Skewness is a term that describes whether, or to what extent, a set of values is not perfectly balanced but rather *trails off* to the left or right of center. I will not discuss how to calculate skew, but it is easy to show. If I use the data from Table 5.6, with some additional adjustment of scores, you can see from Figure 5.11 that the resulting distribution of scores looks imbalanced; in this case, many of the scores trail to the right of the mean. In this case, we can say that the distribution is positively skewed (SK+) since the values trail to the right. The distribution would be negatively skewed (SK−) if the scores trailed away to the left.

Both Excel® and SPSS® provide calculations for Skewness. Figures 5.12 and 5.13 show the output from Excel® and SPSS®, respectively, that includes the skewness figures. In both cases, the skewness value is 0.43.

The interpretation of the skewness value helps to understand the shape of the distribution. A skewness value of zero indicates perfect balance, with the values in the distribution not trailing excessively to the left or right of the center. Positive values indicate values trailing to the right, and negative values indicate left trailing. As you can see in both Figures 5.12 and 5.13, the value of 0.43 is positive and therefore the values will trail to the right (as shown in Figure 5.11).

The important question researchers must ask is, How far from zero is considered excessive? That is, how big can the skewness number get with the distribution still retaining the general shape of a normal distribution? The SPSS® output in

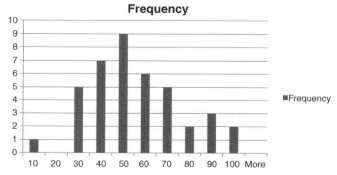

FIGURE 5.11 Illustration of a positively skewed distribution.

	Column1
Mean	52.18
Standard error	3.21
Median	50.00
Mode	50.00
Standard deviation	20.32
Sample variance	413.07
Kurtosis	−0.08
Skewness	0.43
Range	88.00
Minimum	10.00
Maximum	98.00
Sum	2087.00
Count	40.00

FIGURE 5.12 Excel® output showing descriptive statistics including skewness.

Figure 5.13 is more helpful because it provides the "Standard Error of Skewness." We will discuss this value in later chapters, but for now you can establish a general rule of thumb for excessive skew by dividing the skewness value by the Standard Error of Skewness, resulting in a skewness *index*. If the resulting index number does not exceed two or three (positive or negative values), depending on the number of values in the distribution, the distribution is considered normal and balanced. You can use the sign (positive or negative) of the skewness value to indicate which way the skew tends, but the index magnitude indicates whether or not the skewness is excessive.

Statistics

N valid	40
N missing	0
Mean	52.1750
Median	50.0000
Mode	50.00
Skewness	0.430
Standard error of skewness	0.374
Kurtosis	−0.077
Standard error of kurtosis	0.733

FIGURE 5.13 SPSS® output showing descriptive statistics including skewness.

Using the reported figures in Figure 5.13, you can see that this procedure results in a skewness index of 1.15. Therefore, even though the distribution *looks* imbalanced (see Figure 5.11), it can be considered normally distributed for statistical purposes.

I mentioned earlier that the number of values in the distribution affects this skewness index. While this will become clearer later on, I mention here that the Standard Error of Skewness reported by SPSS® will be *smaller* with larger numbers of values in the distribution. So, large datasets (200–400) might have very small Standard Error of Skewness numbers and result in the overall skewness index being very *large* (since dividing by a smaller number yields a larger result). Small datasets (30–50) will typically have large Standard Error of Skewness numbers with resulting small skewness indices.

In light of these issues, the researcher needs to consider the *size of the distribution* as well as the *visual evidence* to make a decision about skewness. In the example in Figures 5.12 and 5.13, a skewness number of 0.43 seems small, since it has a Standard Error of Skewness of .34 and would therefore yield the index of 1.15 (within our ±2 to 3 rule of thumb). However, if we were to add several more cases to the distribution, the Standard Error of Skewness would likely shrink resulting in a larger skewness index. If we use a large dataset, I might view the visual evidence alone as a better measure of overall balance. Smaller datasets are more problematic, even though the skewness indexes are within normal bounds. Use both the visual and numerical evidence to help you decide upon the overall shape of skewness.

There is another way to help assess the extent of skewness using the three measures of central tendency we discussed (mean, median, and mode). If a distribution of scores is "balanced," with most of the scores massing up around a central point, then the mean, median, and mode will all lie on the same point. The mean is typically the most "sensitive" indicator, and it will get pulled toward the direction of the skew more readily than the median and the mode. You can use both the numerical results and the visual inspection to see if this method helps.

The output shown in Figure 5.5 indicates that the mean is 50.63 while both median and mode are 50. While there is a slight discrepancy, the three values are very close to one another. Compare these results with those reported in Figures 5.12 and 5.13 in which the mean is 52.18 and the median and mode are unchanged. This report, along with the visual evidence in Figure 5.11, indicates a likely positive skew to the distribution.

Kurtosis

Kurtosis is another way to help describe a distribution of values. This measure indicates how "peaked" or flat the distribution of values appears. Distributions where all the values cluster tightly around the mean might show a very high point in the distribution since all the scores are pushing together and therefore upward. This is known as a "leptokurtic" distribution. Distributions with the opposite dynamic, those with few scores massing around the mean, are called "platykurtic" and appear flat. "Perfectly" balanced distributions show the characteristic pattern like the distribution in Figure 5.10, being neither too peaked nor too flat.

Making a determination of the extent of kurtosis is similar to the method with evaluating skewness. You can see from Figures 5.12 and Figure 5.13 that both Excel® and SPSS® report kurtosis and standard error of kurtosis figures. Like skewness, dividing the kurtosis by the standard error of kurtosis (creating a kurtosis index) will provide a helpful measure for interpretation. You can use the same ±2 or 3 rule such that resulting kurtosis index values greater than 2 or 3 are considered out of normal bounds. Looking at the kurtosis values indicate which direction this might take: Positive kurtosis values that are excessive are considered leptokurtic, whereas negative kurtosis values that are excessive are platykurtic. It is also important to use visual evidence as it is for determining skewness.

DESCRIPTIVE STATISTICS—USING GRAPHICAL METHODS

Up to now, we have discussed *numerical* ways of deciding upon the shape of distributions. As I mentioned in the discussion of skewness above, it is also important to be able to describe data *visually*. Many students and researchers, as well as consumers of statistical reports, are visual learners. Therefore, it is important to be able to inspect and analyze the distribution of values by visual means in order to understand better the nature of the data and to effectively communicate the results.

The simplest way to visually describe data is simply to rank order it from high to low, or from low to high. Beyond this are several ways of displaying data to see its underlying patterns. The frequency distribution is the most commonly accepted way of showing the array of data.

Frequency Distributions

Frequency distributions are graphs that literally describe the distribution by showing groups of values. We can use the values in Table 5.4 to demonstrate the method of creating a frequency distribution and how to view the results. Table 5.7 shows these values (because of the size of the table, Table 5.7 shows the values in two panels).

Table 5.7 shows the values in three columns. The "Unordered" column shows the values without ranking or other ordering processes. The "Ordered" column shows the same values ranked from smallest to largest values. The third column, "Freq. (10)" shows the values in groups of 10. That is, the numbers in this column indicate how many values in the ordered list fit into groups of values of 10. Thus, there is one value ("10") that lays in the group between 10 and 20; there are four values (24, 25, 27, 28) that lay in the group between 21 and 30, six values between 31 and 40, and so on. Just visually moving from the top to the bottom of the Freq. (10) column gives you an idea of the shape of the distribution: 1, 4, 6, 6, 11, 5, 6, 1. As you can see, there are more values in the middle than on either end, and the values increase up to the middle and decrease thereafter. This is the numerical shape of the distribution.

By the way, Figure 5.8 shows the SPSS® frequency output of a range of values like ours in the current example. As you can see in the output, SPSS® reports the

DESCRIPTIVE STATISTICS—USING GRAPHICAL METHODS

TABLE 5.7 Creating Frequency Groups for Visual Description

Unordered	Ordered	Freq. (10)	Unordered	Ordered	Freq. (10)
38	10	1	49	62	
37	24		28	63	
46	25		50	63	
63	27		79	66	
51	28	4	89	69	5
30	30		50	72	
36	34		42	73	
50	35		73	73	
40	36		56	74	
41	37		10	77	
59	38	6	56	79	6
74	40		27	89	1
50	40				
62	41				
25	42				
53	46				
63	49	6			
72	50				
50	50				
77	50				
35	50				
73	50				
69	51				
66	53				
24	56				
34	56				
40	58				
58	59	11			

percent of the total set of numbers that each value represents. Thus, there is only one value of "10" in the set of values, and it alone represents 2.5% of all the values. Each of the remaining values are shown along with the percent of the total set of cases that equal that particular value.

Histograms

We can use the graphical capabilities of Excel® and SPSS® to show a group of values in a graph. The easiest way is to use the "histogram" or graph of the values we created in the frequency distribution.

Before we do this, we must first decide the size of the groups within which we will place the values, or the group *interval*. *The decision of what size group intervals to use will affect the visual results of the data.* I used intervals of 10 in the example in Table 5.7. Since the data represent percentages, I thought it would be appropriate to use intervals of 10 percentage points. I could just as easily have used intervals of five percentage points, or some other number. While there is no objective standard for which intervals to use, you need to have some rationale before you create the histogram.

Figure 5.14 shows the first step in creating a histogram in Excel®. As you can see, I selected the Data Analysis button from the Data menu. This results in a separate menu in which I can select Histogram. First, however, I added another column called "Bins." This column is where I specify and list the group intervals. As you can see, I created groups in intervals of 10 percentage points: from 10% to 20% to 30%, and so on. The Histogram procedure requires that we provide this to specify the histogram that will be produced. I like to think of these as "buckets," since they are intervals, or receptacles, in which we are "throwing" our values.

Choosing Histogram from the "Data Analysis" menu results in the "Histogram" sub-menu shown in Figure 5.15. As you can see, I specified cells A1:A41 in the "Input Range:" to indicate the table of data and specified cells B1:B11 to indicate the "Bin Range:" Be sure to check the "Labels" box if you include the variable labels (A1 and B1 in the example in Figure 5.14). Otherwise, the procedure will not run; only numbers can be used in the procedure.

Figure 5.16 shows the frequency distribution that Excel® returns (to a separate sheet in the overall spreadsheet). You will see that the values from the data table have been placed in the bins and that the frequency column specifies how many values are in each bin. If you do not specify bins, Excel® will return the values in evenly distributed bins between the highest and lowest values in the dataset.

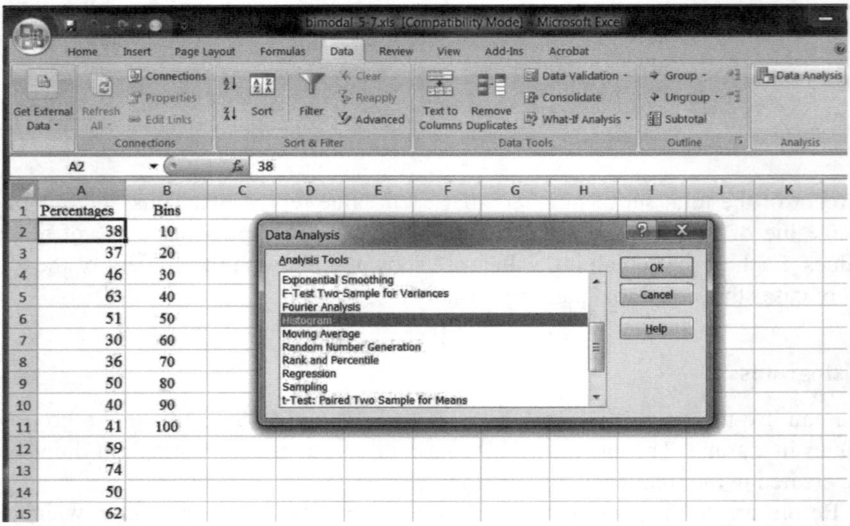

FIGURE 5.14 Excel® output showing the Histogram specification and the data columns.

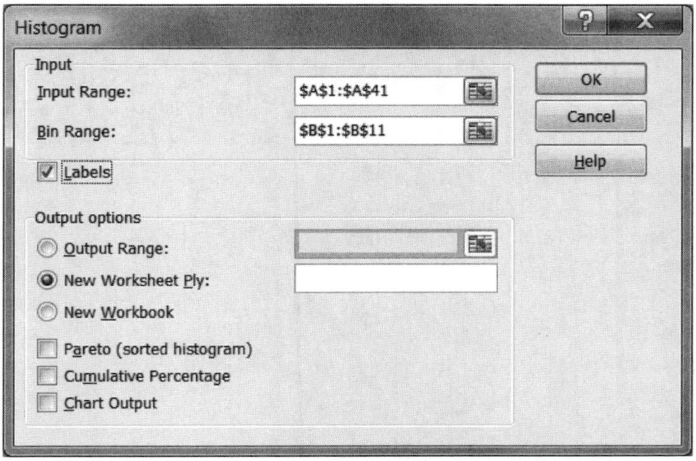

DESCRIPTIVE STATISTICS—USING GRAPHICAL METHODS

FIGURE 5.15 Excel® output showing the Histogram sub-menu.

Note that the frequencies in Figure 5.16 are a bit different than those I specified earlier (1, 4, 6, 6, 11, 5, 6, 1) due to the different method for viewing the intervals. Excel® placed one value in the first bin (1–10), but none in the second (11–20), and so on. I placed one value in the first bin (10–19) and four values in the second bin (20–29), and so on. This difference underscores the necessity for carefully specifying how you want the data treated. Again, there are no formal rules; just make sure you use some criterion for creating the bins. You can adjust the histogram when it is created, but it is easier to do it at the outset by carefully designing the intervals.

Once the frequency distribution is created, you can create the histogram by choosing Insert on the main menu and Column on the submenu as shown in Figure 5.17. This results in a sub-menu of several choices for the style of

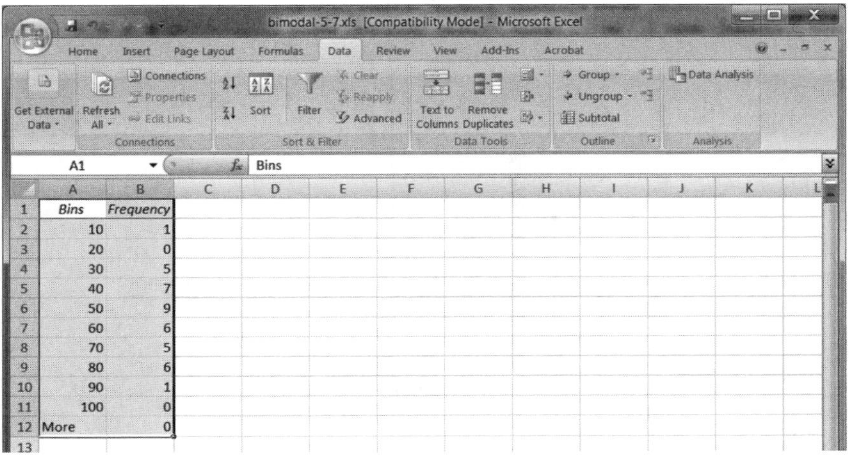

Bins	Frequency
10	1
20	0
30	5
40	7
50	9
60	6
70	5
80	6
90	1
100	0
More	0

FIGURE 5.16 Excel® output showing the frequency distribution.

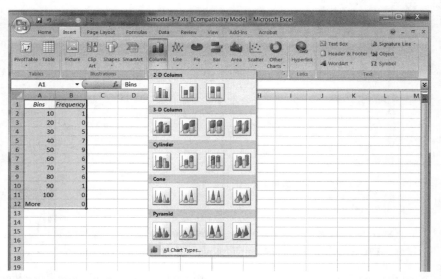

FIGURE 5.17 Excel® output showing how to insert a graph using the frequency distribution.

histogram you wish to create. I ordinarily choose the ordinary bar graph (the first choice under the "2-D Column" group shown in Figure 5.17).

Making this choice results in the histogram shown in Figure 5.18. You can see the general shape of the distribution is roughly normal, with a few minor exceptions. Taken together with the numerical analyses discussed in the earlier sections (skewness and kurtosis values, size and similarity of mean, median, and mode), this indicates that the data in this set are normally distributed.

You can also create histograms in SPSS® by using the Graphs menu. When you select Graphs, SPSS® returns a series of submenus as shown in Figure 5.19. As you can see, Graphs allows you to choose from a series of graph templates, among

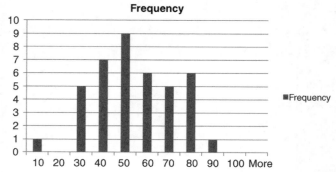

FIGURE 5.18 Excel® histogram generated from the frequency distribution.

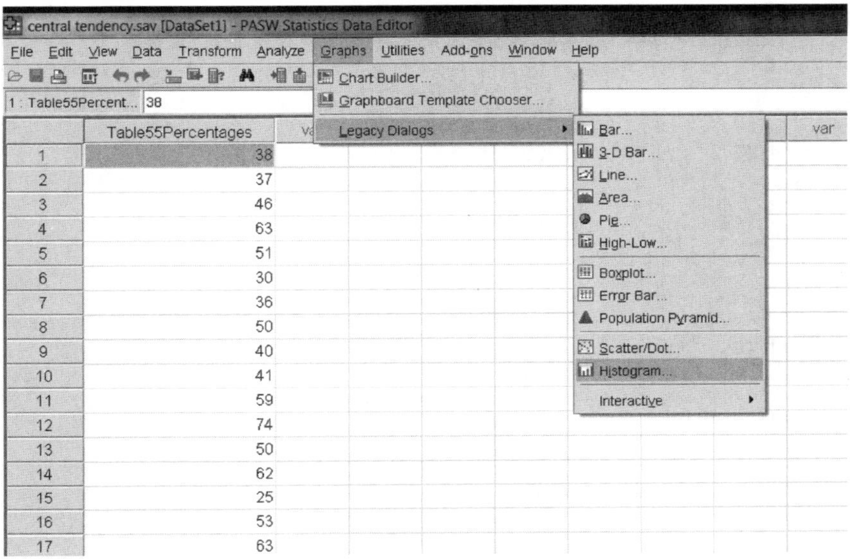

FIGURE 5.19 SPSS® procedure for creating the histogram.

which is "Legacy Dialogs." This gives you a list of classical graph designs including the Histogram. Choosing this option results in the dropbox appearing in which you can specify the features of the histogram, as shown in Figure 5.20. As you can see, you can also call for a line showing the normal curve by checking the box called "Display normal curve."

The final result of these procedures is shown in Figure 5.21, an SPSS® output file including the histogram. You can double-click on the histogram and copy it so that you can include it in a word processing file or other report format. You can also change the format features (axis intervals, titles, etc.) by double-clicking in the graph area. When you do this, you are presented with several dialog boxes that include several tools in which you can produce custom histograms.

TERMS AND CONCEPTS

Bimodal A set of data with two modes.

Central Tendency One of the ways to describe a set of data is to measure the way in which scores "bunch up" around a mean, or central point of a distribution. Mean, median, and mode are typical measures of central tendency.

Descriptive Statistics The branch of statistics that focuses on measuring and describing data in such a way that the researcher can discover patterns that might exist in data not immediately apparent. In these processes, the researcher does

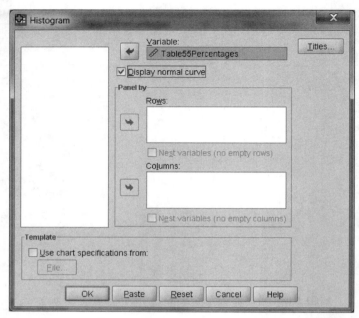

FIGURE 5.20 SPSS® procedure for specifying the features for the histogram.

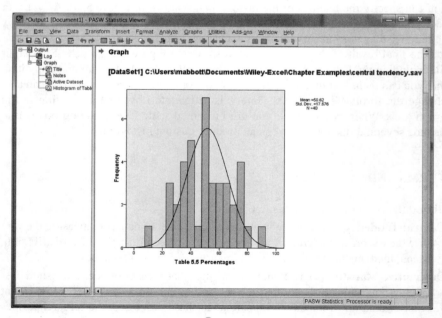

FIGURE 5.21 SPSS® histogram in the output file.

not attempt to use a set of data to refer to populations from which the data may have been derived, but rather to gather insights on the data that exist at hand.

Data Distributions The patterned shape of a set of data. Much of statistics for research uses the "normal distribution," which is a probability distribution consisting of known proportions of the area between the mean and the continuum of values that make it up.

Frequency Distributions Tabular representations of data that show the frequency of values in groupings of data.

Histograms Graphical representations of frequency distributions. Typically, these are in the form of "bar graphs" in which bars represent groups of numerical values.

Inferential Statistics The branch of statistics that uses procedures to estimate and make conclusions about populations based on samples.

Interval Data Data with the qualities of an ordinal scale, but with the assumption of equal distances between the values and without a meaningful zero value. Typically, researchers use standardized test values (e.g., IQ scores) as an example of an interval scale in which the difference between IQ scores of 100 and 101 is the same as the distance between IQ scores of 89 and 90. This assumption allows the researcher to use mathematical properties (i.e., adding, subtracting, multiplying, and dividing) in their statistical procedures. The lack of a meaningful zero (e.g., what does an IQ of "0" mean?) is typically exemplified by the "0" on a Fahrenheit scale not referring to the "absence of heat" but rather simply to a point in the overall temperature scale.

Kurtosis The measurement of a distribution of scores that determines the extent to which the distribution is "peaked" or flat. Data distributions that have excessive kurtosis values may be overly peaked ("leptokurtic") or overly flat ("platykurtic").

Mean Average value in a set of data.

Median Middlemost score in a set of data.

Mode The most commonly occurring value in a set of data.

Nominal Data Data that exist as mutually exclusive categories (e.g., home schooling, public schooling). These data can also refer to "categorical data," "dichotomous variables" (when there are two naturally occurring groups like male/female), or "dichotomized variables" (when two categories are derived from other kinds of data like rich/poor).

Ordinal Data Data that exist in categories that are ranked or related to one another by a "more than/less than" relationship like "strongly agree, agree, disagree, strongly disagree."

Ratio Data Interval data with the assumption of a meaningful zero constitute ratio data. An example might be the amount of money that people have in their wallets at any given time. A zero in this example is meaningful!! The zero allows the researcher to make comparisons between values as "twice than" or "half of" since the zero provides a common benchmark from which to ground the comparisons. Thus, if I have $2 in my pocket and you have $4, I have half of your amount.

Scales of Measurement The descriptive category encompassing different classes of data. Nominal, ordinal, interval, and ratio data differ according to the information contained in their scales of values. Also known as "Levels of Measurement."

Skewness A measurement of a data distribution that determines the extent to which it is "imbalanced" or "leaning" away from a standard bell shape (in the case of a normal distribution).

Spuriousness A condition in which an assumed relationship between two variables is explained by another variable not in the analysis.

REAL-WORLD LAB I: CENTRAL TENDENCY

Understanding educational strategies and outcomes is a challenge for educators and practitioners. What explains academic achievement? What impact does a changing school structure have on student outcomes? These and other questions are routinely investigated in the attempt to clarify why some schools are effective and others ineffective in promoting student learning.

I have worked across the country with different programs designed to improve student outcomes like achievement, retention, and college readiness, among many others. Each of these programs is well-intentioned, but yields differential results. Funding agencies and recent government programs focus on different features of school experience and structure likely to yield positive results for students. I cannot hope to review all of these here, but I refer readers to my earlier cited book (Abbott, 2010) as (a) a general summary of the state of these programs and (b) a recommendation for ways of viewing their success.

This Lab uses data from the state of Washington[1] that identifies several outcomes and characteristics of its schools. Given the large size of the database, we will use a random sample of schools ($N = 40$) to clarify the nature and use of central tendency.

Using the Excel® database in Table 5.8 on 40 schools, conduct the following analyses using Excel® and/or SPSS®:

1. Calculate mean, mode, and median for the following variables:
   Reading percent met standard
   Math percent met standard
2. Report skewness and kurtosis figures.
3. Based on your response to #1 and #2 above, discuss the shape of the distributions of these variables.

[1] The data are used with the permission of the Office of the Superintendent of Public Instruction.

REAL-WORLD LAB I: SOLUTIONS

TABLE 5.8 School Data Sample for Lab 1

Reading Percent Met Standard	Math Percent Met Standard	Reading Percent Met Standard	Math Percent Met Standard
73	38	69	35
69	37	82	73
76	46	81	69
69	63	83	66
66	51	50	24
62	30	54	34
59	36	70	40
81	50	82	58
74	40	85	49
73	41	47	28
80	59	85	50
88	74	90	79
64	50	85	89
83	62	70	50
68	25	70	42
79	53	73	73
81	63	63	56
88	72	37	10
77	50	73	56
85	77	59	27

4. Create a histogram for "Reading Percent Met Standard."
5. Provide a summary or interpretation of the findings.

REAL-WORLD LAB I: SOLUTIONS

1. Calculate mean, mode, and median for the following variables:
   Reading percent met standard
   Math percent met standard
2. Report skewness and kurtosis figures.

Results

Excel® data analysis procedures generate the output shown in Table 5.9, and Table 5.10 shows the SPSS® results. The shaded cells in Table 5.9 show the descriptive statistics required to respond to #1 and #2 above. The mean, median, and modes for reading and math are listed along with skewness and kurtosis figures. Note, however, that the SPSS® results in Table 5.10 indicate multiple modes, whereas the Excel® results in Table 5.9 do not. Recall in our discussion that Excel® does not report multiple modes, so it is important to check both sets of

TABLE 5.9 Excel® Descriptive Statistics for Lab 1

Reading Percent Met Standard		Math Percent Met Standard	
Mean	72.48	Mean	50.61
Standard error	1.90	Standard error	2.79
Median	73.20	Median	50.00
Mode	72.70	Mode	50.00
Standard deviation	12.00	Standard deviation	17.68
Sample variance	144.10	Sample variance	312.44
Kurtosis	0.80	Kurtosis	−0.41
Skewness	−0.94	Skewness	0.01
Range	52.60	Range	78.80
Minimum	37.10	Minimum	9.70
Maximum	89.70	Maximum	88.50
Sum	2899.30	Sum	2024.20
Count	40.00	Count	40.00

results to ensure the variable is not multimodal. Since we are using continuous data, there may be instances of the same values occurring only once or twice, thereby indicating multiple modes, but these may not represent distorted (non-normal distributions). The visual evidence from the graphs will help to confirm this.

3. Based on your response to #1 and #2 above, discuss the shape of the distributions of these variables.

Results

The skewness and kurtosis figures for both variables indicate normally distributed values, although the Excel® output in Table 5.9 does not report the standard error values for each measure. Judging from the values shown in Table 5.9, the variables appear to be "in bounds" of a normal distribution. The mean, median, and mode

TABLE 5.10 SPSS® Descriptive Statistics for Lab 1

	Reading Percent Met Standard	Math Percent Met Standard
N valid	40	40
N missing	0	0
Mean	72.48	50.61
Median	73.20	50.00
Mode	70[a]	50
Skewness	−0.940	0.008
Standard error of skewness	0.374	0.374
Kurtosis	0.797	−0.412
Standard error of kurtosis	0.733	0.733

[a]Multiple modes exist. The smallest value is shown.

REAL-WORLD LAB I: SOLUTIONS

TABLE 5.11 Skewness and Kurtosis Indexes for Reading and Math Data

	Reading	Math
Skewness	−0.940	0.008
Standard error of skewness	0.374	0.374
Index of skewness	2.51	0.021
Kurtosis	0.797	−0.412
Standard error of kurtosis	0.733	0.733
Standard Index of kurtosis	1.09	0.562

values for reading appear to be similar, indicating a balanced distribution. The results for math also appear balanced using the same criterion.

The SPSS® output in Table 5.10 adds the additional information of the standard errors for skewness and kurtosis. As I discussed earlier, if you divide the skewness and kurtosis values by their respective standard errors, you can get a more precise picture of whether or not the variables appear to be normally distributed. Table 5.11 shows the results of creating indexes by making these calculations.

The shaded values show the skewness and kurtosis indexes for reading and math. For reading, the skewness index is large, but it does not exceed 3, which is within the boundaries of our criterion of a normally distributed variable. Note that the skewness value (−0.940) is negative, indicating a tendency toward a skew to the left, but still within normal bounds. The kurtosis value for math (−0.412) is negative, indicating a tendency toward being flat, but the index (0.562) shows that it is within normal bounds. All the other indexes (for both reading and math) are within the boundaries we discussed earlier for normal distributions.

4. Create a histogram for "Reading Percent Met Standard."

Figure 5.22 shows the Excel® histogram for reading, and Figure 5.23 shows the SPSS® histogram. In both cases, the distribution appears to be negatively skewed. This follows the numerical trends described in Table 5.11.

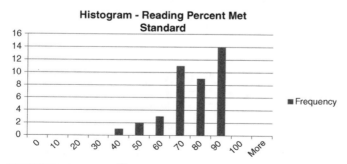

FIGURE 5.22 Excel® histogram for reading percent met standard.

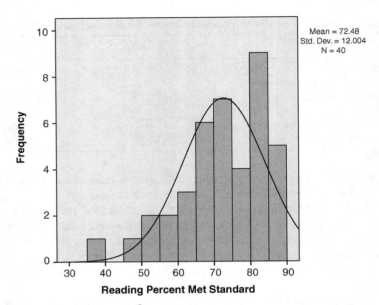

FIGURE 5.23 SPSS® histogram for reading percent met standard.

5. Provide a summary or interpretation of the findings.

Recall that with fewer cases, a distribution of scores may have a large standard error, thus creating a small skewness figure and therefore a small skewness index (i.e., within bounds). In this case, we might combine the evidence of the visual data provided by histograms to help make the determination about whether the variable is normally distributed. Taken together, the numerical and visual evidence suggest that reading percent met standard in this sample of data

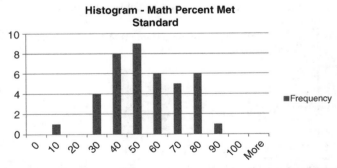

FIGURE 5.24 Excel® histogram for math percent met standard.

REAL-WORLD LAB I: SOLUTIONS

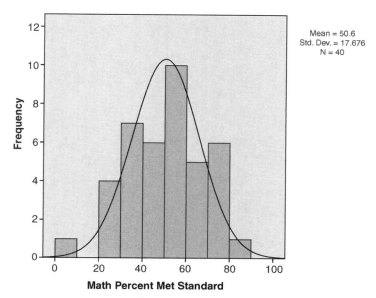

FIGURE 5.25 SPSS® histogram for math percent met standard.

may be slightly negatively skewed, even though the numerical data suggest that it is normally distributed. We should keep this in mind if we use this variable in subsequent statistical analyses.

By contrast with the reading results, Figures 5.24 and 5.25 show the Excel® and SPSS® histograms (respectively) for math percent met standard. You can see how this variable appears normal, which matches the numerical results.

6

DESCRIPTIVE STATISTICS—VARIABLITY

In Chapter 5, we examined several aspects of central tendency in the attempt to describe a set of data. In addition to mean, median, and mode, we discussed skewness and kurtosis as measures of the balance of a distribution of values. Excel® and SPSS® provided visual descriptions as well as numerical results to help make assessments of the likely normal distribution of our data. The frequency distribution and histogram are effective ways of communicating the shape and features of the distribution.

I will continue to explore descriptive statistics in this chapter. This time, we will examine the extent to which scores spread out from the mean of a distribution of values. It is important to understand the characteristic score or value of a distribution, as we saw with central tendency, but it is also important to understand the extent of the *scatter* of scores away from the center. How far away do scores fall, and what is the average distance of a score from the mean? The answers to these and similar questions will help us to complete our description of the distribution of values.

This chapter thus deals with *variability* or *dispersion* of scores. Several measures of variability are important to grasp since we will use them throughout the book in virtually all the statistical procedures we cover.

Understanding Educational Statistics Using Microsoft Excel® and SPSS®. By Martin Lee Abbott.
© 2011 John Wiley & Sons, Inc. Published 2011 by John Wiley & Sons, Inc.

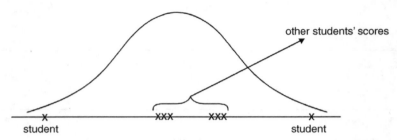

FIGURE 6.1 The characteristics of the range.

RANGE

The first way to measure variability is the simplest. The range is simply the numerical difference between the highest and lowest scores in the distribution and represents a helpful global measure of the spread of the scores. But remember it is a global measure and will not provide extensive information. Look at Figure 6.1. If two students score on the extremes of the distribution of achievement scores, and everyone else scores near the mean, the range will provide a distorted view of the nature of the variation. The range represents the "brackets" around the scores, but it cannot tell you, for example, how far the "typical" score might vary from the mean.

Nevertheless, the range contains important information. It provides a convenient shorthand measure of dispersion and can provide helpful benchmarks for assessing whether or not a distribution is generally distributed normally (I will develop this benchmark later in this chapter).

There are several ways of "dividing up" the distribution of scores in order to make further sense of the dispersion of the scores. Some of these are rarely used, but others are quite important to all the procedures we will explore in this book.

PERCENTILE

The percentile or percentile rank is the point in a distribution of scores below which a given percentage of scores fall. This is an indication of *rank* since it establishes a score that is above the percentage of a set of scores. For example, a student scoring in the 82nd percentile on a math achievement test would score above 82% of the other students who took the test.

Therefore, percentiles describe where a certain score is in relation to the others in the distribution. The usefulness of percentiles for educators is clear since most schools report percentile results for achievement tests. The general public also sees these measures reported in newspaper and website reports of school and district progress.

Statistically, it is important to remember that percentile ranks are ranks and are therefore not interval data. Figure 6.2 shows the uneven scale of percentile scores along the bottom axis of values in the frequency distribution.

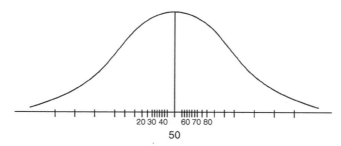

FIGURE 6.2 The uneven scale of percentile scores.

As Figure 6.2 shows, the distance between the 30th and 40th percentiles is not the same as the distance between the 10th and 20th percentiles, for example. The 30th to 40th percentile distance is much shorter than the 10th to 20th percentile distance. That is because the bunching up of the total set of scores around the mean and the tailing off of the scores toward either end of the frequency distribution result in uneven percentages along the scale of the distribution. Thus, a greater percentage of the scores lay closer to the mean, and a lesser percentage of the scores lay in the tails of the distribution.

Many educational researchers have fallen into the trap of assuming that percentiles are interval data and using them in statistical procedures that require interval data. The results are somewhat distorted under these conditions since the scores are actually only ordinal data. The U.S. Department of Education developed the Normal Curve Equivalent (NCE) score as a way of standardizing the percentile scores. This results in a scale of scores along the bottom axis of the frequency distribution that have equal distances between values. This transforms rank scores to interval scores enabling the educational researcher to use the values in more powerful statistical procedures.

SCORES BASED ON PERCENTILES

Education researchers use a variety of measures to help describe how scores relate to other scores and to show rankings within the total set of scores. The following are some of these descriptors based on percentiles:

Quartiles. These are measures that divide the total set of scores into four equal groups. This is accomplished by using three "cutpoints" or values that create the four groups. These correspond to the 25th, 50th, and 75th percentiles.

Deciles. These break a frequency distribution into ten equal groups using nine cutpoints (the 10th, 20th, 30th, 40th, 50th, 60th, 70th, 80th, and 90th percentiles). They are called deciles since they are based on groups of 10 percentiles.

Interquartile Range. These scores represent the middle half of a frequency distribution since they represent the difference between the first and third quartiles (the 75th minus the 25th percentiles). This is a global descriptor for the variability of a set of scores since the other half of the scores would reside in both of the tails of the distribution.

USING EXCEL® AND SPSS® TO IDENTIFY PERCENTILES

Through much of this chapter, I will use a sample of the TAGLIT (Taking a Good Look at Instructional Technology) database.[1] These data come from the 2003 nationwide survey of elementary and secondary teachers, students, and school administrators regarding the potential impacts of instructional technology on classroom outcomes. Originally developed by educators in the University of North Carolina, these data are currently housed and operated by T.E.S.T. Inc.

The TAGLIT measure is comprised of a series of instruments administered to different groups within schools in the attempt to understand the extent of technological literacy and the impact that technology might have on different student, teacher, and school outcomes. I combined some of the original data into "factors" that express the following aspects of middle- and high-school teacher experience with technology:

- The technology skill levels of teachers in secondary schools.
- How often technology applications were used in learning.
- What access to technology did teachers have?
- What technology support was available for teachers?

The database I will use consists of a random sample of teachers from forty schools. There are several variables in this database, but I will only use one, student-to-teacher ratios, to demonstrate how to use Excel® and SPSS® to generate percentiles.

Figure 6.3 shows the Excel® data screen in which I selected the statistical functions from the More Functions menu. You can see that I selected the "PERCENTILE" function, which allows me to specify a given percentile with these student-to-teacher ratios ("ratio"). Note that I selected a cell out of the range of my data ("DK2") so that the percentile I choose will be returned outside the data field.

For this example, I will choose to identify the 72nd percentile, or the score below which 72% of the cases lay. Figure 6.4 shows the percentile function in which I specified the 72nd percentile (0.72 in the "K" window) from the array of data in the DI column ("DI2:DI41"). When I select OK, Excel® returns the 72nd percentile value of 16.67. Therefore, 72% of the student–teacher ratios in this sample are below 16.67. You will note in the screenshot in Figure 6.4 that the percentile value is identified in the dropbox (in the right middle of the box) if you do not wish to paste it in the selected cell in the spreadsheet.

---

[1] The author acknowledges the kind approval of T.E.S.T., Inc., the owner and manager of the 2003 TAGLIT data for the use of TAGLIT databases for this book (http://www.testkids.com/taglit/).

USING EXCEL® AND SPSS® TO IDENTIFY PERCENTILES 85

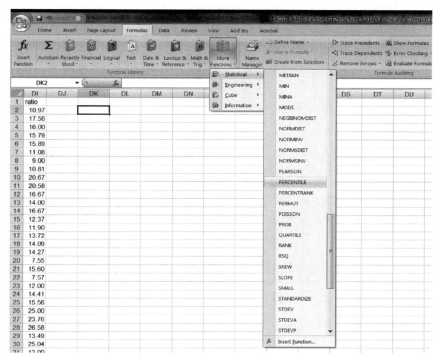

FIGURE 6.3 Using the Excel® functions to create percentiles.

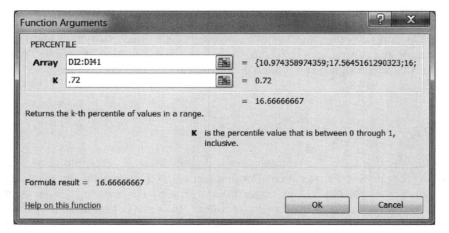

FIGURE 6.4 Specifying a percentile with Excel® functions.

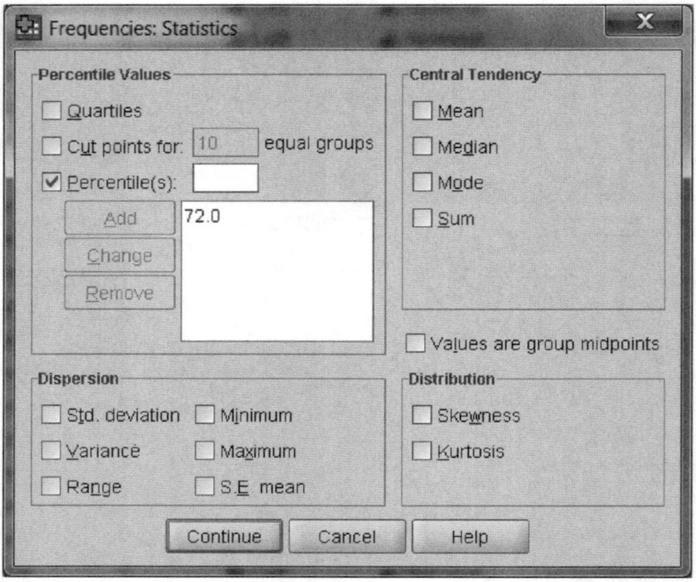

FIGURE 6.5 Specifying a percentile with SPSS® functions.

Ratio		
N	Valid	40
N	Missing	0
Percentiles	72	16.6667

FIGURE 6.6 SPSS® output for percentile calculation.

The procedure is also very simple using SPSS®. Selecting the Frequencies procedure, as I described in Chapter 5 (see Figure 5.6), I can specify the same 72nd percentile and generate the screen shown in Figure 6.5. Selecting "Continue" generates the screen shown in Figure 6.6. As you can see, the same value (16.67) is generated by SPSS®.

NOTE

Please note that Excel® often calculates a slightly different percentile value than SPSS® because of the way it handles tied values. Excel® does this by *assigning the same rank for tied values rather than averaging the tied ranks*. That is, if two values are the same in a string of values, their *ranks* should be averaged; apparently, Excel® does not average the ranks; rather, it simply assigns the same rank to the tied scores. With our example variable, ratio, there happen to be two identical ratios (i.e., schools with the same student–teacher ratio), where

STANDARD DEVIATION AND VARIANCE

both schools have a ratio of 16.67 (cells DI12 and DI14). Excel® assigns both values a rank of 11, since they are the 11th and 12th values in a ranked list. However, the ranks should technically be averaged, so that both values of 16.67 would receive ranks of 11.5 (where $(11 + 12)/2 = 11.5$).

The examples I have discussed above (Figures 6.4 through 6.6) show the same percentile value when calculated in Excel® and SPSS®. This just happens to be because the 72nd percentile of 72 falls at the ratio percentage of 16.67, the two schools with the tied scores. However, if you specify a percentile of 90, Excel® returns "23.38" whereas SPSS® returns "23.72." Researchers occasionally find two or more cases with exactly the same values (especially among small datasets using percentages) among a set of continuous scores in research. You should be aware of this discrepancy in the way Excel® and SPSS® calculate percentiles.

STANDARD DEVIATION AND VARIANCE

To some, these statistical measures appear mysterious; to others, they may seem superfluous. However, both measures are crucial to calculation and understanding of statistical procedures we will cover in this book. Make sure you have a level of comfort with what they represent and how to calculate them before you move on to the further topics we discuss.

The standard deviation (SD) and variance (VAR) are both measures of the dispersion of scores in a distribution. That is, *these measures provide a view of the nature and extent of the scatter of scores around the mean*. So, along with the mean, skewness, and kurtosis, they provide a fourth way of describing the distribution of a set of scores. With these measures, the researcher can decide whether a distribution of scores is normally distributed.

Figure 6.7 shows how scores in a distribution spread out around the mean value. Each score can be thought to have a "deviation amount" or a distance from the mean. Figure 6.7 shows these deviation amounts for four raw scores $(X_1 - X_4)$.

The VAR is by definition the square of the SD. Conceptually, the VAR is a *global measure of the spread of scores* since it represents an average squared deviation. If you summed the squared distances between each score and the mean of a distribution of scores (i.e., if you squared and summed the deviation amounts), you

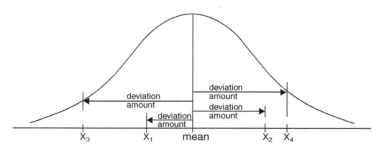

FIGURE 6.7 The components of the SD.

would have a global measure of the total amount of variation among all the scores. If you divided this number by the number of scores, the result would be the VAR, or the *average squared distance from the mean*.

The SD is the square root of the VAR. If you were to take the square root of the average squared distances from the mean, the resulting figure is the *standard deviation*. That is, it represents a *standard* amount of distance between the mean and each score in the distribution (not the average *squared* distance, which is the VAR). We refer to this as *standard* since we created a standardized unit by dividing it by the number of scores, yielding a value that has known properties to statisticians and researchers. We know that if a distribution is perfectly normally distributed, the distribution will contain about six SD units, three on each side of the mean.

Both the SD and the VAR provide an idea of the extent of the spread of scores in a distribution. If the SD is small, the scores will be more alike and have little spread. If it is large, the scores will vary greatly and spread out more extensively. Thus, if a distribution of test scores has an SD of 2, it conceptually indicates that *typically*, the scores were within two points of the mean. In such a case, the overall distribution would probably appear to be quite "scrunched together," in comparison to a distribution of test scores with an SD of 5.

Conceptually, as I mentioned, the SD is a standardized measure of how much all of the scores in a distribution vary from the mean. A related measure is the "average deviation," which is actually the *mean of all the absolute values of the deviation scores* (the distances between each score and the mean). So, in a practical way, the researcher can report that the average achievement test score (irrespective of sign) among a group of students was two points away from the mean, for example.

CALCULATING THE VARIANCE AND STANDARD DEVIATION

I will discuss two ways of calculating the SD, which can then be squared to provide the VAR: the deviation method and the computation method. Both methods have merit; the first is helpful to visualize how the measure is created, and the second is a simpler calculation.

The Deviation Method

The deviation method allows you to see what is happening with the numbers. In this formula, the X values are raw scores, M is the mean of the distribution, and N is the total number of scores in the distribution.

$$SD = \sqrt{\frac{\sum (X - M)^2}{N}}$$

As you can see, the formula uses deviation scores in the calculation $(X - M)$. Thus, the mean is subtracted from each score to obtain the deviation amount. These $X - M$ values in the formula represent the deviation amounts shown in Figure 6.7.

CALCULATING THE VARIANCE AND STANDARD DEVIATION   89

The deviation scores are squared, summed, and divided by the number of scores. Finally, the square root is taken to yield the SD. If the square root were not taken, the resulting value would be the VAR.

One curious aspect of the formula is that the deviation scores are *squared* and then the *square root* is taken of the entire calculation! This appears pointless until you consider some of the features of deviation scores that become important to later statistical procedures.

An example may help to illustrate the features of this formula and the process for calculating SD. Table 6.1 shows the sample TAGLIT values we used earlier in this chapter to discuss percentiles.

As you can see, I placed the ratio variable in a column along with separate columns for the deviation scores $(X-M)$, the squared deviation scores $(X-M)^2$, and an additional column I will discuss in the next section. I included the mean values for each of the first three columns. You can see that the mean value for the deviation scores is "0.00" because in a set of normally distributed values there will be equal numbers of values that lay to the left and to the right of the mean. Adding them up in effect creates a zero total because there will be as many negative as positive values. Therefore, it is necessary to square all the values to avoid the effects of the negative and positive signs. Once the values are squared and divided by N, the formula puts everything under the square root sign to remove the earlier squaring of values. The formula thus provides a neat solution to avoid the negative and positive deviation values adding to zero.

Creating the final calculation shows the SD to be 4.77 for this sample of schools. Therefore, the typical school in this group of 40 schools deviates 4.77 from the mean of 15.13. The largest deviation amount is 11.45 and the smallest deviation amount is (-0.06).

$$SD = \sqrt{\frac{\sum(X-M)^2}{N}} \quad SD = \sqrt{\frac{900.37}{40}} \quad SD = \sqrt{22.51} \quad SD = 4.77$$

The Average Deviation

I mentioned that the average deviation is the mean of the absolute values of the deviation scores. In this example, if we created a mean from the absolute value of each of the school ratios, the average deviation (AD) would equal 3.63. (You can compute this in Excel® by using the Formula menu and then choosing Statistical— AVEDEV.)

Why the discrepancy between the AD and the SD? There is a long answer and a short answer. The short answer is that the standard deviation is based upon calculations of the perfect normal distribution with known mathematical properties of the curve, and based on large sample values. That is why the SD is known as standardized deviation, to make use of these known properties and characteristics. The

TABLE 6.1 Using the Deviation Method to Calculate SD

Ratio (X)	$X - M$	$(X - M)^2$	X^2
10.97	−4.16	17.27	120.44
17.56	2.43	5.93	308.51
16.00	0.87	0.76	256.00
15.79	0.66	0.43	249.19
15.89	0.76	0.58	252.46
11.08	−4.05	16.39	122.79
9.00	−6.13	37.58	81.00
10.81	−4.32	18.70	116.77
20.67	5.54	30.65	427.11
20.58	5.45	29.74	423.67
16.67	1.54	2.36	277.78
14.00	−1.13	1.28	196.00
16.67	1.54	2.36	277.78
12.37	−2.76	7.63	152.98
11.90	−3.23	10.40	141.72
13.72	−1.41	1.99	188.22
14.09	−1.04	1.08	198.55
14.27	−0.86	0.74	203.66
7.55	−7.58	57.53	56.93
15.60	0.47	0.22	243.36
7.57	−7.56	57.13	57.33
12.00	−3.13	9.80	144.00
14.41	−0.72	0.52	207.70
15.56	0.43	0.19	242.26
25.00	9.87	97.42	625.00
23.76	8.63	74.56	564.76
26.58	11.45	131.18	706.67
13.49	−1.64	2.69	181.97
25.04	9.91	98.24	627.09
13.00	−2.13	4.54	169.00
18.00	2.87	8.24	324.00
10.00	−5.13	26.32	100.00
11.15	−3.98	15.81	124.41
14.09	−1.04	1.09	198.48
8.08	−7.05	49.75	65.24
17.00	1.87	3.50	289.00
12.28	−2.85	8.15	150.70
15.07	−0.06	0.00	227.23
14.52	−0.61	0.37	210.83
23.33	8.20	67.29	544.44
Mean 15.13	0.00	22.51	
Sum 605.13		900.37	10055.02

THE COMPUTATION METHOD

average deviation does not make reference to these properties; rather, it is simply the average (absolute) deviation from the mean of a distribution of values. Generally, it yields a slightly different value than the SD to account for the properties of the perfect normal distribution.

THE COMPUTATION METHOD

The second method for calculating the SD is using the computation formula. It looks more complex than the deviation method, but it is much easier computationally because it does not involve creating the deviation amounts. The formula is as follows:

$$SD = \sqrt{\frac{\sum X^2 - \frac{(\sum X)^2}{N}}{N}}$$

Computing the SD with this formula involves only two columns, namely, the column of values and the column of squared values. If you look at Table 6.1, you will see that I included the second column as "(Ratio)^2" indicating that each of the ratio values are squared. The bottom row includes the sums of the two appropriate columns. When you use the values in the formula, you compute the same SD as you did with the deviation formula (with slight differences due to rounding).

$$SD = \sqrt{\frac{\sum X^2 - \frac{(\sum X)^2}{N}}{N}} \qquad SD = \sqrt{\frac{10055.02 - \frac{(605.13)^2}{40}}{40}}$$

$$SD = \sqrt{\frac{900.46}{40}} \qquad SD = 4.75$$

There is one trick to using this formula. The values of $\sum X^2$ and $(\sum X)^2$ are quite different. The first value is obtained by adding up all the values of the raw scores *that have been squared*, yielding 10,055.02 (the sum of the X^2 column). The second value is squaring the *sum of the raw score values themselves*, which is the sum of the ratio (X) values (605.13) squared (605.13^2 or 366,182.32).

The Sum of Squares

The top part of both the deviation and computation formulas, under the radical sign, is known as the "Sum of Squares" since it is a global measure of all the variation in a distribution of scores. These measures will be important for later procedures that analyze the differences between sets of scores by comparing variance amounts.

Remember, these are equivalent formulas, with the second being an algebraically equal (computation) approach.

$$\sum (X - M)^2 \quad \text{or} \quad \sum X^{2'} - \frac{(\sum X)^2}{N}$$

Sample SD and Population SD

I will have much more to say about this difference in later chapters when I discuss inferential statistics. For now, it is important to point out that computing SD for a sample of values, as we did with the ratio data, will yield a different value depending on whether we understand the distribution of data to represent a complete set of scores or merely a sample of a population.

Remember that inferential statistics differs from descriptive statistics primarily in the fact that, with inferential statistics, we are using sample values to make inferences or decisions about the populations from which the samples are thought to come. In descriptive statistics, we make no such attributions; rather, we simply measure the distribution of values at hand and treat all the values we have as the complete set of information (i.e., its own population). When we get to the inferential statistics chapters, you will find that, *in order to make attributions about populations based on sample values, we typically must adjust the sample values since we are making guesses about what the populations look like.* To make better estimates of population values, we adjust the sample values.

Excel® and SPSS® have no way of distinguishing inferential or descriptive computations of SD. Therefore, they present the inferential SD as the default value. I will show how to determine the differences and examine the resulting values with both Excel® and SPSS®.

Figure 6.8 shows the descriptive statistics output for our student–teacher ratio variable. As you can see, the mean value (15.13) is the same as that reported in Table 6.1. However, the SD (the fifth value from the top of the first column) is 4.8

Mean	15.13
Standard error	0.76
Median	14.34
Mode	16.67
Standard deviation	4.80
Sample variance	23.09
Kurtosis	0.18
Skewness	0.74
Range	19.04
Minimum	7.55
Maximum	26.58
Sum	605.13
Count	40.00

FIGURE 6.8 The Excel® descriptive statistics output for ratio.

THE COMPUTATION METHOD

according to this output. In the earlier sections, we calculated the SD to be 4.77. This discrepancy is due to the fact that Excel® reports the inferential SD in Figure 6.8.

You can obtain the "actual" or population SD by using the formula menus of Excel® as shown in Figure 6.9. As you can see, I have highlighted the "STDEVP"

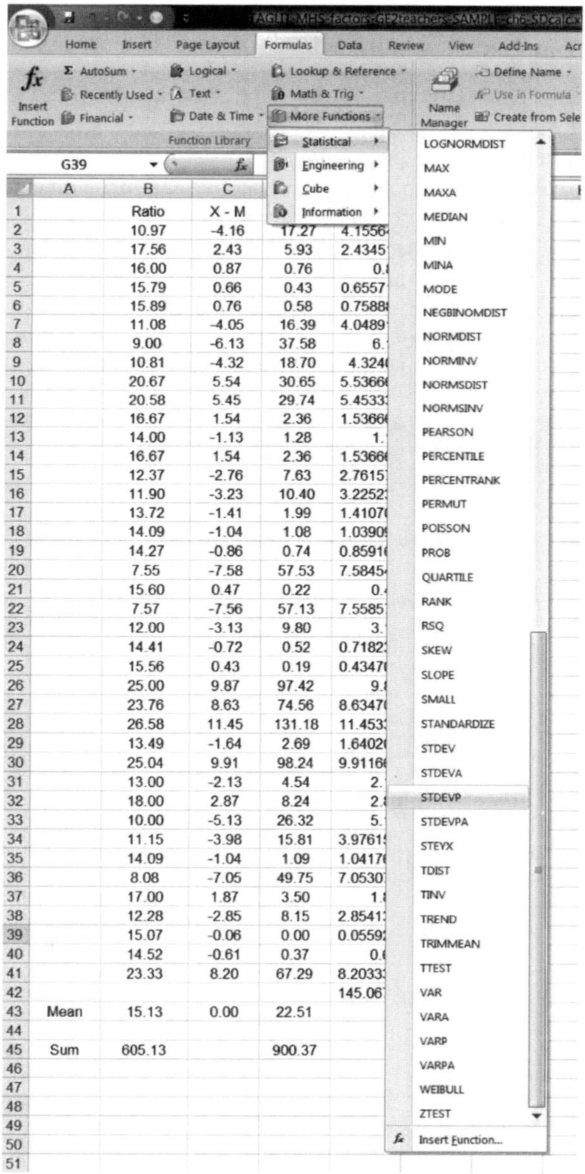

FIGURE 6.9 Using the Excel® functions to calculate the "actual" SD.

choice of functions which is the "standard deviation of the population." This will calculate a SD based on all of a set of scores as if they constituted the entire population. This value will be equivalent to the SD calculations we discussed above. The Excel® default value for the SD (the inferential SD) is also available in the list of functions as "STDEV."

Obtaining SD from Excel® and SPSS®

I demonstrated how to obtain the SD from Excel® in the preceding section. Obtaining the SD from SPSS® is straightforward, but, like Excel®, it returns the inferential SD as the default value. Figure 6.10 shows the menu screens and options for creating descriptive statistics, including standard deviations. You can obtain the SD through the SPSS® menus for descriptive-frequencies that we discussed in the sections above regarding percentiles. Figures 6.5 and 6.6 showed how to use the frequencies menus to obtain percentiles, and they included options for SD and means, among other measures.

Figure 6.11 shows the output from this descriptive–descriptive request. As you can see, the listed SD is 4.80, the same (inferential SD) value reported in Excel®.

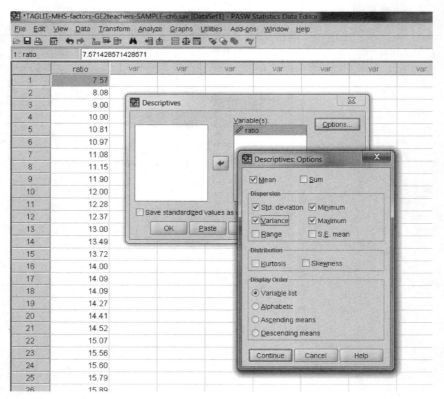

FIGURE 6.10 Using the "Descriptives" menus in SPSS®.

THE COMPUTATION METHOD

Descriptive Statistics

	N	Range	Mean	Standard deviation	Variance	Skewness		Kurtosis	
	Statistic	Statistic	Statistic	Statistic	Statistic	Statistic	Standard error	Statistic	Standard error
Ratio	40	19.04	15.1283	4.80482	23.086	0.736	0.374	0.183	0.733
Valid N (listwise)	40								

FIGURE 6.11 The descriptive statistics output from SPSS®.

Because these are default values, you can use Excel® to calculate the population SD values or you can convert the inferential SD to the population SD using the following formula:

$$SD_{(Population)} = SD_{(Inferential)} \sqrt{\frac{N-1}{N}},$$

$$SD_{(Population)} = 4.80 \sqrt{\frac{39}{40}}, \qquad SD_{(Population)} = 4.74$$

Note the other descriptive statistics reported in Figure 6.11. The skewness and kurtosis values, when divided by their respective standard errors, are within bounds of a normal distribution (1.97 and 0.25, respectively). The visual evidence suggests a slight positive skew to the distribution, however, as seen in Figure 6.12.

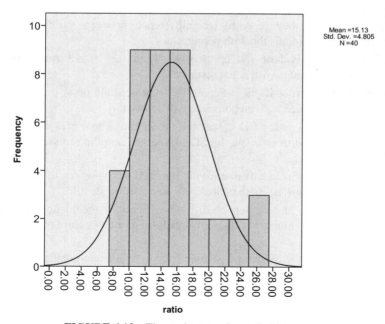

FIGURE 6.12 The student–teacher ratio histogram.

Note also the relationship between the range (19.04) and the SD. From our earlier calculation, the "converted" SD (i.e., the "actual" SD) was 4.74. Because a normally distributed set of data contains just over 6 SDs (± 3 SD), compare the range to 6 SDs (i.e., 28.44). Since these numbers are quite far apart, the distribution may not be normally distributed. This is only a "ballpark" rule because extreme values and sample size will affect both the measures, but it is another piece of "evidence" to use in making the determination of whether the distribution of scores is normally distributed.

Another rule of thumb is that you cannot calculate a standard deviation or variance less than 0 or "negative variability." You can have small variation, or even no variation (where every score is the same, for example), but never less than zero. This is one way to check whether you are calculating the variance and standard deviation correctly.

TERMS AND CONCEPTS

Average Deviation The average deviation represents the mean of the absolute values of the deviation scores in a distribution.

Deciles These measures break a frequency distribution into 10 equal groups using nine cutpoints based on the 10th, 20th, 30th, 40th, 50th, 60th, 70th, 80th, and 90th percentiles.

Interquartile Range These scores represent the middle half of a frequency distribution since they represent the difference between the first and third quartiles (the 75th minus the 25th percentiles).

Normal Curve Equivalent (NCE) Scores These are transformed percentile scores that yield values with equal distances.

Percentile (or Percentile Rank) Percentiles represent the point in a distribution of scores below which a given percentage of scores fall.

Population SD This is the SD calculated on all the scores of a distribution of values and not used to estimate the SD of a larger sampling unit. The "actual" SD of a set of scores.

Quartiles These are measures that divide the total set of scores in a distribution into four equal groups using the 25th, 50th, and 75th percentiles.

Range The range is the numerical difference between the highest and lowest scores in the distribution and represents a helpful global measure of the spread of scores.

Sample SD This is the SD from a sample used to estimate a population SD. This is used in inferential statistics and can be understood to be an inferential SD. This is different from an SD used to represent *only* the values of the sample distribution of values (Population SD).

Standard Deviation (SD) The SD represents a *standard* amount of distance between the mean and each score in the distribution. It is the square root of the variance (VAR).

Variance (VAR) The variance is the average squared distance of the scores in a distribution from the mean. It is the squared SD.

REAL-WORLD LAB II: VARIABILITY

Continue the Real-World Lab from Chapter 5. That lab addressed central tendency whereas we will focus on variability using the data in Table 5.8.

1. What are the SDs (both population and inferential values) for school-based reading and math achievement percentages?
2. What is the range for both variables?
3. Identify the 25th, 50th, and 75th percentiles for each variable.

REAL-WORLD LAB II: SOLUTIONS

1. What are the SDs (both population and inferential values) for school-based reading and math achievement percentages?

Results

Table 5.9 reports the (inferential) SDs as 12.00 and 17.68 for reading and math, respectively. You can use the Excel® formula menus to calculate the (population) SDs, using the "Statistical-STDEVP" formula, as I demonstrated in Figure 6.9. This yields SDs for reading (11.85) and math (17.45) that differ from the (inferential) SDs shown in Figure 5.9. The alternative method is to use the conversion formula I discussed earlier to obtain the (population) SDs from the Excel® and SPSS® output files:

$$SD_{(Population)} = SD_{(Inferential)} \sqrt{\frac{N-1}{N}}$$

This formula yields the same results for reading and math as obtaining them through the Excel® formula menus.

2. What is the Range for both variables?

Table 5.9 also reports the range values for reading (52.60) and math (78.80). Using the relationship between the SD and range, we discussed earlier (range $= 6 * SD$) as

TABLE 6.2 Range and SDs for School-Based Reading and Math Achievement

	Range	SD	6 * SD
Reading	52.6	11.85	71.1
Math	78.8	17.45	104.7

a gauge, we find the figures shown in Table 6.2. In both cases, the range and SD values are discrepant. Recall, however, that smaller distributions (i.e., those with less than 40) may not be strictly normally distributed because there are not enough cases to completely populate the distribution curve. In these cases, rely on both the numerical and visual (see Figures 5.22 through 5.25) evidence to determine the extent to which the data form a normal distribution.

3. Identify the 25th, 50th, and 75th percentiles for each variable.

Figure 6.13 shows the SPSS® output for a frequency analysis in which I specified the 25th, 50th, and 75th percentiles. The output identifies selected values in addition to the percentile figures for reading and math.

	Reading Percent Met Standard	Math PercentMet Standard
N valid	40	40
N missing	0	0
Mean	72.48	50.61
Standard deviation	12.004	17.676
Skewness	−0.940	0.008
Standard error of skewness	0.374	0.374
Kurtosis	0.797	−0.412
Standard error of kurtosis	0.733	0.733
Range	53	79
Percentiles		
25	66.58	37.10
50	73.20	50.00
75	81.95	62.80

FIGURE 6.13 The SPSS® 25th, 50th, and 75th percentile values for reading and math.

TABLE 6.3 The Excel® 25th, 50th, and 75th Percentile Values for Reading and Math

	Reading	Math
25th	67.725	37.5
50th	73.2	50
75th	81.85	62.6

Table 6.3 shows the same percentiles obtained through Excel® (I used the Percentile formula pull-down menus as shown in Figure 6.3). Note the discrepancies between the Excel® and SPSS® output values. As I discussed earlier in this chapter, the discrepancies are due to Excel® method for (not) accounting for tied scores in the same way that SPSS® accounts for them. SPSS® accurately calculates tied ranks making their percentile report a bit more precise.

7

THE NORMAL DISTRIBUTION

Thus far, we have discussed how to describe distributions of raw scores graphically and in terms of central tendency, variability, skewness, and kurtosis. We will continue to perform these calculations because most all of the statistical procedures we will discuss in subsequent chapters require that data be normally distributed. Using what we have learned with calculating these descriptive statistics, we can confirm whether our data are normally distributed or if we must use different procedures. Often, even if the variables are not strictly or exactly normally distributed, we can still use them because many statistical procedures are "robust" or able to provide meaningful and precise results even if there are some violations of the normal distribution assumptions.

THE NATURE OF THE NORMAL CURVE

The normal distribution, as I explained in an earlier chapter, is very common in educational research, so we need to deepen our understanding of some of the properties of the normal *curve*. I call the normal distribution a curve because the histogram forms a curve when the top midpoints of the bars are joined together. Technically, this is called a *frequency polygon*. If you look back to Figure 5.25, you will see the SPSS® histogram for the school-based math achievement variable we discussed. In the figure, SPSS® overlaid the normal curve on top of the histogram so you can see the extent to which the data approximate a normal distribution. As you can see from that figure, if you were to connect the top midpoints with a line, it would not be the smooth line you see, but rather a more jagged line. However, as a

---

Understanding Educational Statistics Using Microsoft Excel® and SPSS®. By Martin Lee Abbott.
© 2011 John Wiley & Sons, Inc. Published 2011 by John Wiley & Sons, Inc.

database increases its size, the histogram approximates the smooth normal curve in variables that are normally distributed; the jagged line becomes filled in as more cases are added.

When we speak of the normal distribution and how our sample dataset is normally distributed, we actually speak about our data *approximating* a normal distribution. We refer to the *perfect* normal distribution as an ideal so that we have a model distribution for comparison to our *actual* data. Thus, the normal curve is a kind of perfect ruler with known features and dimensions. In fact, we can mathematically chart the perfect normal curve and derive a picture of how the areas under the curve are distributed. Because of these features, we refer to the perfect normal distribution as a *standard normal distribution*.

Look at Figure 5.10, which is reproduced here as Figure 7.1. As you can see, the perfect normal curve is represented as having known proportions of the total area between the mean and given standard deviation units. A standard normal curve (also known as a *z distribution*) has a mean of 0 and a standard deviation of 1.0. This is always a bit puzzling until you consider how the mean and standard deviation are calculated. Since a perfect distribution has equal numbers of scores lying to the left and to the right of the mean, calculating the mean is akin to adding positive and negative values resulting in 0. Dividing 0 by N, of whatever size, will always equal 0. Therefore the mean of a perfect, standard normal distribution is equal to 0.

The standard normal distribution has a standard deviation equal to 1 unit. This is simply an easy way to designate the known areas under the curve. Figure 7.1 shows that there are six standard deviation units that capture almost all the cases under the perfect normal curve area. (This is the source of the rule for the range equaling six times the SD in a raw score distribution.) This is how the standard normal curve is "arranged" mathematically. So, for example, 13.59% of the area of the curve lies between the first (+1) and second (+2) standard deviation on the right side of the mean. Because the curve is symmetrical, there is also 13.59% of the area of the curve between the first (−1) and second (−2) standard deviation on the left side of the curve, and so on.

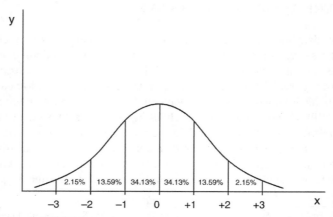

FIGURE 7.1 The normal curve with known properties.

THE STANDARD NORMAL SCORE: z SCORE

Remember that this is an ideal distribution. As such, we can compare our actual data distributions to it as a way of understanding our own raw data better. Also, we can use it to compare two sets of raw score data since we have a perfect measuring stick that relates to both sets of "imperfect" data.

There are other features of the standard distribution we should notice.

- The scores cluster in the middle, and they "thin out" toward either end.
- It is a balanced or symmetrical distribution, with equal numbers of scores on either side of the middle.
- The mean, median, and mode all fall on the same point.
- The curve is "asymptotic" to the x axis. This means that it gets closer and closer to the x axis but never touches because, in theory, there may be a case very far from the other scores—off the chart, so to speak. There has to be room under the curve for these kinds of possibilities.
- The *inflection point* of the standard normal curve is at the point of the (negative and positive) first standard deviation unit. This point is where the steep decline of the curve slows down and widens out. (This is a helpful visual cue to an advanced procedure called *factor analysis*, which uses a *scree plot* to help decide how many factors to use from the results.)

THE STANDARD NORMAL SCORE: z SCORE

When we refer to the standard normal deviation, we speak of the z score, which is a very important measure in statistics. *A z score is a score expressed in standard deviation units*—that is, a complete standard deviation score of 1, -1 and so on. That is the way scores in the standard normal distribution are expressed. Thus, a score of 0.67 would be a score that is two-thirds of one standard deviation to the right of the mean. These scores are shown on the x axis of Figure 7.1, and represent the standard deviation values that define the areas of the distribution. Thus, $+1$ standard deviation unit to the right of the mean contains 34.13% of the area under the standard normal curve, and we can refer to this point as a z score of $+1$. So, if a school's student–teacher ratio had a z score of 3.5, we would recognize immediately that this school would have an inordinately high ratio, relative to the other values, since it would fall three and one-half standard deviations above the mean where there is only an extremely small percent of curve represented.

Because it has standardized meaning, the z score allows us to understand where each score resides compared to the entire set of scores in the distribution. It also allows us to compare one individual's performance on two different sets of (normally distributed) scores. It is important to note that z scores are expressed not just in whole numbers but as decimal values, as I used in the example above. Thus, a z score of -1.96 would indicate that this score is slightly less than two standard deviations below the mean on a standard normal curve as shown in Figure 7.2.

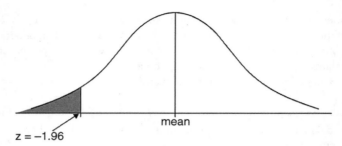

FIGURE 7.2 The location of $z = (-)\,1.96$.

THE z-SCORE TABLE OF VALUES

Statisticians have prepared a table of values to help researchers understand how various scores in the standard normal distribution relate to the total area under the curve. Consider table in Figure 7.3. This is a part (the second half) of the z-score table you will find in the Appendices.

z	0.05		0.06		0.07		0.08		0.09	
0.0	0.0199	0.4801	0.0239	0.4761	0.0279	0.4721	0.0319	0.4681	0.0359	0.4641
0.1	0.0596	0.4404	0.0636	0.4364	0.0675	0.4325	0.0714	0.4286	0.0753	0.4247
0.2	0.0987	0.4013	0.1026	0.3974	0.1064	0.3936	0.1103	0.3897	0.1141	0.3859
0.3	0.1368	0.3632	0.1406	0.3594	0.1443	0.3557	0.1480	0.3520	0.1517	0.3483
0.4	0.1736	0.3264	0.1772	0.3228	0.1808	0.3192	0.1844	0.3156	0.1879	0.3121
0.5	0.2088	0.2912	0.2123	0.2877	0.2157	0.2843	0.2190	0.2810	0.2224	0.2776
0.6	0.2422	0.2578	0.2454	0.2546	0.2486	0.2514	0.2517	0.2483	0.2549	0.2451
0.7	0.2734	0.2266	0.2764	0.2236	0.2794	0.2206	0.2823	0.2177	0.2852	0.2148
0.8	0.3023	0.1977	0.3051	0.1949	0.3078	0.1922	0.3106	0.1894	0.3133	0.1867
0.9	0.3289	0.1711	0.3315	0.1685	0.3340	0.1660	0.3365	0.1635	0.3389	0.1611
1.0	0.3531	0.1469	0.3554	0.1446	0.3577	0.1423	0.3599	0.1401	0.3621	0.1379
1.1	0.3749	0.1251	0.3770	0.1230	0.3790	0.1210	0.3810	0.1190	0.3830	0.1170
1.2	0.3944	0.1056	0.3962	0.1038	0.3980	0.1020	0.3997	0.1003	0.4015	0.0985
1.3	0.4115	0.0885	0.4131	0.0869	0.4147	0.0853	0.4162	0.0838	0.4177	0.0823
1.4	0.4265	0.0735	0.4279	0.0721	0.4292	0.0708	0.4306	0.0694	0.4319	0.0681
1.5	0.4394	0.0606	0.4406	0.0594	0.4418	0.0582	0.4429	0.0571	0.4441	0.0559
1.6	0.4505	0.0495	0.4515	0.0485	0.4525	0.0475	0.4535	0.0465	0.4545	0.0455
1.7	0.4599	0.0401	0.4608	0.0392	0.4616	0.0384	0.4625	0.0375	0.4633	0.0367
1.8	0.4678	0.0322	0.4686	0.0314	0.4693	0.0307	0.4699	0.0301	0.4706	0.0294
1.9	0.4744	0.0256	0.4750	0.0250	0.4756	0.0244	0.4761	0.0239	0.4767	0.0233
2.0	0.4798	0.0202	0.4803	0.0197	0.4808	0.0192	0.4812	0.0188	0.4817	0.0183
2.1	0.4842	0.0158	0.4846	0.0154	0.4850	0.0150	0.4854	0.0146	0.4857	0.0143
2.2	0.4878	0.0122	0.4881	0.0119	0.4884	0.0116	0.4887	0.0113	0.4890	0.0110
2.3	0.4906	0.0094	0.4909	0.0091	0.4911	0.0089	0.4913	0.0087	0.4916	0.0084
2.4	0.4929	0.0071	0.4931	0.0069	0.4932	0.0068	0.4934	0.0066	0.4936	0.0064
2.5	0.4946	0.0054	0.4948	0.0052	0.4949	0.0051	0.4951	0.0049	0.4952	0.0048
2.6	0.4960	0.0040	0.4961	0.0039	0.4962	0.0038	0.4963	0.0037	0.4964	0.0036
2.7	0.4970	0.0030	0.4971	0.0029	0.4972	0.0028	0.4973	0.0027	0.4974	0.0026
2.8	0.4978	0.0022	0.4979	0.0021	0.4979	0.0021	0.4980	0.0020	0.4981	0.0019
2.9	0.4984	0.0016	0.4985	0.0015	0.4985	0.0015	0.4986	0.0014	0.4986	0.0014
3.0	0.4989	0.0011	0.4989	0.0011	0.4989	0.0011	0.4990	0.0010	0.4990	0.0010

FIGURE 7.3 The z-score table of values.

This table of values is based on *proportions of the area under the normal curve between the mean and various z scores*. Since the total proportion of the area under the normal curve is 1.0 and the total area is 100%, I have multiplied the proportions by 100 to show that the proportions represent *percentages* of the area within the distribution. Note also that the percentages shown are calculated in reference to the mean. The shaded graphs at the top of the columns indicate the proportion of the curve in relationship to specific z-score values.

As you can see, each z-score value (to two decimal places) is represented in the table in such a way that you can identify the percent of the area in the normal curve that is between a given z score and the mean of the distribution (in the first of the two data columns at each z-score column). The second of the two data columns shows the percent of the area of the curve that lies in the "tail" of the distribution, or the percent of the area that lies beyond the given z score.

To take the example in Figure 7.2 (-1.96), you would find the "tenths" part of the score (1.9) in the first column titled "z" and then follow that row across until you found the "hundredths" part of the score (in the "0.06" column). When combined, the score of -1.96 indicates a value of "47.50" (or 47.50%) which you will find if you look at the value in the 1.9 row and the 0.06 column. This number indicates the percent of the curve between the mean and the z score of -1.96. Since -1.96 is almost 2 SDs below the mean, you can do a quick mental check to see if this is reasonable by looking at Figure 7.1. Because there is 34.13% between the mean and the first SD, and another 13.59% between the first and second SDs, then the combined total of 47.72% (adding the percentages together) shows the percent of the curve between the mean and the second SD. This value is very close to the table value (47.50%) for $z = -1.96$.

Note that the sign of negative or positive does not affect you locating the value in the table because the values in the table are symmetrical; they apply to both "sides" of the distribution equally. The sign is crucial to remember however, since it indicates the *direction of the score in relation to the mean*. Negative scores are located to the left of the mean, as shown in Figure 7.2, and positive scores are located to the right of the mean.

NAVIGATING THE z-SCORE DISTRIBUTION

It is a good idea to familiarize yourself with the z-score table because you may need to visualize where certain scores are in relation to one another in the z distribution. You can also use the table to create "cumulative proportions" of the normal curve at certain z-score values. Cumulative proportions are simply the summed percentages or proportions of the area in the normal distribution. Percentiles represent one such proportion measure because they are by definition the percent of the scores below a given score. You can also calculate cumulative proportions that exist *above* given z scores in the same fashion using the z-score table. Here is an example.

If you are interested in the percentage of scores that lie below a given score (as you would in the calculation of a percentile), you can use the z-score table to help you. *What percentage of the standard normal curve lies below a z score of 1.96?*

Before you consider your answer in detail, try to create a "ballpark" solution:

- We are looking for a score almost two SDs above the mean (1.96).
- The percentage distribution in the standard normal curve (see Figure 7.1) is 34.13% between the mean and SD 1 and is 13.59% between SD 1 and SD 2.
- Therefore, about 47.72% of the curve (34.13% + 13.59%) lies between the mean and a z score of 1.96.
- Since the score is to the right of the mean, you will need to add 50% (the other half of the distribution) to 47.72% to get an approximate 97.72% of the curve lying below $z = 1.96$.

Now, compare this ballpark answer to a more precise method of using the z-score table.

- Locate 1.96 in the table of values as we did earlier using table in Figure 7.3. The value is 47.50% of the distribution lie between the mean and the score of 1.96. Remember, 47.50% *is not the entire amount of the curve below* 1.96. It is only the amount of the curve that lies between the mean and the z score of 1.96.
- Add 50% (the other half of the curve) to get 97.50% of the distribution lying below a z score of 1.96. In order to get the total percentage of the curve that lies below the z score of 1.96, you would need to add 50.00 to this amount. Why? Because the curve is symmetrical and each side contains 50.00 (50%). Therefore, add the 50.00% from the left half of the curve to the 47.50% from the right half. This would yield a total of 97.50% of the curve below a z score of 1.96.
- This percentage (97.50%) is slightly less than the estimated percentage (97.72%) because the target score is slightly less than 2 SDs.
- Figure 7.4 shows this identification.

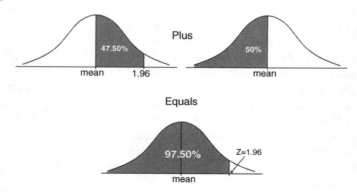

FIGURE 7.4 Using the z-score table to identify the percent of the distribution below a z score.

NAVIGATING THE z-SCORE DISTRIBUTION

The process is the same for negative z-score values. Take the example of −1.96. *What percent of the curve lies below a z score of −1.96?* Using the ballpark estimate method:

- We know from the standard normal curve that a score 2 SDs from the mean (either above or below) contains about 47.72% of the distribution (i.e., 34.13% between the mean and SD 1 plus 13.59% between SD 1 and SD 2).
- Because we need the percent of the distribution below −1.96, we need to subtract 47.72% from 50%, the total area in the left half of the distribution.
- This results in approximately 2.28% of the distribution below $z = -1.96$.

Now, using the table of values, create a more precise calculation of the percent of the distribution below $z = -1.96$.

- Locate 1.96 in the table of values using the table in Figure 7.3. The value is 47.50% of the distribution lying between the mean and the score of 1.96 (because the table identifies the percentages for either positive or negative values).
- Subtract this amount from 50% (the left half of the curve) to get 2.50% of the distribution lying below a z score of −1.96. Because you are not interested in the percentage of the distribution that lies above the score of −1.96, you must "subtract it out" of the entire half of the distribution (50%) to see what percentage lies below the given scores.
- This percentage (2.50%) is slightly greater than the estimated percentage (2.28%) because the target score is slightly closer to the mean, leaving a greater area to the left of the target score of −1.96.
- Figure 7.5 shows this identification.
- Note that you can also identify this same amount (2.50%) using the z-score table because this proportion in the "tail" is also identified in the same set of columns you used to obtain 47.50%. Because the table is symmetrical, you can see that the percent of the area in the right (positive) tail would be equivalent

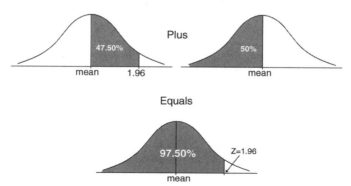

FIGURE 7.5 Using the z-score table to identify the percent of the distribution below $z = -1.96$.

to the area in the left (negative) tail, which is the area you sought to solve the problem.

Calculating Percentiles

You will note that the two examples in Figures 7.4 and 7.5 show how to identify a *percentile*. This is a special feature of using the cumulative proportions in the standard normal distribution and one that we will revisit in subsequent sections.

Creating Rules for Locating z Scores

Some statisticians and researchers like to create rules for the operations we discussed above. Thus, for example, we might create the following rules:

1. For locating the percent of the distribution lying below a positive z score, add the tabled value to the 50% of the other half of the distribution.
2. For locating the percent of the distribution lying below a negative z score, subtract the tabled value from the 50% of the left half of the distribution.

While these may be helpful to some students, they can also be confusing because there are so many different research questions that could be asked. Generating a rule for each question would present an additional burden for the student to remember!! Here is another example that illustrates this point. *What percent of the standard normal curve lies between a z score of −1.96 and −1.35?*

Look at Figure 7.6, which illustrates the solution to this problem.

Visualizing the distribution is helpful so that you can keep the "order of operations" straight.

- Using the table of values, we find that 47.50% of the curve lies between the mean and −1.96.
- The table of values identifies 41.15% of the area between the mean and −1.35.
- Subtracting these areas identifies 6.35% as the total area of the distribution that lies between z scores of −1.96 and −1.35.

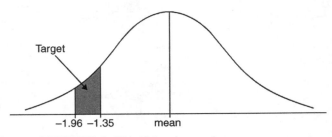

FIGURE 7.6 Identifying the area between z scores.

NAVIGATING THE z-SCORE DISTRIBUTION

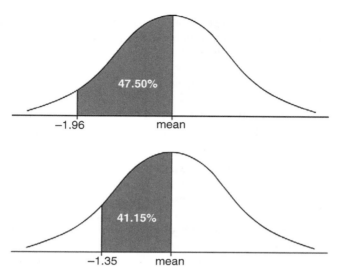

FIGURE 7.7 Identifying the tabled values of z scores of −1.96 and −1.35.

Figures 7.7 and 7.8 show the visual progression that will enable you to answer this question. Figure 7.7 shows the areas between the mean and the different z scores (i.e., the tabled values), and Figure 7.8 shows how to treat the two areas to respond to the question.

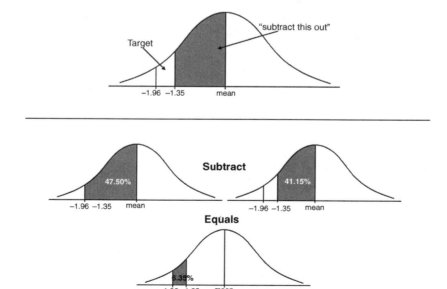

FIGURE 7.8 Subtracting the areas to identify the given area of the distribution.

If we created a rule for this operation, it would be something like, "in order to identify the area between two negative z scores, subtract the tabled values of the two scores." Again, this might be useful, but adds to the list of rules we already generated above.

Other situations for which we would need to generate rules would be as follows:

- Identifying the area between two positive z scores.
- Identifying the area between one positive and one negative z score.
- Identifying the area that lies above a negative z score.
- Identifying the area that lies above a positive z score.

There are many other potential rules as well. My overall point is that I think it is better to *visualize* the distribution and then highlight which area percentage you need to identify. In this method, there is only one rule: *Draw a picture of the curve and shade in the portion of the curve that you need to identify.*

Here is an example for you to visualize. *What percentage of the distribution falls between the z scores of -1.96 and $+1.96$?* This one might be easy now that we have used these values before, but if you use the "visualization" method, simply freehand draw a normal distribution and shade in the target area (i.e., the percentage of the distribution you need to answer the question). It might look like the drawings in Figure 7.9.

Visualization provides the student with the means to answer any such question. Essentially, after drawing the figures, you can use the table of values and then manipulate them in whatever way you need to identify the shaded areas you identified. Memorizing all the possible rules for identification of percentages of areas seems to me to be more burdensome and complex.

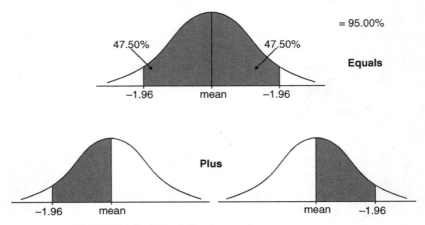

FIGURE 7.9 Visualizing the area between two z scores.

CALCULATING z SCORES

A z score is very important score because it provides a perfect standard of measurement that we can use for comparison to raw score distributions that may not be perfectly normally distributed. Remember, the standard normal curve is a perfect curve. When we create distributions in real life, we often make assumptions that the raw score data *approximate* a normal distribution. Raw score distributions, like math achievement of our sample of 40 schools we saw earlier, are near normally distributed, but still not perfectly so. Look again at Figure 5.25 to see the difference between the raw score of the histogram and the perfect normal curve that is superimposed on the histogram.

Because raw score distributions are not always perfectly distributed, we must perform descriptive analyses to see if they are within normal boundaries. Thus, by looking at the skewness and kurtosis of a distribution, we can see if the raw score distribution is balanced and close to a normal shape; if the mean, median, and mode are on the same point (or close), that is another indication that the data approximate a normal distribution. Finally, we can use the visual evidence of the histogram and frequency polygon to help us understand the shape of the distribution.

Suppose we have a set of students' test scores that have a mean of 100 and an SD of 15. Let's say that one student's score on the test is 120 and his mother calls to ask how he did on the test. What would you tell her? Look at Figure 7.10.

The distribution of raw scores is not perfect because it is based on scores derived from real life, and probably not on a large group. For the purposes of the research and after assessing skewness, kurtosis, and so on, we might assume that the data are normally distributed. The difficulty is that, since it is not the standard normal curve, we cannot use the method we just described for computing a percentile (the percent of the scores that lie below the student's score). We can tell the mother that her son

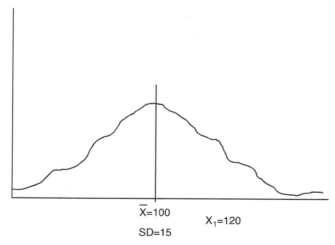

FIGURE 7.10 The raw score distribution.

scored more than one standard deviation above the mean, if that would be of value to her. What we really need, however, is to translate the student's raw score of 120 into a z score so that we can calculate a meaningful percentile.

Before we do this, however, we might try our estimation skills just by looking at the raw score information. The student's score is 120, which is 20 points above the mean. Because the SD is 15, that means that the student's score of 120 is 1 and 1/3 SD above the mean (or approximately 1.33 standard deviations). This means a ballpark estimate of the percentile as follows:

50% (left half of distribution)
+ 34.13% (distance between mean and SD 1)
+ 4.48% (approximately 0.33 of the distance between SD 1 and SD 2)
Total estimate = cumulative proportion of 88.61% or the 89th percentile.

We can *transform* raw scores to z scores in order to understand where the scores in a raw score distribution fall in relation to the other scores by making use of the standard normal distribution. This process involves visualizing and calculating where a certain raw score would fall on a standard normal distribution. That is, we would "translate" the x (raw score) to a z (standard score) to enable us to understand where the raw score is in relation to its own mean, and so on. (I like to think of this as "ecstasy" because we are transforming x values to z values: x to z!) In doing so, we are using the standard normal curve as a kind of yardstick to help us compare information that we create in real life.

As an example, suppose that I want to calculate a certain school's math achievement score in relation to the other schools' scores in the sample of 40 schools. Using the data presented in Figure 7.11 (that I reproduced from Figure 5.25), consider the "percentile" of a school with a math achievement score of 69. You can identify where 69 falls on the x axis, but how many "standard deviation units" above the mean does this score represent?

The formula for transforming raw score values (x) to standard normal scores (z) is as follows, where X is the raw score, M is the mean, and SD is the (population) SD:

$$Z = \frac{X - M}{SD}$$

Using this formula and the data reported from Figure 7.12, we can calculate the value of the z score:

$$Z = \frac{X - M}{SD}, \quad Z = \frac{69 - 50.6}{17.68}, \quad Z = 1.03$$

Mathematically, this formula transforms raw scores into standard deviation units, which is the definition of a z score. Remember that z scores in the standard

CALCULATING z SCORES

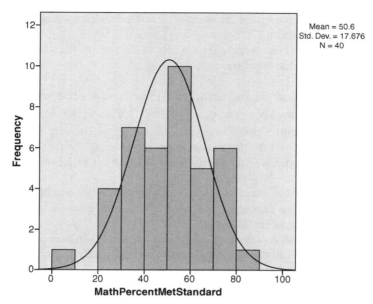

FIGURE 7.11 The histogram of math achievement values in a sample of 40 schools.

normal curve are expressed in standard deviation units. The numerator $(X - M)$ identifies how far the raw score is from its mean, or the deviation amount we discussed in Chapter 6. When the deviation amount is divided by SD, the denominator, it transforms the deviation amount into standard deviation units, since dividing something by a number creates a solution with the character of that number. (It is

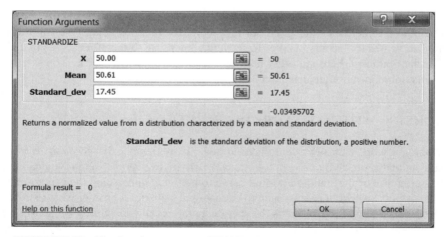

FIGURE 7.12 The function argument STANDARDIZE with math achievement data.

like miles per hour created from the number of miles divided by time; the result is distance per unit of time.)

The raw score of 69 in our example translates to a z score of 1.03. Just by looking at the z score, you can identify that:

- The raw score lies above the mean, and is therefore a positive value.
- The score is slightly above one standard deviation above the mean.
- The score represents a cumulative proportion of approximately 84.13%, or the 84th percentile (since the percent between the mean and the first SD is 34.13% and adding 50% yields 84.13%, which translates to the 84th percentile).

Consulting the z table of values only changes this value slightly to the 85th percentile (since the area between a z value of 1.03 and the mean is 34.85%, which, when added to 50%, becomes 84.85%).

Now, recall the mother who called in about her son's achievement score. We provided an estimate of her son's performance (approximately the 89th percentile), but let's use the z-score formula to get a precise figure.

$$Z = \frac{X - M}{SD}, \quad Z = \frac{120 - 100}{15}, \quad Z = 1.33$$

Consulting the z table of values, we find that the percent of the distribution between the mean and the z score of 1.33 is 40.82%. When we add the left half of the distribution (50%), we arrive at a cumulative percentage of 90.82% (50% + 40.82% = 90.82%, or the 91st percentile). Our estimate of the 89th percentile was close to the actual 91st percentile but not exact because we did not take into account the uneven areas of the curve represented by the first 1/3 of the second SD.

The bottom line of all these examples is that we can use the standard normal distribution to help us understand our raw score distributions better. We can translate raw scores to z scores in order to see how they relate to the other scores in their own distribution. There are other uses of the standard normal distribution that we will examine in subsequent sections.

WORKING WITH RAW SCORE DISTRIBUTIONS

When you work with raw score distributions, just remember that in order to compare student scores, you need first to transform them to z scores so that you can use the perfection of the standard normal curve to help you understand the original raw scores. As is critical in an approach to educational statistics, consider what research question is being asked: What is the specific nature of the question? What ultimate outcome is required? Using the visualization and estimation processes we discussed above, you can proceed with z transformations and calculation of percentiles, if that is the desired outcome.

USING EXCEL® TO CREATE z SCORES AND CUMULATIVE PROPORTIONS

There are a variety of ways to use Excel® with the normal distribution features we have discussed in this chapter. In fact, in Chapter 6 we have already discussed one way, namely, the percentile function (see Figures 6.3 and 6.4). This function returns the percentile of a point in the distribution of raw scores that you specify. However, as we discussed earlier, this function may be problematic with tied values.

There are three additional functions we can use that employ the z-score formula. I will use the school math achievement distribution to illustrate these (see Table 5.4).

STANDARDIZE Function

This function, accessible in the same way as the other Excel® functions (refer to the discussions in Chapter 6) makes use of the z-score transformation formula we discussed above:

$$Z = \frac{X - M}{SD}$$

Figure 7.12 shows the specification of the values we are using in STANDARDIZE to create a z score. Note that I chose a raw score (X) value of 50.00 and entered the mean (50.61) and the SD (17.45). I chose to use the population SD (which I created using the STDEVP function) rather than the default inferential SD used by Excel®. If I chose the OK button, the z-score value will be pasted into the cell I chose in the spreadsheet. However, you can see that it is listed below the three input values as $z = -0.03495702$. You can calculate the same value using a calculator as shown below:

$$Z = \frac{X - M}{SD}, \quad Z = \frac{50.00 - 50.61}{17.45}, \quad Z = -0.034957$$

The STANDARDIZE function therefore uses the z-score transformation formula to return z scores from specified raw scores when you provide the mean and SD.

By the way, you can get the same outcome by *directly* entering the formula into a spreadsheet cell. If you look at Figure 7.13, you will see that I have entered the formula directly into the cell (B2) adjacent to the value I want to transform to a z score (37.70). You can see the formula, with the appropriate values (raw score, mean, and SD) located in the formula band directly above the label row in the spreadsheet. I entered the formula into a cell I selected outside the column of data (in this case, I chose cell B2), hit the "Enter" key, and Excel® returned the value of (-0.74), the z-score transformation of the raw score value of 37.70.

As I discussed, using the equals sign notifies Excel® that you are entering a formula. I did so in B2, but it is not shown because, for this example, I wanted to show how the value is returned and how Excel® places the formula in the formula bar. If I make an error, I can make corrections in that band more easily than in the

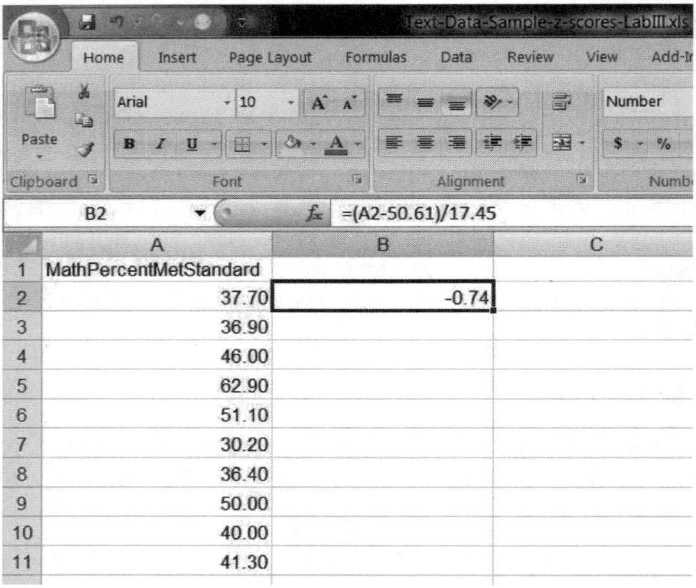

FIGURE 7.13 Entering formulas directly in Excel® using the Enter key.

cell where I input the formula. For this example, I hit return after I created the formula in cell B2 and Excel® returned the z score for $x = 37.70$.

I created this example for another reason. You can use Excel® to "replicate" actions by selecting and dragging cells. So, in the formula I created, I specified the cell location (A2) in the formula rather than inputting the raw score value of 37.70. This is so I could "drag" the formula in cell B2 down the B column. You can do this by selecting the cell with the formula (B2 in this example); when you move your cursor around the cell, you will see a variety of cursor styles. By hovering over the bottom right corner of the cell, you get a cursor style that looks like a "+" sign. By clicking and holding this cursor, drag the formula in cell B2 down the B column and Excel® will replicate the formula for each value in the adjacent data column (column A in this example). Figure 7.14 shows the results of this operation for several of the values of the data sample.

You can see the z scores in column B that correspond to the raw scores in Column A. For example, row 10, column A shows a raw score value of 40.00, and column B, row 10 shows this to be a z score of -0.61. (Confirm this with your calculator.) Likewise, the raw score value of 73.90 (in column A, row 13) is transformed to a z score of 1.33 (in column B, row 13). This procedure is a quick way to generate z-score values from a set of raw scores in Excel®.

Excel® has other ways of helping you to calculate z scores and cumulative proportions. I will review two of these below. The names of the functions are similar, so they may be confusing, but they are quite different. The NORMSDIST function calculates the cumulative proportion of the normal distribution below a given z

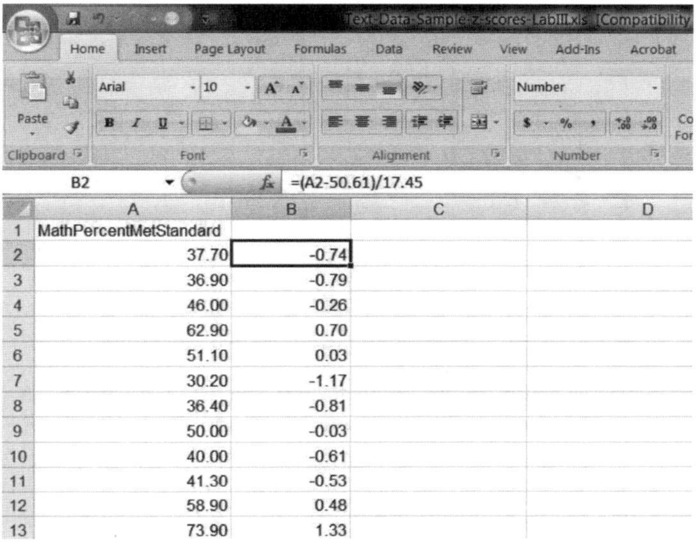

USING EXCEL® TO CREATE z SCORES AND CUMULATIVE PROPORTIONS

[Excel spreadsheet showing formula =(A2-50.61)/17.45 in cell B2, with column A "MathPercentMetStandard" values: 37.70, 36.90, 46.00, 62.90, 51.10, 30.20, 36.40, 50.00, 40.00, 41.30, 58.90, 73.90 and column B values: -0.74, -0.79, -0.26, 0.70, 0.03, -1.17, -0.81, -0.03, -0.61, -0.53, 0.48, 1.33]

FIGURE 7.14 "Dragging" the formula in Excel®.

score that you enter. The NORMDIST function calculates the cumulative proportion of the normal distribution below a point that is calculated from the raw score, mean, and standard deviation values (i.e., the data upon which a z score is calculated) which you enter.

NORMSDIST Function

This Excel® function is quite helpful because it uses an embedded z-score table of values. It is quite simple to use. Figure 7.15 shows the NORMSDIST function sub-menu deriving from the statistics formula menus. The user simply inputs a z score and Excel® returns the proportion of the standard normal

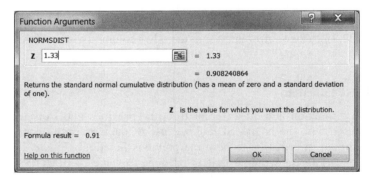

FIGURE 7.15 Using the Excel® NORMSDIST function.

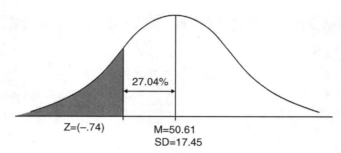

FIGURE 7.16 The z-score distribution identifying the area below $z = -0.74$.

distribution that lies below this value (essentially, the percentile). As you can see from Figure 7.15, I entered a z score of 1.33 and the value "0.908240864" is returned immediately below the 1.33 number. This 0.908 value (rounded to three decimals) is the percent of the area below the z score of 1.33. Thus, 90.8% of the area of the standard normal distribution lies below $z = 1.33$. It is therefore the 91st percentile. I chose this value purposely to confirm our analysis earlier for the mother who wanted to know her son's achievement score. If you recall, we calculated his z score to be 1.33, in the 91st percentile.

The NORMSDIST function works with negative z scores as well. Simply specify a negative value and it returns the area of the distribution below the value. For example, we created the z score of -0.74 using the STANDARDIZE function (see Figure 7.16). If we enter this value in the NORMSDIST function, it returns the value .2296. Because this is a negative value, it lies to the left of the mean, so you must use the visualization I discussed earlier to identify the value of .2296.

Figure 7.16 shows a picture of the curve with the appropriate data. Because we want to identify the area below the z score of -0.74, we must use a z table of values to identify the percent of the standard normal distribution between the mean and the z score. We find this to be 27.04%. In order to calculate the area below the z score, we need to subtract out the 27.04% from 50% to get 22.96%. This is the same value identified in Excel® using the NORMSDIST function.

Once again, note that you can get the same value from the z-score table by looking in the "tail" column of the row corresponding to $z = -0.74$. This value is 0.2297 or 22.97%, the same value using the other column (with a slight rounding difference).

NORMDIST Function

This Excel® function is a bit different from the previous two I discussed. It is often confused with the NORMSDIST because of the similarity of the name. It produces similar information to both the previous functions. Figure 7.17 shows the function argument window that is produced by the NORMDIST formula sub-menu. As you can see, I specified the first raw score value in our dataset

USING SPSS® TO CREATE z SCORES

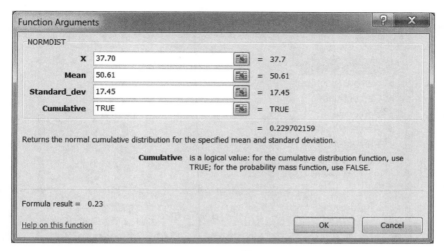

FIGURE 7.17 Using the Excel® NORMDIST function.

(37.70) so you can reference the output to our earlier example in Figure 7.14. As you can see in Figure 7.17, NORMDIST returns the value "0.229702159" below the right-hand "TRUE" entry. You will recognize this as the area of the standard normal distribution that lies below the raw score value of 37.70 and its z score equivalent of −0.74. Both STANDARDIZE and NORMDIST create z scores from raw scores, and NORMSDIST calculates the cumulative proportion from the z score. NORMDIST skips the step of reporting the z score by calculating the cumulative proportion directly from the raw score. The "Cumulative" information provided by the NORMDIST function output can be either "TRUE" or "FALSE," depending on your needs. I input TRUE because I wanted the percentile (or the "cumulative distribution function"). I can use FALSE to calculate the probability of the given score, but that is the subject of a later chapter. We will return to this function.

USING SPSS® TO CREATE z SCORES

Using SPSS® to create z scores is very easy using the "Descriptives–Descriptives" selection from the "Analyze" menu. Figure 7.18 shows the specification window in which I have identified the math achievement variable ("MathPercentMetStandard"). I can use the "Options" button to further specify mean, skewness, and a variety of other statistical procedures. However, I can create z scores by simply checking the box at the lower left corner of the Descriptives menu window. It is shown in Figure 7.18 as the box "Save Standardized values as variables." When I check this box, a new variable is added to the dataset, one consisting of the z-score values corresponding to the raw score values

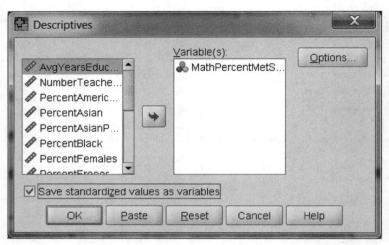

FIGURE 7.18 Using SPSS® to create z scores.

of the (math achievement) variable selected. It is called "ZMathPercentMetStandard," which is simply the name of the original variable with a "Z" on the front to indicate that the values are z scores. Figure 7.19 shows the new variable beside the original.

Please note that the z scores created in SPSS® use the inferential SD, so they will be slightly different if you used the population SD, as I did, with Excel®. Compare the same school value to see the differences—they will be very slight.

	MathPercentMetStandard	ZMathPercentMetStandard	var
1	50.000	-.03423	
2	42.200	-.47550	
3	37.700	-.73008	
4	41.300	-.52642	
5	62.500	.67294	
6	69.200	1.05199	
7	9.700	-2.31415	
8	27.900	-1.28450	
9	24.300	-1.48817	
10	34.200	-.92809	
11	26.500	-1.36371	
12	36.400	-.80363	
13	30.200	-1.15439	

FIGURE 7.19 Creating a z-score variable using an SPSS® descriptive menu.

TERMS AND CONCEPTS

Frequency Polygon A graph that is formed by joining the midpoints of the bars of a histogram by a line.

Standard Normal Distribution A normal distribution that is "perfectly" shaped such that the percentages of the area under the curve are distributed in known and standard amounts around the mean. The mean has a value of 0 and the SD = 1. Also known as the "z distribution."

z score A raw score expressed in standard deviation units. Also known as a "standard score" when viewed as scores of a standard normal distribution.

REAL-WORLD LAB III: THE NORMAL CURVE AND z SCORES

In Chapter 6 I introduced you to the TAGLIT database that consists of school-level data from school leaders and students regarding the role of technology in teaching and learning. The earlier example focused on student to teacher ratios as a way of demonstrating how to use Excel® and SPSS® to generate percentiles. In this Real-World Lab, we will use similar data but focus on student–teacher ratios from the Washington State database. Use the data in Table 7.1 to complete this Real-World Lab.

TABLE 7.1 Student–Teacher Data from Washington State

StudentsPerClassroomTeacher	
12	27
16	12
14	24
16	17
15	15
21	17
12	12
15	18
14	15
20	15
21	14
18	16
10	17
17	15
17	18
18	9
20	12
21	16
19	17
15	13

1. What is the descriptive data for the "StudentsPerClassroomTeacher" variable from the Washington School database?
2. Create z scores for each of the student–teacher ratios.
3. Which, if any, of the z scores would you consider "extreme"?
4. Comment on your findings.

REAL-WORLD LAB III: SOLUTIONS

1. What is the descriptive data for the "StudentsPerClassroomTeacher" variable from the Washington School database?

   Figures 7.20 and 7.21 show the descriptive analyses for Excel® and SPSS®, respectively. Both reports suggest that the data are fairly normally distributed. The mean, median, and mode lie close to the same point, and the skewness and kurtosis data are within normal boundaries. This observation is confirmed visually by the histogram in Figure 7.22. Although there are a couple of observations on the upper end of the distribution that may be pulling the mean up a bit, the distribution appears to be normal.

2. Create z scores for each of the student–teacher ratios.

   Table 7.2 shows the z scores of the student–teacher ratio data from Table 7.1. The z scores in Table 7.2 are from Excel®, in which I used the Standardize formula to calculate the first raw score (12); I then created the other z scores by using the method of dragging the cell formula down across all raw scores. The z scores generated from SPSS®, using the descriptive–descriptive command, reveal identical values as the Excel® values.

3. Which, if any, of the z scores would you consider "extreme"?

   Table 7.3 is identical to Table 7.2 except that I sorted the z-score values in order to make it easier to identify extreme values. How large or small does a

StudentsPerClassroomTeacher	
Mean	16.25
Standard error	0.57
Median	16.00
Mode	15.00
Standard deviation	3.63
Sample variance	13.17
Kurtosis	1.09
Skewness	0.61
Range	18.00
Minimum	9.00
Maximum	27.00
Sum	650.00
Count	40.00

FIGURE 7.20 Student–teacher ratio descriptive data from Excel®.

REAL-WORLD LAB III: SOLUTIONS

StudentsPerClassroomTeacher	
N valid	40
N missing	0
Mean	16.25
Median	16.00
Mode	15
Standard deviation	3.629
Variance	13.167
Skewness	.615
Standard error of skewness	.374
Kurtosis	1.086
Standard error of kurtosis	.733
Range	18

FIGURE 7.21 Student–teacher ratio descriptive data from SPSS®

value have to be in order to be called "extreme"? The answer to this question is somewhat subjective, although we have examined some aspects of the normal distribution that can help us answer this question. If you recall, the middle 4 standard deviations of the standard normal curve defined by the 2 SDs to the left of the mean (negative side) and the 2 SDs to the right of the mean (positive side) comprises about 95% of the total area under the curve (47.72% on each side).

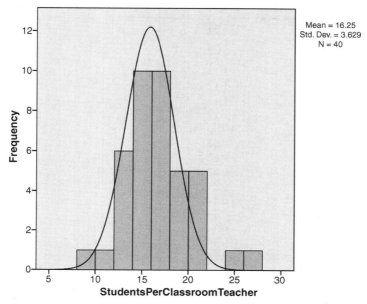

FIGURE 7.22 Student–teacher ratio histogram from SPSS®.

TABLE 7.2 z Scores Using Excel® Standardize Formula

StudentsPerClassroomTeacher	z Scores	StudentsPerClassroomTeacher	z Scores
12	−1.17	27	2.96
16	−0.07	12	−1.17
14	−0.62	24	2.13
16	−0.07	17	0.21
15	−0.34	15	−0.34
21	1.31	17	0.21
12	−1.17	12	−1.17
15	−0.34	18	0.48
14	−0.62	15	−0.34
20	1.03	15	−0.34
21	1.31	14	−0.62
18	0.48	16	−0.07
10	−1.72	17	0.21
17	0.21	15	−0.34
17	0.21	18	0.48
18	0.48	9	−2.00
20	1.03	12	−1.17
21	1.31	16	−0.07
19	0.76	17	0.21
15	−0.34	13	−0.90

TABLE 7.3 Sorted z Scores for Identifying Extreme Values

StudentsPerClassroomTeacher	z Scores	StudentsPerClassroomTeacher	z Scores
9	−2.00	16	−0.07
10	−1.72	16	−0.07
12	−1.17	17	0.21
12	−1.17	17	0.21
12	−1.17	17	0.21
12	−1.17	17	0.21
12	−1.17	17	0.21
13	−0.90	17	0.21
14	−0.62	18	0.48
14	−0.62	18	0.48
14	−0.62	18	0.48
15	−0.34	18	0.48
15	−0.34	19	0.76
15	−0.34	20	1.03
15	−0.34	20	1.03
15	−0.34	21	1.31
15	−0.34	21	1.31
15	−0.34	21	1.31
16	−0.07	24	2.13
16	−0.07	27	2.96

While there is nothing magical about the "mean ± 2 SD" value, it does represent the great majority of the area of the distribution, and statisticians use it to help establish a common benchmark for various aspects of statistical procedures. Using this criterion, examine the data in Table 7.3. You will find three scores that lie at or above these amounts. I shaded three values: z scores of -2.00, $+2.13$, and $+2.96$. The highest z-score value of 2.96 is nearly 3 SDs above the mean, which represents a very small percentage of the distribution (approximately 2.15% of the distribution falls between 2 and 3 SDs).

4. Comment on your findings.

The Washington State sample ($N = 40$) of student–teacher classroom ratios appear to be approximately normally distributed with a mean of 16.25 and SD of 3.63 (inferential SD). Schools in this sample range from student–teacher ratios of 9 to ratios of 27. Mean, median, and mode values are close to one another, and skewness and kurtosis values are within normal bounds. The histogram also confirms normal tendency with a slight positive skew possibly due to one extreme score. The student–teacher ratios are slightly higher than the national data sample (TAGLIT) we discussed in Chapter 6, but not markedly so.

The z scores in the Washington State sample identify perhaps three schools that are somewhat extreme, using a ± 2 SD criterion. Of the three, one score is clearly extreme, representing a school with a student–teacher ratio of 27:1.

8

THE Z DISTRIBUTION AND PROBABILITY

The z distribution is very important to statistics because it provides an ideal model that we can use to understand the raw data distributions we create. In fact, in this book, the z distribution straddles both descriptive and inferential statistics, the "branches" of statistics I introduced earlier. We have seen how we can use the z distribution to understand raw scores by transforming them to z scores, and we have learned how to create and manage cumulative percentages. We will continue this discussion to complete our descriptive statistics section, and we will then consider how the standard normal distribution can help us to understand inferential statistics.

Chapter 7 discussed both z scores and cumulative percentages, focusing on how to use the z table to transform z scores. As you recall, the z-score formula is as follows:

$$z = \frac{X - M}{SD}$$

There are many uses for this formula, as you will see throughout the book. For now it is enough to know that sometimes you may have different kinds of information available and you can use the formula to help you with various analyses of your data. You can actually use simple algebra to solve for the different parts of the formula. We will consider one such formula because it will become very important to our later statistical procedures.

Understanding Educational Statistics Using Microsoft Excel® and SPSS®. By Martin Lee Abbott.
© 2011 John Wiley & Sons, Inc. Published 2011 by John Wiley & Sons, Inc.

TRANSFORMING A z SCORE TO A RAW SCORE

In Chapter 7, we used the hypothetical example of a mother trying to understand the test score of her child. Using this same scenario, let us suppose that the student's mother was informed that on a certain test with a mean of 100 and a standard deviation of 15, her son got a z score of -1.64. Not being sophisticated in the interpretation of z scores, she may be interested in what the raw score was on the test. We can use the z-score formula to calculate the raw score from the z-score data and the descriptive information about the distribution. First, we should use the information we have to estimate what a solution might be.

What can we observe generally? Certainly, because z scores are expressed in standard deviation units on the standard normal distribution, the student probably did not perform highly on the test (i.e., because negative scores are on the left of the distribution—in this case more than 1.5 SDs to the left of the mean). Consider the following formula, which is derived from the z-score formula listed above:

$$X = Z(\text{SD}) + M$$

In this formula, the z score we need to transform to a raw score (X) is known, along with the SD and mean of the raw score distribution. Substituting the values we listed above, we obtain

$$X = -1.64(15) + 100$$
$$X = 75.40$$

So, we can inform the student's mother that her child received 75.40 on the test. Although we don't know how that might translate into a teacher's grade, it probably has more meaning to a parent who does not normally see z scores. To the researcher, however, the z score contains more information.

TRANSFORMING CUMULATIVE PROPORTIONS TO z SCORES

Another situation may arise in which we have cumulative proportions or percentiles available and wish to transform them to z scores. This is a fairly easy step because both are based on z scores.

In our previous example, the student's inquisitive mother would probably have been given a percentile rather than a z score because the educational system uses percentiles extensively as the means to make comparisons among scores. Here is a brief example, again using the mother. Suppose the student's mother was told that her son received a score that was at the 60th percentile. What would be the z-score equivalent?

TRANSFORMING CUMULATIVE PROPORTIONS TO z SCORES

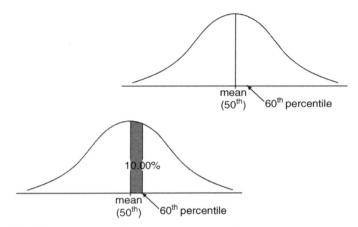

FIGURE 8.1 Visualizing the transformation of a percentile to a z score.

As we learned in previous sections, it might be good to first try to visualize the solution. Figure 8.1 shows how you might visualize this problem. In the top panel of Figure 8.1, we can see that a percentile score of 60 (or a cumulative percentage of 60%) would fall to the right of the mean by definition (i.e., because the score surpasses 60% of all the scores, which is above the 50% mark).

Because the 60th percentile is 10% away from the mean (or the 50th percentile), we know that the percent of the distribution that lies between the mean and the target score is 10%. We know that the distance between the mean and the first SD on a z distribution contains about 34.13%, so our percentile score of 60 should represent an approximate z score of 0.29 (10% divided by 34.13%), using a ballpark estimate.

If you recall, knowing that the area of the distribution between the percentile score and the mean is 10% gives us additional information since this is what the z-score table provides. We can now use the z-score table *in reverse* to "finish" our visualization of the answer. If you locate the closest percentage to 10.00% that you can find in the table (among the first of the column "pairs" of percentages), you find 9.87% and 10.26%, corresponding to z scores of 0.25 and 0.26, respectively. Because 10.00% is closest to 9.87%, we can conclude that a percentile score with this distance from the mean represents a z score of 0.25. (Locate 9.87% in the table, note the 0.2 in the first z column which is on the same row, add the 0.05 from the z figure above in the z-score numbered columns for a total of 0.25: that is, $0.2 + 0.05 = 0.25$.)

A second method of arriving at the same z score of 0.25 is to use the second of the column pairs of percentages, those that are listed in the "tails" of the distribution. Knowing that the percentile score of 60 also creates a cumulative proportion of 40% in the tail, look for the closest value to 40.00% among the "tail" columns and you will find that the closest value is $z = 0.25$.

DERIVING SAMPLE SCORES FROM CUMULATIVE PERCENTAGES

This is another process that involves an understanding of the area in the standard normal distribution and how to transform scores. In essence, it simply combines the procedures we have already covered. I will use an example from the school database we have been using from the state of Washington.[1]

Administrators in schools with fourth-grade students recorded the 2009 results of math and reading achievement tests, as we have see with earlier examples. The results were "aggregated"; that is, all individual (fourth grade) student achievement test scores were averaged across the entire school. We have examined the reading and math achievement scores at the school level from a sample of the schools. We also examined other aggregate variables, like the student–teacher ratio. We turn now to another aggregate variable that describes the overall school, "Percent Free or Reduced Price Meals" (F/R). This demographic variable aggregates the percent of students in a school who are eligible for free or reduced price meals because of their family income level. This is not a perfect measure of family income for various reasons. However, in education research, it is often the only measure available to the researcher.[2]

Our sample schools ($N = 40$) show a mean of 50.16 and a (population) SD of 26.54 on the F/R variable, and the data appear normal according to the descriptive procedures we discussed earlier. Using this descriptive information, which sample (school) score would represent a cumulative percentage of 67% (i.e., fall at the 67th percentile) of this sample of data?

Using the admonition to try visualization before we use a formula, how can we represent the question? Figure 8.2 shows how you can "see" this with a graph of the distribution. As you can see, the 67th percentile, the "target" school score, lies well above the mean. Because 34.13% of the distribution lies between the mean and SD

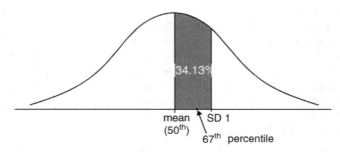

FIGURE 8.2 Visualizing the 67th percentile of the F/R distribution.

---

[1] Data were downloaded from the website of the Office of the Superintendent of Public Instruction at http://www.k12.wa.us/ and were used by permission.
[2] A further dynamic of the data is that F/R measures all the students in the schools that have fourth-grade scores in reading and math, not just the percent of fourth graders who took the test.

1, our target score should be approximately at the middle of this area [because 17%, the distance beyond 50% (the mean of the distribution), is about half of 34.13%], although a bit closer to the mean given the shape of the curve. Therefore, we can visualize a z score of about 0.5. With the z score, we can say the school score would be approximately 63, determined as follows:

1. Our visual estimate of 0.5 z scores represents SD units; therefore the target score represents approximately one-half of the sample score SD of 26.54.
2. This amount plus the mean of 50.16 equals approximately 63.

Checking this visual calculation with the formula I presented above, we can see how close we were to the actual sample score value. The 67th percentile falls at the z score of 0.44 (17% above the mean or 37% in the tail), so, substituting the values in the formula, we find the sample score to be 61.84%. Therefore, schools with 61.84% F/R lunch qualification are at the 67th percentile of our data sample. Our estimate (63%) was not perfect, but it helped us to check our analyses and try to see the relationship among all the various pieces of data.

$$X = 0.44(26.54) + 50.16$$
$$X = 61.84\%$$

ADDITIONAL TRANSFORMATIONS USING THE STANDARD NORMAL DISTRIBUTION

In education, researchers often come across a variety of scores and need to understand how they are derived. Knowing about these helps to understand students' performance, but also how to interpret the information more completely.

Normal Curve Equivalent

We have already discussed one such transformation, the *normal curve equivalent* (NCE), in Chapter 6. As you recall, these scores are important because they transform percentile scores into interval data by making the distance between percentiles equal. While this is a more important feature to researchers than to parents or teachers, it nevertheless shows how we can use the standard normal distribution to understand features of our derived (raw score) data.

Stanine

Stanine (standard nine) scores are scores created by transformations using the standard normal distribution. Used in education, these scores divide the standard normal distribution into nine parts. (The z scores we have discussed divide the distribution into about 6 parts by comparison.) The stanine has a mean of 5 and an SD of 2. So, by changing our formula a bit, we can calculate the stanine, rather than

a raw score, as follows:
$$X = Z(SD) + M$$
$$\text{Stanine} = Z(2) + 5$$

Thus, from our example, we have
$$\text{Stanine} = 0.44(2) + 5$$
$$\text{Stanine} = 5.88$$

T Score

The *T score* is another transformation used in education because it transforms z scores into more intuitively understandable scores; Z scores are both positive and negative, and they have a narrow range (from about -3.00 to $+3.00$). T scores transform z scores into scores that are always positive, and they have a larger range. T scores have a mean of 50 and an SD of 10. We can use the same formula to transform z scores to T scores:
$$X = Z(SD) + M$$
$$T \text{ score} = Z(10) + 50$$

Using our previous example:
$$T \text{ score} = 0.44(10) + 50$$
$$T \text{ score} = 54.40$$

Using the standard normal distribution and the z score, you can therefore make any sort of transformation. (If you wanted to create your own score, you can use the same process.) We will use these extensively in our study.

Grade Equivalent Score

Another score used by educators is the *grade equivalent score* (GES). While these can have some utility for comparison purposes, you should see that they are somewhat subjective and depend on the composition of a comparison group. The scores are expressed in grades and months based on the typical score of a comparison group of students during a nine-month school year. Thus, if my child has a GES of 3.5, this represents her estimate of performance representative of a comparison group of students at the third-grade level in the 5th month (or January). Because of their derivation, researchers should use great caution when using GESs in statistical procedures.

USING EXCEL® AND SPSS® TO TRANSFORM SCORES

There are no established formulas or menus in Excel® or SPSS® to make the transformations we discussed earlier in this chapter. Aside from the z-score

USING EXCEL® AND SPSS® TO TRANSFORM SCORES   133

transformations from raw score distributions that we reviewed in Chapter 7, we are on our own to use both of these statistical programs to create appropriate transformations.

These are not difficult to do using the programs as elaborate "calculators." I already demonstrated how to create z scores using Excel®. (Review the "STANDARDIZE Function" section of Chapter 7.) You can easily use SPSS® to do the same thing by using the "Transform" menu and then selecting "Compute." Review the section in Chapter 7, "Using SPSS® to Create z Scores." In that section, I showed how to create z scores by simply checking a box on the "Analyze–Descriptive Statistics –Descriptives" menu choice (see Figure 7.18).

You can use SPSS® to compute a new variable by entering the z-score formula, or other relevant formulas. To use the example of a z-score transformation, consider Figure 8.3, in which I use the "Transform" menu at the top of the page; then from that menu choose "Compute Variable," which will yield the sub-menu shown in Figure 8.3. As you can see, I created a new variable called "zFreeReduced" in the upper left window. Then, in the "Numeric Expression" window, I created the formula for calculating a z score from a raw score. I had to first include the existing raw score variable in the equation by selecting it from the "Type & Label" window and then moving it to the Numeric Expression window using the arrow button.

It is hoped that you will recognize this formula as simply the z-score formula from above:

$$Z = \frac{X - M}{SD}$$

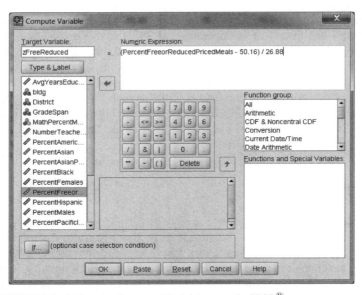

FIGURE 8.3 Using the Compute Variable menu in SPSS® to create z scores.

THE Z DISTRIBUTION AND PROBABILITY

	PercentFreeorReducedPricedMeals	zFreeReduced	ZPercentFreeorReducedPricedMeals	var
1	42.94	-.27	-.27	
2	13.27	-1.37	-1.37	
3	42.37	-.29	-.29	
4	53.80	.14	.14	
5	24.62	-.95	-.95	
6	28.86	-.79	-.79	
7	84.75	1.29	1.29	
8	91.59	1.54	1.54	
9	86.09	1.34	1.34	
10	68.15	.67	.67	
11	73.90	.88	.88	
12	65.36	.57	.57	
13	99.76	1.85	1.85	
14	76.52	.98	.98	
15	60.87	.40	.40	
16	81.29	1.16	1.16	
17	87.20	1.38	1.38	
18	49.38	-.03	-.03	
19	58.19	.30	.30	
20	32.07	-.67	-.67	
21	38.92	-.42	-.42	

FIGURE 8.4 SPSS® data file with two z-score variables.

In SPSS®, it looks like the Numeric Expression in Figure 8.3:

$$\text{zFreeReduced} = (\text{PercentFreeorReducedPricedMeals} - 50.16)/26.88$$

I obtained the mean and (inferential) SD from the Descriptives procedure I have used before. Thus, I simply indicated in SPSS® that I wanted to create a new variable that represents the raw score variable (X) minus the mean (50.16) divided by the SD (26.88). When I perform this operation, SPSS® makes the calculations that are identical to the z scores I obtained using the descriptive procedure. Figure 8.4 shows the SPSS® datafile that now contains two z-score variables.

The first variable shown in Figure 8.4 is the original raw score variable. The second is the new variable I created using the Transform–Compute process. The third variable is the variable SPSS® created when I chose to keep standard scores in the Descriptives–Descriptives menu. In both of the latter cases, all the transformed (z scores) scores are identical.

PROBABILITY

When I teach statistics at any level, I always shade in a certain date in the syllabus and call it the "headache day" because it is the day I begin to discuss inferential statistics. Really, I do this to make sure I have my students' attention! This is an interesting benchmark in statistics because it requires a shift of thinking from working with *individual raw scores* to working with *samples of raw scores and their relationship to the populations from which the samples supposedly came*.

Up to now, we have discussed and used examples of sets of raw scores to examine descriptive statistics. We concentrated on how to use statistical procedures to better describe data and to do it in such a way that we will gain a fresh perspective on what the data may mean. *Now, we turn our attention to a different level of inquiry.*

Determinism Versus Probability

In this chapter, we will only briefly look at the topic of probability from a mathematical perspective. To be sure, we could spend a great deal of time on this topic; and if this were a mathematics book, we would do just that. I would like to focus on the *application* of probability to the kinds of problems we often encounter in conducting research and, in so doing, begin to shift our attention to inferential statistics.

Human actions are rarely, if ever, determined. However, it is also the case that human action is fairly predictable. One has only to consider the many ways in which the things we (think) we choose to do are really those things that are expected. Marketing specialists have made billions of dollars on this principle by targeting the "baby boomer" generation alone! Insurance providers have known this for years. Sociologically, when people repeatedly act in society, they create patterns or ruts that are both helpful (in economizing action and energy) and potentially problematic (because it becomes difficult to act differently than the rut allows). The result is that human behavior can be characterized by "predictability."

What does this have to do with statistics? Plenty. We have already seen that behaviors, attitudes, and beliefs have a great deal of variability. Why is there such variability? Why do people not always believe the same thing, act the same way, and so on? In descriptive statistics, we learn the ways to understand the extent of the variability and whether the resultant distribution of behaviors and beliefs conforms to a normal distribution. But that does not explain the *why*.

Human actions and beliefs have many causes. We cannot understand all of them. The fact that variance exists may be due to our inability to understand the full range of forces acting on the individual at any particular moment. But it may also exist because we cannot fully explicate individual choice or action.

In trying to understand this complexity, we must recognize the importance of probability. *Probability involves the realm of expectation.* By observing and measuring actions and behaviors over time, we develop expectations that can help us better predict outcomes. If we observe students struggling with a certain subject area among schools over time, we might eventually predict the same outcome on future occasions, if there are no changes in conditions. Our expectation, being based in observation, will help us predict more accurately. This still does not explain *why* the students struggle, but it does turn the scrutiny toward the potential conditions for the struggle. We may never discover all the reasons, but the study of probability gets us closer to a more comprehensive understanding.

Elements of Probability

At this point, we could discuss several aspects of probability theory and measurement. However, focusing on the *mathematical* properties of probability takes us beyond the primary purpose of this book: to understand how we can use statistical procedures, along with Excel® and SPSS®, to make statistical decisions. We will discuss the nature of probability and how to measure it in the context of actual research data.

You may wish to explore the formal properties of probability beyond this book. To do so, you might examine the following topics:

- *The Binomial Distribution.* This is a distribution of probabilities for a variable with two outcomes like "success–failure," "true–false," or "heads–tails." Observing repeated occurrences of these variables, like tossing a coin repeatedly, yields a known distribution of probabilities.
- *Combining Probabilities.* There are different rules for how to combine the probability of events according to your interest in determining (1) whether one event occurs *or* another event occurs and (2) whether one event occurs *and* another occurs.
- *Combinations and Permutations.* These are categories of probability that focus on ordered sequences of events. In a group of 10 students, a *combination* is whether Jim and Suzie winning a spelling bee. They can place as "Jim first, Suzie second" or "Suzie first, Jim second." Either way, we are considering these as a combination. However, this combination contains two *permutations*; the order of finish is the focus. Thus, "Suzie first, Jim second" is a distinct permutation from "Jim first, Suzie second." Thus, the combination considers all possible ways a set of events can occur, whereas a permutation considers the order of the finish among the events.

Probability and the Normal Curve

If you think about the normal curve, you will realize that human actions can take a number of different courses. Most responses tend to be clustered together, but there will be some responses that fall in different directions away from the main cluster. Therefore, we can think of the normal curve as a visual representation of the fact that we do not have certainty, but have probability in matters of such things as attitudes and buying behavior, test scores, and aptitude.

In inferential statistics, statisticians think of the normal curve in terms of probability. Because approximately 68% of the area (or a proportion of 0.68) of the normal curve lies between one standard deviation, positive or negative, for example, we can think of any given case having a 0.68 probability of falling between one standard deviation on either side of the mean.

Empirical probability is simply the number of occurrences of a specific event divided by the total number of possible occurrences. If a student announces "I am here" at the top of their lungs when they enter the classroom, you can calculate the empirical probability of this event by observing the number of times the student makes the declaration divided by the number of days they enter the classroom. Thus, if they declare their presence 10 times out of 18 days of class, the probability of their making the declaration is

$$\text{Probability} = \frac{\text{Occurrences}}{\text{Possible occurrences}}$$

$$\text{Probability} = \frac{10\ (\text{Occurrences})}{18\ (\text{Possible occurrences})}$$

$$\text{Probability} = 0.556$$

Therefore, if we wanted to predict whether they would make the declaration on the next possible occasion, we would have slightly better than equal chance (0.556 versus 0.50) of predicting their behavior (i.e., that they would make the declaration). In this example, you can see how probability is measured: as a number between 0 (no likelihood) and 1.00 (complete likelihood). We can indicate these events as follows:

p = the probability of observing the outcome,
q = the probability of not observing the outcome

Therefore,

$$p + q = 1 \quad \text{and} \quad p = (1 - q)$$

Relationship of z Score and Probability

The primary issue in this book is to recognize that the z score and probability are very much related, since they both can be characterized by the proportion of the area under the standard normal curve. *We can therefore think about certain kinds of problems as probability statements.* Knowing what we do of the distribution of area in the standard normal curve, we can observe that possible scores "beyond" 2 SDs (in either a positive or negative direction) are in areas of the distribution where there are very few cases. Thus, randomly selecting a score from a distribution in these small areas would have a much smaller probability than randomly selecting a score nearer the mean of the distribution. In a normally distributed variable, only 5% of cases or so will fall outside the ± 2 SD area. Therefore, with 100 cases, selecting a case randomly from this area would represent a probability of 0.05 (since $5/100 = 0.05$).

Since the standard normal distribution represents percentages (as proportions of the entire area in the distribution) and probabilities, it is easy to convert one to the other. You can convert percentages to probability simply by dividing the percentage by 100 (%) as follows:

$$\text{Probability}\ (p) = \frac{68\%}{100\%} = \frac{68}{100} = 0.68$$

For example, consider a principal asking a teacher to select, at random, a student from a class to represent the entire class on an all-school council of "technology IQ." Furthermore, suppose that the teacher has just administered a test of technology knowledge to the class and the distribution of results has a mean of 95 and a standard deviation of 12. What would be the probability of selecting at random from the class a student who scored between 89 and 101 to represent the class on the council?

Using the procedures we learned in previous chapters, we can visualize the result and then calculate it specifically with the z-score formulas.

1. Remember that probability statements are based on the same information as z scores.
2. Consider where you need to end; that is, what information must you have and how can you work backwards to get it?
3. Draw or visualize a picture of the normal distribution with a shaded target area.

Figure 8.5 shows how you might visualize this problem. As you can see, with a mean of 95 and an SD of 12, the scores of 89 and 101 fall 0.5 a SD below and above the mean, respectively. Thus, both the lower and upper numbers are 0.5 z scores from the mean. With the distance from the mean to SD 1 (both positive and negative) at about 34.13%, the distance between 89 and 101 would occupy about 34% (since 17% and 17% = 34%). The probability would therefore be approximately 0.34 (34%/100%).

We can calculate the distances using the z-score formula:

$$Z = \frac{X - M}{SD}$$

Thus, the raw scores of 89 (X_1) and 101 (X_2) would have z scores (Z_1 and Z_2) calculated as follows:

$$Z_1 = \frac{89 - 95}{12}, \quad Z_1 = \frac{-6}{12}, \quad Z_1 = -0.5$$

$$Z_2 = \frac{101 - 95}{12}, \quad Z_2 = \frac{6}{12}, \quad Z_2 = 0.5$$

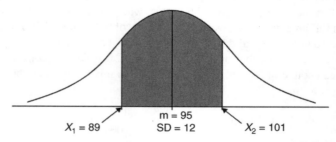

FIGURE 8.5 Visualizing the probabilities as a preliminary solution.

From the z-score table, we find that the area of the curve between the mean and z scores of 0.5 (positive and negative) is 19.15%. Therefore, 38.30% of the area of the distribution falls between scores of 89 and 101. This means that there is a probability of 0.38 of selecting at random a student whose score lies between 89 and 101 on our test of technology knowledge.

Our visualization ($p = 0.34$) was lower than the calculated probability (0.38) due to the shape of the curve. However, this exercise does point out how visualization can help you to "see" the dynamics and approximate solution to questions similar to this.

"Inside" and "Outside" Areas of the Standard Normal Distribution

Oftentimes, in statistics and in education, students are asked to identify probabilities of cases that lie inside or outside sets of scores in a distribution. For example, we may be asked to identify the middle 90% of the students who take a specific test; or, conversely, we may be asked to identify the 10% that lie outside, in the extremes of the distribution (the two "tails" of the curve). In both cases, you can use the information you have already learned about the normal distribution to solve these problems.

Calculating these areas is very helpful for you to understand inferential statistics. Statisticians and researchers have created conventions about which z scores create "exclusion areas" that can be used to compare with the results of actual studies. When calculated values from statistical studies fall within or beyond these exclusion areas (defined by specific z scores), researchers can make statistical decisions about whether the findings are "likely" or "beyond chance." We will look at two kinds of examples in which we can establish these exclusion areas using the F/R variable from our sample of schools.

Inside Area Example. Recall that our sample schools ($N = 40$) show a mean of 50.16 and a (population) SD of 26.54 on the F/R variable. What two raw scores "capture" the middle 90% of the distribution?

Using visualization, we are identifying the middle portion of the distribution encompassed by 45% to the left of the mean and 45% to the right of the mean (since we want to identify the raw scores on both sides of the 90%). We know that the first 2 SDs contain about 47.72% of the distribution on either side of the mean (for a total of 95.44%), which is close to our 45% target area on either side of the mean. Therefore, our raw scores should fall inside the -2 SD mark and the $+2$ SD mark. Figure 8.6 shows these areas.

Now, we calculate the raw score defining the area to the left (45%) of the distribution: $Z = 1.645$ (or 1.65, from the z table). Both Z values lie the same distance from the mean, but on different sides. Therefore, $Z_1 = -1.65$ and $Z_2 = 1.65$.

$$X_1 = Z_1(\text{SD}) + M, \quad X_1 = -1.65(26.54) + 50.16, \quad X_1 = 6.37$$

The raw score defining the right (45%) of the mean is

$$X_2 = 1.65(26.54) + 50.16, \quad X_2 = 93.95$$

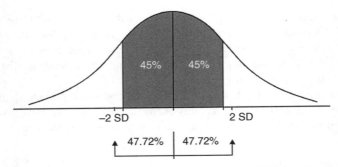

FIGURE 8.6 Visualizing the middle 90% area of the F/R sample values.

Therefore, the middle 90% of the sample distribution of schools' F/R values is bracketed by the two raw scores of 6.37 and 93.95. Typically, the middle 90% of the schools will have F/R values between these two values.

Outside Area Example

We can use a similar process to figure the outside portion of the distribution. This is the portion of the distribution that is in the two "tails" of the distribution. As you will see, this is an important identification because much of how we will use probability will involve the "exclusion area" of scores in the tails of the distribution.

Using the same data, identify the two raw scores that cut off 5% of the distribution and isolate it into the two tails. Recall that our sample schools ($N = 40$) show a mean of 50.16 and a (population) SD of 26.54 on the F/R variable.

Figure 8.7 shows how this looks so you can practice the visualization of the solution. Picture the distribution with small shaded areas (5%) distributed in the two tails of the distribution. We need to identify the raw score values that define this area. First, however, we need to use the z-score table and calculations to help us.

Like the last example, we can use our knowledge about the standard normal distribution to help us with this visualization.

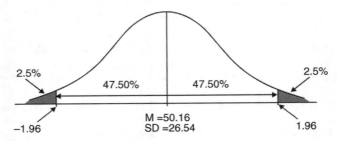

FIGURE 8.7 Visualizing the excluded 5% of the F/R sample values.

1. Recall that 2 SDs (above and below the mean) contain about 95% of the distribution (47.72% on either side of the mean).
2. This is very close to our target area because 2.5% of the 5% to be distributed between the two tails lies on each side of the distribution (i.e., one-half of 5% is 2.5%), and this leaves 47.50 in the middle of both sides of the distribution (i.e., 50% − 2.5% = 47.50%);
3. Therefore, our raw score is going to lie close to a z score of ±1.96 (the tabled value for 47.50% of the area between the mean and the tail).

Because the distribution is symmetrical, the z-score table identifies the same z score on the left (negative value) and the right (positive value) of the distribution. Calculating the raw score defining the 2.5% in the left of the distribution, we obtain

$$Z_1 = -1.96 \quad \text{and} \quad Z_2 = 1.96$$
$$X_1 = Z_1(\text{SD}) + M, \quad X_1 = -1.96(26.54) + 50.16, \quad X_1 = -1.86$$

The raw score defining the right (2.5%) tail is

$$X_2 = 1.96(26.54) + 50.16, \quad X_2 = 102.18$$

Therefore the scores of −1.86 and 102.18 identify the 5% area distributed in the two tails of the F/R sample distribution! This is interesting because the scores range from 0 to 100. Why the discrepancy? Because we are working with sample values to estimate population values.

Perhaps an easier way to identify the z score in this example using our z table of values is to find the z score that "creates" a tail of 2.50%. This value would be identified as the z score closest to a tail value of 2.50% from the tail columns of the z-score table (i.e., 1.96).

"EXACT" PROBABILITY

Thus far, we have discussed probability in terms of the area under the standard normal distribution. We have seen how to translate given areas (e.g., the percent of the distribution between the mean and SD 1) into probability statements (e.g., $p = 0.34$, from the example above). We have thus converted a *range of scores* into a probability amount. Excel® and SPSS® provide this information, but they also report the *probability of exact values occurring* among a set of values. Thus, the exact probability of a value of 1.00 occurring in the standard normal distribution is approximately 0.013. This is the value lying at the SD 1 mark because the standard normal distribution has mean = 0 and SD = 1. Thus, a value that lies directly at the SD 1 mark has a probability of occurring of $p = 0.013$.

To show an example of this, recall our discussion in Chapter 7 regarding the NORMDIST function in Excel®. Figure 7.17 shows the example of the math

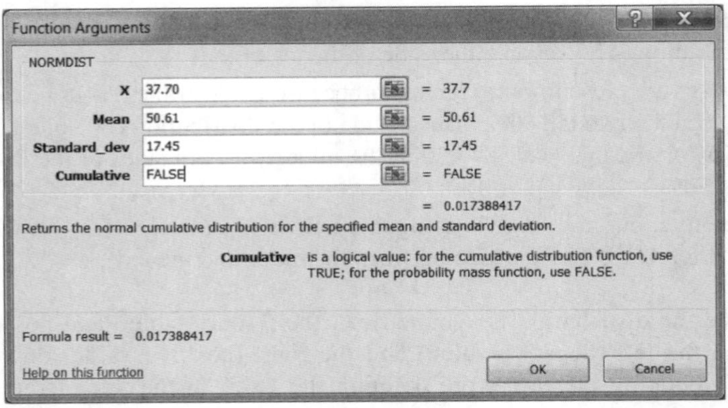

142 THE Z DISTRIBUTION AND PROBABILITY

FIGURE 8.8 Using the Excel® NORMDIST function for exact probabilities.

achievement variable for our sample of 40 schools. As you can see, we used the "TRUE" argument in the "Cumulative" window of the function. This returned the percent of the normal distribution below the raw score of 37.70 (i.e., 0.2297 or 22.97%).

Figure 8.8 shows the same function window from Excel®, except this time I specified "FALSE" in the "Cumulative" window. As you can see, this returns the value of 0.017388417. This is the *exact probability* of the raw score of 37.70 occurring (known as the probability mass function) in this distribution of values. We will see these specific probabilities reported by both Excel® and SPSS® in our discussion of statistical procedures in the chapters ahead.

You can also *estimate* this value using the z-score table. Figure 8.9 shows how to do this using the example in Figure 8.8. As you can see, I identified the target point of 37.7, but I also identified two values very close to it. Creating a raw score value 0.5% to the left of the target (37.20%) and 0.5% to the right (38.20%), I can create a rough estimate of the target point within one raw score percentage "band."

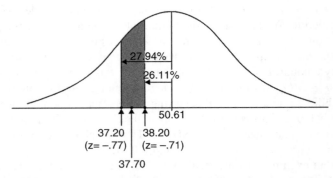

FIGURE 8.9 Calculation of an exact probability using the z-score table.

Calculating the z scores for these points (−0.77 and −0.71, respectively), you can calculate a percentage of 1.83 using the z-score table as we have in past exercises. Transforming this percentage to a probability yields $p = 0.0183$. This is close to the probability identified in Figure 8.8 (0.017388417), but not exact. The probability of $p = 0.0183$ is the probability of a score occurring *between* raw scores of 37.20 and 38.20, not the exact probability of the raw score of 37.70. Nevertheless, you may see from this example what we mean by the calculation of an exact probability.

FROM SAMPLE VALUES TO SAMPLE DISTRIBUTIONS

As we end this chapter, let me make some comments about the nature of inferential statistics that will propel us into the next chapters. Thus far, we have examined raw score distributions made up of individual raw score values. For example, our set of 40 schools each had student–teacher ratios, math and reading achievement scores, and F/R levels. We tried to understand the nature of these variables: whether they were normally distributed and how the percentages were distributed compared to the standard normal distribution.

In the real world of research and statistics, practitioners almost always deal with sample values because they very rarely have access to population information. Thus, for example, we might want to understand our students' math achievement scores compared to the math achievement scores of all similar students in the state. Perhaps we implemented a new way of teaching mathematics and we want to see if our methods were effective with our students compared to the students in the state. (If so, we might want to market the approach!) We can gain access to our own student's scores quite easily, but it is another matter to gain access to all the students' scores in the state!

Inferential statistics are methods to help us make decisions about how real-world data indicate whether dynamics at the sample level are likely to be related to dynamics at the population level. That is to say, we need to start thinking about our *entire sample distribution* (perhaps our group of 40 students math scores) as being one possible sample of the overall population of students (math scores) in the state.

If we derive a sample that is unbiased, the sample values (mean and SD, for instance) should be similar to the population values. If, however, we intentionally make a change to see what will happen to our sample, we should observe that our sample values differ from population values. Or, to take another example, we might observe that another teacher has been using a "traditional" method of teaching math while I have used my new method and I want to see which is more effective. In any case, what we are doing is:

1. Assuming at the outset that our sample reflects the population values from which it supposedly comes.
2. Changing our sample somehow, or observing different conditions between our sample and another similar sample.

3. Then seeing if our changed sample is now different from before, or different from the other sample.

In all these cases, we are comparing a sample to a population, not examining individual scores within a sample. We no longer think of our sample values individually, but as *one set that could be derived from a population along with many more such sets*; our sample set of values are now seen as simply one possible set of values alongside many other possible sample sets. That is the difference between inferential and descriptive statistics. We therefore change the nature of our research question:

1. Are our sample values normally distributed? (Descriptive statistics)
2. Do our sample values likely reflect the known (or unknown) population values from which our sample supposedly came? (Inferential statistics)

TERMS AND CONCEPTS

Grade Equivalent Scores   These are scores typically used by educators to express grades and months based on the typical score of a comparison group of students during a nine month school year.

Probability   This is the field of mathematics that studies the likelihood of certain events happening out of the total number of possible events.

Stanines   Stanines (standard nine) scores are created by transformations using the standard normal distribution. Widely used in education, these scores divide the standard normal distribution into nine parts. The stanine has a mean of 5 and an SD of 2.

***T* Scores**   The *T score* is a transformed score used in education because it recalculates z scores into more intuitively understandable scores. Z scores are both positive and negative, and they have a narrow range (from about -3.00 to $+3.00$). *T* scores transform z scores into values that are always positive, and with a larger range.

REAL-WORLD LAB IV

1. Using the data from Table 7.1, identify the raw scores that contain the middle 92% of the distribution.
2. What is the probability of obtaining a student–teacher ratio between 10 and 12?
3. What is the exact probability of obtaining a student–teacher ratio of 20?

REAL-WORLD LAB IV: SOLUTIONS

1. Using the data from Table 7.1, identify the raw scores that contain the middle 92% of the distribution.

Figure 8.10 shows the pertinent information so that you can visualize the results. The calculations are in accordance with the figure.

The middle 92% of the distribution identifies an area that lays 46% to the left of the mean to 46% to the right of the mean. The resulting z scores are therefore located at -1.75 and $+1.75$.

$$Z_1 = -1.75 \quad \text{and} \quad Z_2 = 1.75$$
$$X_1 = Z_1(\text{SD}) + M$$
$$X_1 = -1.75(3.58) + 16.25, \quad X_1 = 9.99$$
$$X_1 = 1.75(3.58) + 16.25, \quad X_2 = 22.52$$

Therefore, the middle 92% of the distribution falls between the raw score values of 9.99 and 22.52. Given the nature of the sample distribution, this encompasses most all of the scores.

2. What is the probability of obtaining a student–teacher ratio between 10 and 12?

Figure 8.11 shows the target area for visualization. As you can see, the z scores for these raw scores calculate to be -1.75 and -1.19, respectively.

The target area occupies about 7.69% of the distribution. Therefore, the probability of obtaining a student–teacher ratio between the raw scores of 10 and 12 is $p = 0.077$.

3. What is the exact probability of obtaining a student–teacher ratio of 20?

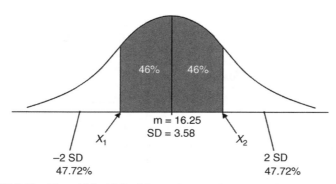

FIGURE 8.10 The middle 92% of the student–teacher ratios of the sample schools.

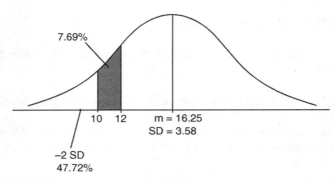

FIGURE 8.11 Calculating the probability between raw scores of 10 and 12.

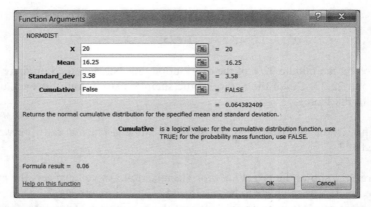

FIGURE 8.12 Calculating the exact probability of obtaining a raw score value of 20.

We can use the NORMDIST function to calculate this probability. Figure 8.12 shows the Excel® function with the result of $p = .064$.

You can estimate this using the z-score table by calculating the area between scores of 19.5 and 20.5 as follows:

X	z	% of Distribution Between Mean and z Score
19.5	0.91	31.86
20.5	1.19	38.30

Subtracting the relevant percentages yields 6.44% (38.30 − 31.86). Converting this to a probability, we obtain $p = 0.0644$. Compare this to the value yielded by the NORMDIST function in Excel® (in Figure 8.12). The values are almost identical.

9

THE NATURE OF RESEARCH DESIGN AND INFERENTIAL STATISTICS

Does class size have an effect on learning? Education researchers are just beginning to take a closer look at this research question because it is such a widely held opinion and it has resulted in funding opportunities for schools that have low student achievement. It has been fashionable in recent years not only to reduce class size, but also to create smaller schools or "schools-within-schools" as a way to limit size. The idea underlying all of these notions is that fewer students means that teachers can more effectively communicate with students and help them learn.

I have evaluated many educational reform efforts that took this tack and have written in several places about similar strategies. One of the main issues in reform efforts, it seems to me, is not whether the class is smaller, but the approach to learning taken by teachers, students, and educational administrations.

The question of class size and achievement is persistent. This being the case, we can use the example to discuss the nature of research and inferential statistics. To these issues we now turn.

As I mentioned in Chapter 8, my students anxiously await "headache day" because it means we must take a different approach to our subject. This is indeed the case, although it does not have to be overly complex. The main requirement for understanding inferential statistics is to learn to think abstractly. We have dealt with descriptive statistics, which, in a sense, are procedures to measure what you see. Inferential statistics looks at data on a different level of abstraction. We must learn to understand the connection between what data we see before us and the statistical world that lies outside and beyond what we see.

Understanding Educational Statistics Using Microsoft Excel® and SPSS®. By Martin Lee Abbott.
© 2011 John Wiley & Sons, Inc. Published 2011 by John Wiley & Sons, Inc.

RESEARCH DESIGN

Before we look in depth at inferential statistics, however, we need to cover some essential matters of research design. Statistics and research design are companion topics that need to be understood together. Often, the two subjects are taught together in college curricula, or they are prerequisites for one another. There is no best way to sequence the ideas, so I simply try to introduce research design elements at the brink of discussing inferential statistics because they have mutual reliance.

This is a book on statistics, so we cannot hope to cover all the complexities of research design. We can only attempt to provide a research "primer." In what follows, I will outline some of the basics, but for a comprehensive understanding of research design, you need to consult standard authorities in the field. You might start with Earl Babbie's excellent work (Babbie, 2010) that has served for many years to provide an excellent examination of social research practice.

Social research is a field of inquiry in which we devise standardized methods for examining available data to answer research questions. Typically, this involves collecting data from subjects or existing sources and subjecting the data to the methods of statistics to provide illumination. We may start with a research question and then figure out the best way to proceed to solve it. This procedure is research design. How can we structure our inquiry so that we can find and use data in the most defensible way to answer a research question? Figure 9.1 shows a process we might envision that will help us negotiate a research question.

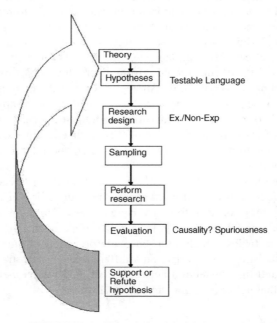

FIGURE 9.1 The process of social research.

Theory

As you can see, the top element in Figure 9.1 is *"Theory,"* an abstract idea in which we state the conceptual nature of the relationship among our ideas of inquiry. For instance, we might state our earlier question as a theoretical question: What is the relationship between size of learning environment and student learning?

Hypothesis

Because a theory cannot be directly assessed (it exists on an abstract and conceptual level), we must find a way to empirically restate the theory so that it can be assessed. That is the role of the *hypothesis*, a statement that captures the nature of the theoretical question in such a way that it can be directly verified. As you can see in Figure 9.1, I note that the hypothesis is written in "testable language," or written in a way that will provide empirical evidence in support of, or contrary to, a theoretical question.

Let me add here that I am introducing a scientific process that underlies most all scientific attempts to understand the world. The physical sciences and social sciences alike use the methods I am describing to generate and verify knowledge. The theory testing process shown in Figure 9.1 is the heart of this process. By following the process, we can support or refute a theoretical position, but we can never "prove" it directly. If a hypothesis statement, being constructed in empirical, and therefore limited, language is borne out, we simply add to our confidence in the theory. There are many hypotheses that can be generated to test any one theory since the empirical world cannot completely capture the essence of the abstract conceptual world.

Here is an example of what I mean. I might generate the following hypothesis regarding the theoretical statement above about size and learning: "Students in classroom A (smaller student to teacher ratio) will evidence higher standard test scores in mathematics than those in classroom B (larger student to teacher ratio)." Do you see the difference in language between theory and hypothesis? Theory is abstract while hypothesis is concrete. Theory is more "general," while hypothesis is more "specific."

Theory cannot be captured by a single hypothesis statement; there are simply too many empirical possibilities that can be created to 'test' the theory. For example, I might suggest another hypothesis: "Schools' student to teacher ratios will be negatively correlated with their aggregated test scores in math." As you can see, this is another restatement of the theory in language that is testable. I can easily (well, perhaps easier!) collect school data on a sample of schools and statistically assess the relationship between their student to teacher ratios and their aggregated test scores. If there is a relationship, as predicted by the hypothesis, this would lend support to the theoretical tie between size of learning environment and learning, but it would not "prove" it.

TYPES OF RESEARCH DESIGNS

Research designs are simply the "housing" within which we carry out our analyses to test a theory. You can see in Figure 9.1 that the research design enables the researcher to *situate* the hypothesis. How can we create the analysis so that we can use statistical tools to their best advantage to provide evidence for the theory? While there are many different possibilities, we can note three:

1. Experiment
2. *Post Facto*—Correlational
3. *Post Facto*—Comparative

Experiment

There are two "classes" of designs; experimental and nonexperimental (*post facto*). *Experiments are designs in which the researcher consciously changes the values of a study variable under controlled conditions and observes the effects on an outcome variable.* Post facto designs are those that involve measuring the relationships among variables using data that have already been collected.

Of the two examples of hypothesis I listed above, the first is closer to an experiment, depending on how I can control the conditions. Thus, if a principal allows me to randomly select students and randomly assign them to two different classrooms (A and B) with different student to teacher ratios, and then after a period of time assess the differences in student test scores, I would be performing an experiment. I consciously change the values of a study variable (student to teacher ratio) and assess the effects on an outcome variable (student achievement).

Control Groups. Of course, the particular way in which I control all the influences other than the two research variables will have a bearing on the strength and validity of my results. The key to a powerful experimental design is limiting these influences. One way to do so is to create a *"control group,"* which is typically a group similar in every way to a *"treatment group"* except for the research variable of interest. In our example, classroom A may have substantially lower student to teacher ratios (e.g., 12:1) than the "normal" classroom B (e.g., 16:1). In this case, the treatment variable is ratio size, and students in classroom B are the control group. Table 9.1 shows these groups. The only difference they have from students in classroom A is that there are more students per teacher in their classroom. Therefore, if students in classroom A get superior test scores, the experimenter will

TABLE 9.1 The Experimental Groups

Research Treatment Variable: Student–Teacher Ratio		Outcome Variable
Classroom A (low ratio)	Experimental group	Test scores
Classroom B (typical ratio)	Control group	Test scores

attribute the test score increase (the outcome) to a lower student–teacher ratio. Theoretically, there are no other differences present in the design that could account for the difference in test scores.

As you might imagine, there are a host of potentially "confounding" conditions, or ways that the two groups cannot be called comparable. Perhaps the experimenter cannot truly choose students randomly and assign them randomly to different classrooms. If so, then there are differences being "built in" to the experiment: Were similar students chosen for the different classes? Were there students of "equal" aptitudes, genders, and personality types represented in both groups, for example?

Randomization. Experimenters use *randomization* methods to ensure comparability of experimental groups. By randomization, I mean (1) selecting students randomly and (2) randomly assigning them to different conditions. The power of randomness is that it results in individual differences between students to be equated across groups. If every student has an equal chance of being chosen for an experiment, along with an equal chance of being assigned to classroom A or B, then the resulting groups should be as equal as possible; there should be very little bias that would normally influence some students to be chosen for one class and other students to be chosen for the other class.

Quasi-Experimental Design. Experiments can be either "strongly" constructed or "weakly" constructed according to how well the experimenter can control the differences between both groups. Often, an experimenter cannot control all the conditions that lead to inequality of groups, but they still implement the study. The *quasi-experimental design* is just such a design. Here, the experimenter may be forced, because of the practicalities of the situation, to use a design that does not include all of the controls that would make it an ideal experiment. Perhaps they do not have the ability to create a control group and must rely on a similar, "comparable group," or they may be confronted with using existing classes of students rather than being able to create the classes themselves.

In the experimental design we discussed above, the experimenter may not be able to randomly select students from the student body and then randomly assign them to different conditions. Perhaps the experimenter can only assign the students randomly to different classrooms. In this case, we cannot be assured that the students in the two classrooms are equal, since we could not assure complete randomness. However, we might proceed with the experiment and analyze how this potential inequality might affect our conclusions. This is shown in Table 9.2, in which the difference from the experimental design shown in Table 9.1 is the absence of randomization and the lack of a true control group.

There are a great many variations of experimental and quasi-experimental designs. The key differences usually focus on the lack of randomization and/or true comparison groups in the latter. For a comprehensive understanding of experimental design and the attendant challenges of each variation, consult Campbell and Stanley (1963) for the definitive statement. In this authoritative discussion, the authors discuss different types of designs and how each can address problems of

TABLE 9.2 A Quasi-Experimental Design

Research Treatment Variable: Student–Teacher Ratio		Outcome Variable
Classroom A (low ratio)	Experimental group (not randomly selected and assigned)	Test scores
Classroom B (typical ratio)	Comparison group (not randomly selected and assigned, but chosen to be comparable to the experimental group)	Test scores

internal validity (whether the conditions of the experiment were present to control extraneous forces) and external validity (including generalizability).

It is probably best to think of experimental designs as being stronger or weaker rather than as specific "types" that can be employed in certain situations. Many research design books list and describe several specific (experimental and quasi-experimental) designs, noting the features that limit problems of internal and external validity. Figure 9.2 shows how research designs exist on a continuum in which they can approach 'true' experimental designs on one end that limit all problems and can make causal attributions to those on the other end of the continuum that are beset with problems which limit their ability to produce meaningful experimental conclusions.

Variables. By now, you will recognize that I have used the language of "variables" in my explanation of experimental design. Before we proceed to discuss other designs, we need to note different kinds of variables. Variables, by definition, are the quantification of concepts (like the student to teacher ratios or test scores) used in research that can take different values (i.e., vary). Thus, math achievement is a quantified set of test scores that vary by individual student.

Independent Variables. In research design, we often refer to certain types of variables. The "*independent variable*" is understood to be a variable whose measure does not relate to or depend upon other variables. Thus, in our experimental design

Design Strong	**Few problems** of internal
Full randomization;	validity and generalizability;
Comparable Control Group	can make causal attributions.

Design Weak	**Many problems** of internal
No randomization;	validity and generalizability;
Lack of Comparable Control Group	cannot make causal attributions.

FIGURE 9.2 The nature of experimental designs.

example, student–teacher ratio is such a variable because in our research problem we assume that this is the influence that will lead to an impact on other variables. It is assumed to be a "cause" of some research action.

There are a host of problems with the independent variable designation. Typically, we refer to a variable as independent only in the context of an experiment because we are framing it as leading to certain effects. In nonexperimental contexts, I prefer to use the designation "*predictor variable*," which does not evoke the language of causality. A variable can be a predictor of an outcome without being an independent variable.

In research designs, independent study variables can be "manipulated" or "nonmanipulated," depending on their nature. *Manipulated independent variables* are those the experimenter consciously changes, or manipulates, in order to create the conditions for observing differential effects of treatment groups on the outcome variable. In our example, student–teacher ratio is the manipulated independent variable because the researcher could assign students to two different levels or conditions of this variable: low or high ratios. Another example could be group size, if a researcher wanted to compare different reading groups according to the numbers of students in the group; perhaps one group would consist of "few" students and another "many" students. In this example, group size is manipulated (i.e., consciously changed by the researcher) by creating two reading groups of different sizes.

Nonmanipulated independent variables are those that cannot change or cannot be manipulated by the researcher. Typically, they are characteristics, traits, or attributes of individuals. For example, gender or age can be independent variables in a study, but they cannot be changed, only measured. When these types of variables are used in a research study, the researcher cannot make causal conclusions. The essence of a true experiment is to observe the effects of changing the conditions of a variable differentially for different groups and then observing the effects on the outcome. If nonmanipulated variables are used, by definition the research design cannot be experimental. For example, if the researcher was interested in the effects of gender on achievement, the research design can only group the subjects by their already designated gender; no causal conclusions can be made.

Dependent Variables. *Dependent variables* are those thought to be the "receivers of action" in a research study; their value depends upon (is tied to) a previously occurring variable. Where independent variables are causes, dependent variables are "effects" or results. In nonexperimental contexts, I like to think of these as "*outcome variables*" that are linked to predictors.

Post Facto **Research Designs**

The second hypothesis example I presented above ("schools' student–teacher ratios will be negatively correlated with their aggregated test scores in math") is a *post facto correlational* design. Here, I am simply using data that already exist (on schools' student–teacher ratios and their aggregate test scores)—hence, *post facto*, which means "after the fact." I do not consciously change anything; rather I use what data I can gather *from what is already generated* to see if the two set of scores are correlated. This design uses the statistical process of correlation to measure the association of two sets of existing scores. [We will discuss this process at length in the correlation chapter. (Chapter 14).]

A *post facto* design can also *compare* conditions rather than *correlate* conditions. The *post facto comparative* design seeks to understand *difference*. Thus, for example, I might compare two *already existing classes* of students to see if they have different test scores. (Perhaps we are interested in whether one class, composed of girls, has different test scores in math than another class composed of boys.) Statistically, I will assess whether there is a difference between the means of the test scores, for example. This type of approach uses methods of difference like the t test, ANOVA, and others. It is *post facto*, since we are using data that already exist (i.e., I did not randomly select and assign students to different classes; I used classes already operative), but it is not correlational, since we see to assess *difference* rather than *association*.

THE NATURE OF RESEARCH DESIGN

I cannot hope to discuss the nuances of each type of design. However, I will introduce the different designs in the context of discussing different statistical procedures in the chapters ahead. For now, it is enough to know that there are different ways to assess theories. We devise hypotheses according to the nature of our interests and inquiry, and we thereby validate or question theories by empirical (statistical) processes.

I should mention here some important aspects of research designs that I will develop in later chapters. In brief, each design has strengths and limitations. The experiment can be a powerful way of making "causal statements" because, if only one thing changes (the main treatment) while everything else is similar between the groups being tested, we can attribute any effects or changes in outcomes to the thing that was changed. Using the first example again, if we chose and assigned students appropriately and if the only difference between the two groups was the student to teacher ratio, then we could attribute any resultant difference in test scores primarily to the ratios. (Of course, as we will learn, we have to take great care to control all other influences beside the ratios in order to make a causal conclusion.)

Post facto designs cannot lead to causal attributions. Because the data are already collected, a number of different influences are already "contained in the data." In this event, any two groups we compare have differences other than the research interest (student to teacher ratio) that will intrude upon differences in test outcomes. It is a matter of controlling these influences that is the difference between an experiment and a *post facto* design.

Research Design Varieties

There is another dimension to research design, namely, the variety of ways in which it is carried out to collect data. Experiments can take place in the laboratory or in the field, for example. *Post facto* designs can use self-report instruments (i.e., questionnaires) or researcher observation to generate outcome data. More generally, research can be quantitative (focusing on the statistical analysis of numerical measures for the concepts studied) or qualitative (focusing on the aspects of research assumed to be irreducible to numbers, like the emergent meaning of a concept among a group of people.)

Sampling

I have already alluded to the importance of sampling. By now, sampling is generally understood to be the process by which a small group of elements is chosen from a larger (population) group so that the small group chosen is representative of the larger group. Thus, for example, in my first hypothesis above, sampling might involve choosing a small number of students from a school in order to create two different classes rather than including everyone at the school. The purpose is to create a representative group for study; the conclusions of the study are then thought to characterize the entire population. Sampling thus follows the hypothesis in Figure 9.1.

As you might imagine, the sampling process is vulnerable to problems. How do you choose the sample so that you can be assured it is representative of the entire population from which it was drawn? There is no way to be entirely certain! But, if we choose a sample randomly, where each element has an equal probability of being chosen, we can rest assured that we will have the most representative sample possible. What we are saying here is that we are using a *probability sampling process*; that is, we are using the methods of probability to arrive at a sample group in which the variability of the sample reflects the variability of the population. Our sample should have the same gender, age, and other characteristic mix as the population.

Sampling is more complex than I am making it out; however, at its heart, it really is a simple principle. The situation I described above is the *simple random sample*. There are other types of sampling that recognize the complexity of the process and the nature of a research problem. *Stratified random sampling*, for example, allows the researcher to build in *levels or categories* so that we can ensure each of the crucial components of a population is taken into account. If our school has unequal numbers of boys and girls, for example, we might want to sample randomly within sex categories to ensure we have a representative sample group.

What makes sampling difficult is that we often cannot control certain features of the process that would result in a representative sample. Often, we do not have the ability to use probability sampling procedures. Perhaps we are studying a phenomenon in which it is difficult or impossible to identify the population from which to sample. In such cases, we need to use whatever procedures are available. I did a study a number of years ago on street-corner preachers. It would be impossible to identify the population of street-corner preachers, so I interviewed and observed those street-corner preachers that were convenient. The *convenience sample* is obviously not representative, but it may be a researcher's only option.

Another nonrepresentative sampling process is *snowball sampling*. Often, because a study population is secretive or difficult to identify, the researcher might gain additional subject interviews from an existing interview. Then, the additional interviews might uncover other potential subjects, like a snowball gathering momentum rolling down a hill. I was forced to resort to this process when conducting a special census some years ago. The population was resistant and secretive, so I had to carefully construct an interview list through existing contacts. This and other nonprobabilistic sampling methods are available to researchers, but the limitations need to be identified in the study. The sampling problems will be registered in the conclusions.

INFERENTIAL STATISTICS

Now that we have covered some of the essentials of research design, we can return to the topic of inferential statistics. The two topics are highly intertwined. As I mentioned in Chapter 8, inferential statistics involves a shift in thinking from individual scores to sets of scores (or samples).

One Sample from Many Possible Samples

We need to begin thinking of our data as a sample that we have drawn from some larger population, rather than a set of data that is a sample unto itself. Or, stated differently, we must move from measuring distributions of raw scores to measuring the probability of sample distributions belonging to certain populations.

In order to pursue the matter of whether size of learning environment affects learning, we can use inferential statistics to help us understand whether our observed changes from a sample study likely apply to the population of all students. Figure 9.3 shows how this works.

When we conduct a research study, we typically select a sample that we try to ensure is representative of a population of interest. Figure 9.3 shows that this sampling process, if it is random, can result in a sample group drawn from any area of a population distribution. There are four samples shown in Figure 9.3 to show that most samples will be selected from the area close to the mean of the population if probabilistic methods are used. We talked about this in the section entitled "Probability and the Normal Curve" in Chapter 8. The greatest likelihood in sample selection is that it will come from the area massed most closely to the mean. There are probabilities that the sample can be drawn from out in the tails, but that is not as likely.

In our size–learning example, we can understand the population of interest to be the school, and our sample to be the set of students that we will randomly assign to two classes, each with a different student–teacher ratio. I must note an issue of generalizability here. If our population is limited to a school, then the conclusions of our study can only extend to the population from which it is drawn. The conclusion, in this case, would be quite limited; it would only apply to the dynamics of our

INFERENTIAL STATISTICS

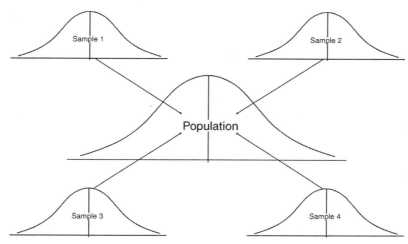

FIGURE 9.3 The sampling process for inferential statistics.

particular school and would constitute a *case study*. If we were to conduct a national study and selected a larger group of students from the national population, then our conclusions could extend to the national population.

For discussion purposes, Figure 9.3 shows that, if we were to create a sample four times, our samples would fall within the large area close to the mean and not in the tails. If we were to select 1000 samples, we would likely get most of the samples from around the population mean but several in the tails as well. That is the nature of the normal distribution.

Of course, we do not need to create four samples for our study. The figure is just to underscore the fact that when we create samples, they are likely to come from the area closer to the mean than in the tails. We assume, when we take a sample, that it represents a population by coming from the area close to the mean. Even though our sample mean will not likely be exactly the same as the population mean, it will likely be close.

Sampling error is the difference between the sample and population means, among other aspects of the distribution. Whenever we take a sample, we are not likely to come away with a small group with exactly the same mean and standard deviation as the overall population. This doesn't make the sample problematic or unrepresentative unless it is widely divergent from the population mean. Some error is expected in sampling. The extent of the error is the subject matter of inferential statistics.

Central Limit Theorem and Sampling Distributions

Now, here is a curious fact. Let's say we did gather 1000 samples from a large population of students. *If we used only the sample means of each sample to represent their entire sample groups*, we could create a new distribution just made up only of these sample means. In such a process, most of the sample means in this

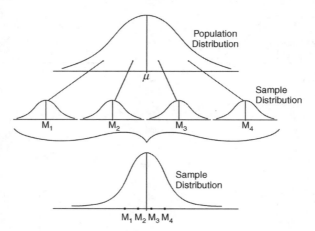

FIGURE 9.4 Creating a sampling distribution.

new "sampling distribution" would lay close to the overall population mean with some spreading out into the tails. In fact, the mean of this sampling distribution of means would be normally distributed, and its mean would be equal to the population mean.

Researchers and statisticians refer to this process as resulting from the assumptions of the *Central Limit Theorem*. This theorem states that means of repeated samples taken from a population will form a standard normal distribution (assuming a large sample size) even if the population was not normally distributed. The sampling distribution that results will have important properties to researchers conducting inferential studies.

As you can see in Figure 9.4, the four hypothetical samples are taken from the population, and their individual means make up a separate distribution called the *sampling distribution*. You can see from Figure 9.4 that the individual means, which represent their sample distributions (M_1, M_2, M_3, M_4), lay close to the population mean in the new distribution. We can say that the sampling error is smaller as we get closer to the population mean. Figure 9.4 shows this process using only four samples for illustration.

There are other important features of the sampling distribution created by using the means of repeated samples. As you can see in Figure 9.4, the following are true:

1. The sampling distribution will be normally distributed.
2. The mean of the sampling distribution will be equal to the population mean. That is, if you added up all the sample means and divided by the number of sample means, the resulting average of sample means would equal the population mean.
3. The standard deviation of the sampling distribution will be smaller than the standard deviation of the population because we are only using the individual mean scores from each sample distribution to represent the entire set of

sample scores. Using only the mean of each sample group results in "lopping off" most of the variability of individual scores around their contributing group means, with the result that the sampling distribution will have a smaller "spread."

The importance of the sampling distribution is that it will be normally distributed. As I noted earlier, if we take repeated random samples and make a separate distribution from just the sample means, the resulting distribution will be normally distributed even if the original population is not normally distributed. You can see this feature of the Central Limit Theorem in Figure 9.5. Because it is normally distributed, you can see how a single sample mean relates to all other possible sample means. If you think of your study being one such sample, is your study mean close to the population mean? How close or far away is it?

You are probably asking yourself, "so what? What does this have to do with anything because I would only select one sample for my study?" This is the crux of inferential statistics, so let me answer this (anticipated) question.

Sampling distributions are not empirical; they provide the framework within which you can understand your sample as one of several possible samples. You don't actually select multiple samples for a study such as I described, so the sampling distribution is not empirical. But you should understand that, when you obtain the results from studying your sample, the results will reflect two things: (1) the impact of the independent (predictor) variable on the dependent (outcome) variable and (2) *the distance of the sample mean from the population mean.* If your sample is drawn randomly, it should be close to the population characteristics and this source of "error" should be minimized. But the fact of the matter is, you could select a sample from the tails of a distribution just by chance. It certainly happens; that is the nature of probability.

The important question is, How far away from the population mean can I select a sample and still assume it is representative? This is the operative question at the

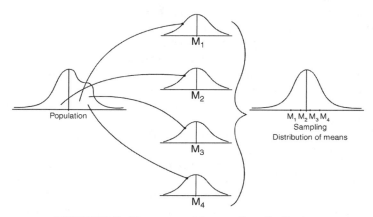

FIGURE 9.5 The nature of the sampling distribution.

center of inferential statistics and the reason we need to envision the sampling distribution. *Even though we will never actually create a sampling distribution by repeatedly sampling from a population, the known features of the sampling distribution will help us to answer the question of representativeness of values.*

The Sampling Distribution and Research

In research, you draw a sample from a population and use the sample to reflect the population characteristics. If you select the sample using probabilistic methods, it will be representative. That is, the sample mean will likely not fall far from the population mean.

Next, you "use" the sample to make a careful study. Perhaps, as a simple example, we perform an "experiment" by selecting a sample of students and introduce some treatment that we think will affect an outcome measure. In our example of student–teacher ratios and student learning, we might take our sample of students and form a class with a small student–teacher ratio. We would then let the class run for a period of time and then measure the outcome of the study (student test scores). Do our students have a larger or smaller test average than "normal" classes with larger student–teacher ratio? Here, we are assuming that the overall population of students has a certain average ratio that is known, and we are purposely decreasing it for our students to see if it makes a difference in learning.

Since we are hypothesizing that smaller student–teacher ratios are helpful to learning, we would anticipate that the average student test scores in our sample group would be higher than the test scores of the general population of students. That is, over the duration of the class, the lower student–teacher ratio would affect the students in various ways and result in higher achievement (test scores) than for students in other classes. If we now compared our class achievement average to the average achievement of the population of students, it might be quite higher. But how much higher would it have to be for us to conclude that the lower student–teacher ratio (that we created) had a *significant impact* on student learning? This is the importance of the sampling distribution.

When we compare a sample mean to the population mean *after we change the sample in some way* (e.g., by an experimental treatment), it may no longer be close to the population mean. In effect, our treatment may have "moved it away" from the population mean a certain distance. We use the sampling distribution of means to "measure" this distance. How far away from the population mean does the sample mean now fall after the treatment?

The answer to this question is found in probability theory. If we assume that we will get a sample mean close to the population mean by choosing it randomly, then we assume that a sample mean, *once changed by an experimental treatment*, will be moved further away from the population mean if the treatment is effective. If the (changed) sample mean now falls in the tail area (exclusion area) of the sampling distribution, that is tantamount to saying that it is no longer representative of the population; the sample mean falls too far away from the sample mean.

How far is too far? Statisticians and researchers have assumed that if a sample mean falls into the tails of a distribution, it is not representative; but how far into the tails? The consensus is generally the 5% exclusion area; that is, the area in the tails (both sides) that represents the extreme 5% of the area under the standard normal distribution (i.e., 2.5% on either side).

Therefore, to go back to our example, if our sample students' test scores are substantially higher than those of the population, we would say that the lower student–teacher ratios moved the sample test mean away from the population mean so far into the tails of the distribution that our students are no longer representative of the students in the population.

Figure 9.6 shows how you might visualize this. The sampling distribution is drawn (in theory) from the population of all students. After the experiment, the students' scores (as represented by the sample mean) are now ahead of the other students' scores as represented on the sampling distribution (the sample mean value is on the far right of the test score distribution of all students).

The sampling distribution thus becomes a kind of "ruler" that we can use to apply our findings. It is (theoretically) created according to the Central Limit Theorem and therefore reflects a "perfect" distribution—that is, a standard normal distribution. *It now stands in the place of the population as our comparison figure.* We can now see where our sample mean falls on this perfect comparison distribution. If it falls in the tails, as it does in the example in Figure 9.6, we would say that the treatment changed the group to such an extent that it no longer is the same as the group of all other students (i.e., population); it is too far away from the population mean.

Another way of saying this is that, after the experiment, our one sample of many possible samples is now not in the area of a "normal" sample. All possible samples are represented by the sampling distribution. Our *changed* sample is now moved into the tails, and we can say it is now different from the others.

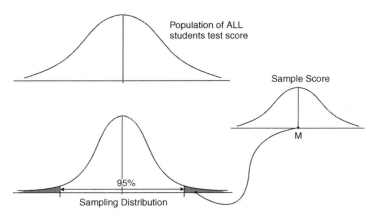

FIGURE 9.6 Using the sampling distribution to 'locate' the sample mean.

Populations and Samples

Parameters refer to measures of entire populations and population distributions. This is distinguished from *statistics, which refer to measures of sample data taken from populations.* We need to distinguish these measures because inferential statistics is quite specific about the measures available for analysis. One way they are distinguished is by their symbols; population parameters are typically represented by Greek symbols.

This is a good place to review some symbols that will distinguish population measures from sample measures. Because we are now beginning to think of different levels of data, we need to be able to be precise in how we speak of both. Table 9.3 is a chart showing the symbols for population and sample sets of data.

The Standard Error of the Mean

This is a new term for a measure we have already discussed: the standard deviation of the sampling distribution of means. The designation "standard error of the means" is used because it is much shorter! I hope you can see why we give it this name. Both are simply ways of saying that σ_M is a standard deviation of a standard normal distribution of sample means.

The standard error of the mean is a standard deviation. But recall that it is a standard deviation of a distribution that is "narrower" than the population standard deviation because we only use the mean scores from repeated sampling to make it up. You can see this by looking again at Figure 9.4.

Because it is now different from the population standard deviation (σ_X) we need to be able to *estimate* its value. One way of doing this is by the following formula:

$$\sigma_M = \frac{\sigma_X}{\sqrt{N}}$$

TABLE 9.3 Population and Sample Symbols for Inferential Statistics

M	The mean of the sample.
SD	The standard deviation of the sample (assumes the sample is its own population)
μ	The Greek letter "mu" is the symbol for the population mean.
M_M	This is the symbol for the mean of the sampling distribution. You can see how this works by observing that it is a mean (M), but a mean of the sampling distribution indicated by the subscript. Thus, it is the *mean of the distribution of means*, or the "mean of the means." Because it is (theoretically) created by all possible samples, it is a parameter.
σ_X	"Sigma X" is the standard deviation of all the population raw scores. This differs from standard deviation SD in that it does not refer to a sample, but to the entire population of individual raw scores.
σ_M	The standard deviation of the sampling distribution of means; also called the *standard error of the mean.*

INFERENTIAL STATISTICS 163

We use this formula because it includes the sample size (N) as a way of helping us make the estimate. The sample size will ultimately determine the size of the σ_M since the group size is registered in the sample means that make it up. As N increases, the σ_M will decrease; you can see this from the formula. When you divide a number by a large number, the result will be a smaller number. Conceptually, however, this simply refers to the fact that larger sample sizes are better estimates and therefore the standard deviation of the sample means will likely be smaller.

The rule of thumb to remember is: $\sigma_M < \sigma_X$.

"Transforming" the Sample Mean to the Sampling Distribution

Remember that "sampling error" is the distance of a sample mean from the population mean. Remember also that the sampling distribution of means is perfectly normally distributed, and its mean is equal to the population mean ($M_M = \mu$). When we place a *sample* mean (M) in the sampling distribution as we did in our example in Figure 9.6, the distance of our sample mean from M_M can be expressed as a "standard distance" by referencing it to the standard deviation of this (perfect) distribution (σ_M).

The resulting distance is therefore like a z score that is expressed in standard deviation units. In effect, we are transforming a raw score mean to a standard value in the sampling distribution so that we can compare it to all possible sample means that could be taken from a population. It therefore helps us to answer the question, How good of an estimate of the population mean is our sample mean?

Example

Let's take an example from our school database. Using our question about the effect of student–teacher ratio upon learning, here are some relevant data to consider. For purposes of the example, I deliberately chose a sample of schools from among those with small student–teacher ratios. All the schools in the state (among schools with fourth-grade test scores) had an average student–teacher ratio of 16.24. The "sample" of schools I chose ($N = 61$) had an average student to teacher ratio of 13. Remember, these were not randomly selected; I chose schools on the basis of having low student–teacher ratios to show the dynamics of our problem.

1. The mean aggregate math achievement of the population of schools: $\mu = 51.96$. (This represents the average percent of students in every school in the state who passed the math assessment.)
2. The population standard deviation (σ_X) of math passing rates is 18.63.
3. The mean math achievement of a sample ($N = 61$) is $M = 44.04$.
4. The standard deviation of math passing rates of the sample (SD) is 19.52.

5. The standard error of the mean (σ_M) is 2.39.

$$\sigma_M = \frac{\sigma_X}{\sqrt{N}}, \qquad \sigma_M = \frac{18.63}{\sqrt{61}}, \qquad \sigma_M = 2.39$$

Figure 9.7 shows how these are 'placed' so you can visualize the various measures. The research question is: Does the mean of our sample (of smaller student–teacher ratio classrooms) have higher or lower math passing rates than the population passing rates? What we need to do is to see where our sample mean falls in the sampling distribution of means so that we can compare it to the population mean.

We can do this by transforming the sample mean to a standardized value, just as we did when we created a z score. Here is the formula we used when we were dealing only with single sample scores:

$$Z = \frac{X - M}{SD}$$

We can use the same formula with some changes to reflect that we are using population rather than only sample values:

$$Z = \frac{M - \mu}{\sigma_M}$$

Compare the two formulas. They look alike, because they do the same thing. The z-score formula (top) is used to *transform a raw score to a standard score*. These are both at the sample distribution "level."

The bottom formula is used to *transform a sample mean to a standard score in the distribution of means*. These are measures used when we are dealing with both population and sample values.

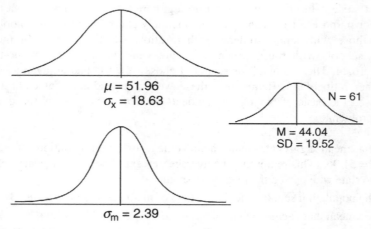

FIGURE 9.7 Using the sampling distribution and the standard error of the mean.

INFERENTIAL STATISTICS

If we calculate the *standardized sample mean score*, Z, we obtain

$$Z = \frac{M - \mu}{\sigma_M}, \quad Z = \frac{44.04 - 51.96}{2.39}, \quad Z = -3.31$$

Findings. What does this figure of $Z = -3.31$ mean? Consider the following findings based on our calculations.

1. It means that the difference between the sample mean and the population mean, which is the top part of the formula $(M - \mu) = -7.92$. This indicates that our sample math achievement mean lies *below* the population math achievement mean by almost 8%.
2. When we divide this distance (-7.92) by the standard error of the mean (2.39), it *transforms* the sample-population distance into a standard measure, like we did with a z score. The resulting figure $Z = -3.31$ means that our sample mean falls about $3\frac{1}{3}$ standard deviations below the population mean!
3. If you recall your z-score table, this means that our sample mean falls well into the left (negative) tail of the standard normal comparison distribution.
4. Using the NORMSDIST function from Excel®, you can calculate the area of the standard normal distribution falling below -3.31 to be 0.00047. In terms of probability, that would mean that our sample mean is extremely atypical; it is nowhere close to the population mean.
5. This probability is far below the benchmark 5% (or 2.5% on each side of the standard normal distribution) that researchers use to conclude that the sample mean is not typical.
6. Our conclusion is that the sample math achievement for this sample is not typical of samples drawn from the population; it is so different that is can be considered belonging to a totally different population of schools.

Figure 9.8 shows how this result looks using the findings from our former figure.

Discussion. We need to remember that these findings are not generalizable because I did not choose the sample randomly. I deliberately chose it from schools with smaller student to teacher ratios to demonstrate the process of making a statistical decision. But consider the findings.

1. We discovered a finding exactly opposite what we expected! Despite the fact that the "sample" was contrived, real research often results in unexpected findings such as this. Researchers should be aware of this possibility. In a later chapter, I will have more to say about ways to address the potential direction of findings.

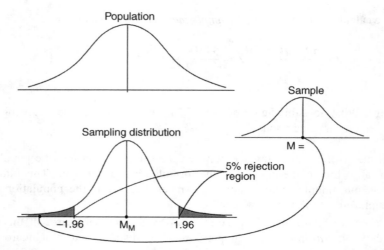

FIGURE 9.8 Using the sampling distribution to make a statistical decision.

2. This was not an experimental design because I used existing school values. To constitute an experiment, I would have had to randomly select a group of schools, and then convince the principals to allow me to deliberately create lower student–teacher ratios so that I could see if it made a difference to math achievement! Although this sounds farfetched, some educational programs have been created for a similar purpose. This is behind many efforts to lower class size by means of programs like creating smaller schools, schools within schools, and so on, because of the assumption that these actions will automatically lead to improved student achievement.

3. This set of schools (i.e., deliberate "sample") is a real set of schools with low average student–teacher ratios. Even though they were not randomly sampled, and therefore the findings not generalizable, they might indicate a trend we can observe (and will) in later studies. It is unclear why this particular set of schools had lower math achievement than the population, but it would indicate that we need to investigate the matter further.

Z TEST

Congratulations! Without being aware of it, you just performed your first inferential statistical test, the Z test. As you can see, it is not really that difficult to understand. We simply transformed a sample mean so that it could be compared to all possible sample means. By doing so, we can see how it "falls" on a standard normal distribution of values and calculate a probability for this score occurring by chance. If it falls too far into the tails (i.e., beyond the extreme 5% area), we can conclude that it is not representative of the population.

The Hypothesis Test

What we have done is to perform a hypothesis test. This is the formal, logical process established to make a scientific decision. If you look back at Figure 9.1, the bottom step in the process is to support or refute a hypothesis and thereby inform a theory.

If we identify the steps we used in our example, you will find that these steps form the general procedure for a hypothesis test that we will follow on all our remaining statistical tests (with some variations for each procedure). Here are the steps with our (contrived) results applied so you can see how it works:

1. The Null Hypothesis or (H_0): $\mu_1 = \mu$. Researchers begin by considering a statement that can be measured and then verified or refuted. They begin with the assumption that there will be no difference between the mean of the study sample (μ_1) and the mean of the population (μ). The object of the research process it to see if this is an accurate assumption, or if our sample violates that assumption by being either too large or too small. Our null hypothesis was that a sample of schools with low student–teacher ratio would have the same math achievement scores as the population of schools.

2. The Alternative Hypothesis or (H_A): $\mu_1 \neq \mu$. This statement is created in order to present the finding that would negate the null hypothesis—thus, the alternate finding. In our study, we proposed that a sample of schools with lower ratios would show substantially higher achievement than the population. Technically, our alternate hypothesis allows findings to be *not equal to*, and thus either higher or lower than, the population values (which came into play in our example).

3. The Critical Value: $\pm 1.96\ z$ Values (5% Exclusion Area). Recall that we need to have a benchmark to help us decide whether our actual, calculated results are considered typical or atypical. Actually, we are using this benchmark to help us decide which hypothesis (null or alternate) is more accurate. As I discussed before, for this particular situation researchers use a 5% benchmark. That is, if a calculated/transformed sample mean falls into the 5% exclusion area (the 2.5% in each tail) of a standard normal distribution, then it would be considered atypical. In probability terms, this would represent a probability of occurrence of ($p < 0.05$, either positive or negative). Stated differently, it would be considered not likely to occur just by chance; rather some reason other than chance would create a finding this extreme.

4. The Calculated Value (-3.31). This is the Z-test value that we calculated from the values we had available. It represents the results of the Z-test formula that transformed the sample value into a standard score so that we can compare it to other possible sample outcomes.

5. Statistical Decision: Reject Null Hypothesis? This step asks us to compare what we calculated (step 4) to the benchmark (step 3) in order to see which hypothesis (null or alternate) is more likely. In our study, the sample schools with lower ratios had much lower math achievement than the population. If we had simply chosen a sample of schools by chance (rather than intentionally chosen a group with low ratios), the sample mean would likely not have been this extreme because most chance selections would be much closer to the population mean.

The lower ratios in this contrived study "pushed" the sample math achievement mean far down into the left tail of the distribution. Given this result, that there would be an extremely small probability ($p < 0.00047$) of a Z-test value this small falling into the tails simply by chance alone, we would "reject the null hypothesis." That is, we could conclude that the null hypothesis is not supported by our findings; the alternate hypothesis is supported by our findings.

6. Interpretation. Researchers must make statistical decisions through the steps above. However, they must *place the findings in the language of the question* so that it has meaning to the audience. We obtained an extremely atypical finding. We would need to capture this in an interpretive discussion. We might say something like: "Our sample group of schools ($N = 61$) had a much lower mean math achievement (44.04%) than the mean of the population of schools (51.96%) as evidenced by a statistical hypothesis test ($p < 0.05$)." (Reporting the $p < 0.05$ finding simply is a general statement that the probability of our sample mean had a much smaller probability of a chance finding—that is, in the extreme 5% of the tails of the distribution.)

Statistical Significance

In probability terms, any finding of $p < 0.05$ is considered "statistically significant." Researchers and statisticians have a specific definition for statistical significance: It refers to the likelihood that a finding we observe in a sample is too far away from the population parameter (in this case the population mean) *by chance alone* to belong to the same population.

Practical Significance: Effect Size

I will develop this measure much further in the following statistical procedures, but a word is in order here. Researchers and statisticians have relied extensively on statistical significance to help make statistical decisions. You can see how this language (i.e., using p values) permeates much of the research literature; it is even widespread among practitioners and those not familiar with statistical procedures.

The emphasis in statistics and research now is on *effect size*, which refers to the "impact" of a finding, regardless of its statistical p value. The two issues are related to be sure. However, effect size addresses the issue of the extent to which a difference or treatment "pushes a sample value away from a parameter." That is, how much "impact" does a research variable have to move a sample value?

Consider our example problem shown in Figure 9.8. We performed a statistical significance test with these findings and concluded that the sample mean of (low) student–teacher ratio was too far into the tails of the sampling distribution to be considered a chance finding. *The effect size consideration is a completely different issue.* It does not concern itself with probability, but, rather, how far away from the population mean our sample mean has been driven as result of a lower ratio. According to Figure 9.8, the sample mean has been pushed far into the left tail of the sampling distribution as a result of having a lower ratio than the population, a distance of about $3\frac{1}{3}$ standard deviations!

Z-Test Elements

In order to use the Z test, we must know the population parameters and the sample statistics. This is often not possible because little may be known about the population. In our example, we did have population information (mean and standard deviation) if we define the population as all the schools in the state. We did not have data on all schools in the United States, so our definition of the population was somewhat restricted.

We will discuss similar procedures in the next chapter, but with the assumption that we do not have population information. As you will see, this makes an interesting situation because we must "estimate" the population parameters in order to use the process we learned in the Z test.

TERMS AND CONCEPTS

Alternative Hypothesis The research assumption that is stated in contrast to the null hypothesis.

Case Study A study that focuses entirely on one setting rather than making inferences from one study setting to another.

Central Limit Theorem The statistical notion that means of repeated samples taken from a population will form a standard normal distribution (assuming a large sample size) even if the population was not normally distributed.

Control Group An experimental group in which the treatment is not applied or administered so that the results can be compared with the "treatment group."

Convenience Sample A nonprobabilistic sample selected from available elements. Usually, this method is used when the researcher has no opportunity to use random sampling methods as in studying secretive groups or groups with difficult-to-identify populations.

Dependent Variables Study variables thought to be "receivers" of the action of independent variables or influenced by predictor variables. Often referred to as "outcome variables" in nonexperimental contexts.

Effect Size The meaningfulness of a study finding. In contrast to statistical significance, which deals with chance or nonchance as a basis for judging a finding, effect size measures the "impact" of a (predictor) study variable to affect the change in an outcome variable. Also known as practical significance.

Experiment   A research design in which the researcher consciously changes the values of a study variable under controlled conditions and observes the effects on an outcome variable.

Hypothesis   A statement that captures the nature of the theoretical question in such a way that it can be quantified and directly verified.

Hypothesis Test   The formal process of assessing whether or not a test statistic is judged to be similar to the population elements from which it was drawn or statistically different from those elements.

Independent Variables   A designation in experimental research for treatment variables or those study variables that are consciously changed to observe effects on outcome variables.

Manipulated Independent Variables   Independent variables the experimenter consciously changes, or manipulates, in order to create the conditions for observing differential effects on the outcome variable.

Nonmanipulated Independent Variables   Independent variables that cannot be changed by the researcher. Typically, they are characteristics, traits, or attributes of individuals.

Null Hypothesis   The assumption in an hypothesis test that there is no difference between the study population yielding a particular sample statistic and the population from which the sample supposedly came.

Parameters   Characteristics or measures of entire populations.

Post facto Research   These study designs are those that involve measuring the relationships among variables using data that have already been collected. The focus of the study may be to determine differences among study variables ("*post facto* comparative") or correlations ("*post facto* correlational").

Predictor Variables   Study variables that are considered prior to and influential upon other study variables. In experiments, these are typically called independent variables; in *post facto* research, they are simply the alleged variable of influence upon an outcome variable.

Probability Sampling   The process of using probability methods in sampling to ensure that the sample elements are representative of the elements in the population.

Quasi-Experimental Design   Experimental designs in which there are elements missing or extraneous influences cannot be controlled sufficiently to ensure that the researcher can make causal conclusions. Ordinarily in program evaluation, this takes the form of working with "intact groups" or groups that already exist, or, more generally, the inability to ensure full randomization in designing the experiment.

Randomization   Experimental research procedures in which the study elements (i.e., sample) are chosen and assigned on the basis of equal chance or probability.

Sampling   The process of selecting elements from a population.

Sampling Distribution The distribution formed from repeatedly drawn means from a population.

Sampling Error The difference between the sample mean in a study and the mean of the population from which the sample was drawn.

Simple Random Sample A sampling process in which each element chosen from a population has an equal chance of being selected.

Snowball Sampling A method of (nonprobabilistic) sampling in which available sample elements are identified from previous sample elements. Oftentimes, in studying secretive or difficult groups, additional contacts can be identified as the researcher proceeds with interviews, for example.

Standard Error of the Mean The standard deviation of the sampling distribution of means.

Statistical Significance This designation is stated differently for different statistical procedures. Essentially, it refers to the likelihood that study results are not obtained by chance or happenstance.

Statistics Characteristics or measures of samples.

Stratified Random Sampling A sampling process that recognizes inherent levels or strata in the population that can be sampled by probabilistic means.

Theory An abstract idea or statement that links the conceptual relationships within an inquiry.

Treatment Group An experimental group in which the values of the independent (or treatment) variable have been changed.

Variables Concepts that have been quantified and used in research. In contrast to "constants" that cannot take different values, variables consist of elements that can vary or change.

Z Test The statistical test of the likelihood that a sample mean comes from a population with known parameters.

REAL-WORLD LAB V

In this lab, we will explore another aspect of the theory relating to the relationship between size of learning environment and achievement. Our chapter example dealt with class size, which led to the conclusion that smaller class size is not related to higher achievement; in fact, it is related to lower achievement.

In this lab, we will look at another hypothesis to see whether the size of learning environment affects achievement. This time, we will explore whether a sample of schools ($N = 40$) with enrollments of less than 200 will have higher average reading passing rates than the population of schools in the state ($N = 1054$).

The mean (fourth grade) reading passing percentages for our sample ($N = 40$) of schools with less than 200 enrollment is 69.29%. The population of schools in the state ($N = 1054$) yields a mean (μ) reading passing rate of 73.33% and a standard deviation (σ_X) of 13.28.

Using this information, perform a Z test to determine whether smaller school size affects reading passing rates. Respond to the following questions:

1. What is the standard error of the mean?
2. What is the calculated Z-test ratio?
3. Perform an hypothesis test.
4. Write a brief summary of your findings.
5. Discuss effect size.

REAL-WORLD LAB V: SOLUTIONS

Using the information provided, perform a Z test to determine whether smaller school size affects reading passing rates.

1. What is the standard error of the mean?

Answer: The standard error of the mean is 2.10. This is the standard deviation of the sampling distribution of means.

$$\sigma_M = \frac{\sigma_X}{\sqrt{N}}, \quad \sigma_M = \frac{13.28}{\sqrt{40}}, \quad \sigma_M = 2.10$$

2. What is the calculated Z-test ratio?

Answer: The Z-test ratio is -1.92. This represents the number of standard deviations that the sample mean is removed from the population mean in the sampling distribution of means. Figure 9.9 shows that the sample mean falls to the left of the sampling distribution about two standard deviations away from the (population) mean.

$$Z = \frac{M - \mu}{\sigma_M}, \quad Z = \frac{69.29 - 73.33}{2.10}, \quad Z = -1.92$$

3. Perform a hypothesis test.

The hypothesis test follows the same steps we used in our example above:

The Null Hypothesis (H_0): $\mu_1 = \mu$. Our null hypothesis was that a sample of low enrollment schools will have the same reading achievement scores as the population of schools.

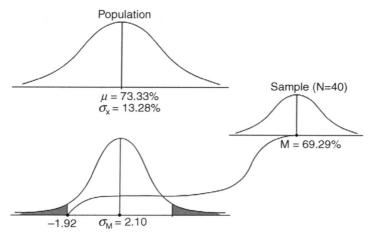

FIGURE 9.9 Using the sampling distribution in the Z test.

The Alternative Hypothesis (H_A): $\mu_1 \neq \mu$. The sample of schools with lower enrollments will not have reading achievement percentages equal to all schools in the state.

The Critical Value: ± 1.96 z Values (5% Exclusion Area). Remember, if a calculated Z ratio falls into the 5% exclusion area (the 2.5% in each tail) of a standard normal distribution, then it would be considered unlikely to occur by chance.

The Calculated Z Ratio $= (-1.92)$. The calculated Z value falls into the left side of the distribution, but not quite in the 5% exclusion area.

Statistical Decision: Do not Reject Null Hypothesis. The calculated Z value was close to the critical value of rejection, but not beyond it. Strictly speaking, the sample mean is therefore still in the "acceptance" region and we can conclude that it is likely a chance finding. That is, our sample of schools has reading passing rates just below the population mean value by chance; the school size does not result in a "significant" finding. Therefore, we do not reject the null hypothesis; we must assume there are no differences between sample and population values.

Interpretation. On the basis of a Z test ($Z = -1.92$) there are no statistically significant differences in the average reading passing rates between a sample ($N = 40$) of schools with low enrollments (<200) and the population of schools in the state.

4. Write a brief summary of your findings.

The interpretation section of the hypothesis test can serve as our summary of findings.

5. Discuss effect size.

Despite the fact that we did not reject the null hypothesis, there does appear to be an effect size of some magnitude in this study. The sample mean falls well below the population mean of reading passing rates. The sample schools' low enrollment size appears to have had the result of pushing the reading passing rates below the population reading passing rates.

10

THE *T* TEST FOR SINGLE SAMPLES

I mentioned earlier that I shade in a portion of my syllabus and call it "headache day" so that students are aware that this is a topic that will take some thinking about. A number of years ago, one of my students told me their story in which the headache became real! A graduate student and teacher, the student was taking my statistics course to get a master's degree to jump into a higher pay scale (as well as to understand statistics).

I noticed on several occasions that the student looked increasingly dour, so I asked the student one day before class began if all was well. The student replied, "Every day I go home after this class and get overwhelmed!" After I pleaded for the student not to take this route to understand statistics, and as we began to discuss the topic, the student began to appear livelier and engaged with the topics. I am happy to report that the student received an "A" grade in the class and, as far as I know, is no longer overwhelmed at the end of the day!

I use this example for students who feel at this point in the class that the material is overwhelming. As my student learned, the material is not unknowable; it simply takes a different way of looking at it to see the direction we are taking in inferential statistics.

Z VERSUS *T*: MAKING ACCOMMODATIONS

Up to now, we have been dealing with population values (parameters) that are known to the researcher. As I mentioned in Chapter 9, however, this is fairly rare. If you do have access to parameter values, you can use the *Z* test to help you make a

---

Understanding Educational Statistics Using Microsoft Excel® and SPSS®. By Martin Lee Abbott.
© 2011 John Wiley & Sons, Inc. Published 2011 by John Wiley & Sons, Inc.

statistical decision about whether your sample values are likely to be taken from that known population.

If you do not know the population values, we will learn in this chapter how to estimate them so that we can use them in our statistical decision-making process. This will seem a little strange at first, since we will use sample values to help estimate the population values, but we are going to make use of our sampling distribution as well. In addition, we will learn to make small "adjustments" based on the sample size as a way to better understand the population values.

As we have discussed, the Z distribution consists of the known areas in the standard normal distribution. We learned in Chapter 9 that we could use the features of this distribution to help us understand whether a sample value could likely come from a distribution with known population parameters (Z test). Now we turn to a related distribution, the T distribution, to help us understand whether a sample value could likely come from a distribution with *unknown population parameters*. This is typically the situation researchers encounter in real-life problem solving because the knowledge of parameters is unlikely in most situations.

The T test for a single sample is a statistical procedure similar to the Z test, but with some limitations:

1. Population parameters are unknown. Typically, the population mean may be known as a general estimate based on similar research, or on the basis of some other reason. However, the standard deviation of the population is not known. Therefore, a T test uses estimates of population parameters based on sample values.
2. Sample size is small. Sample size is very important in statistics because it is used as a denominator in many calculations. Large sample sizes typically result in better estimates of population means. According to the Central Limit Theorem, repeated large samples will more closely approximate a normal distribution. But how large is large? Researchers and statisticians vary on that score. Typically, a sample size of 30 is considered large for statistical procedures by many researchers. Other researchers suggest higher values. I use $N=40$ as my operational definition for "large" samples. As you will see, there is probably no best answer to this question, however.

RESEARCH DESIGN

Does place and income affect technology skills? This is a question that arose during an evaluation of the TAGLIT data I introduced in Chapter 3. As I noted then, TAGLIT is a data evaluation system that addresses the nature of technology in teaching and learning. In 2003, I used the TAGLIT instruments to help evaluate the progress of certain school districts in using technology in the classroom.

TAGLIT uses several instruments to assess different subject groups on the question of technology knowledge and use. Teachers from middle/high schools and elementary schools, students at different levels, and technology leaders in schools were assessed by an electronic data-gathering instrument to explore various aspects of the overall question of instructional technology. The resulting (national-level) database revealed some interesting and informative insights into the primary questions.

As part of my evaluation, I created four "factors" that summarized much of the middle/high teachers' responses using a statistical procedure called "factor analysis" that "reduces" a series of questionnaire items to related groups. I created four factors that "clustered" the similar responses to sets of questionnaire items—a sort of index. I named one of these factors "mhskills" because it represented a set of items that measured the extent of the technology skills of middle/high teachers. This is one good way of combining similar information so that you can operationally define a concept like technology skill. As we make our way through this chapter, I will present a sample of these data and show how to use the T test to make a statistical decision about the "theory" or research question I posed at the outset of this section.

We discussed research design in Chapter 9; I mentioned then that I would discuss various designs in more detail in subsequent chapters. The design appropriate for the subject of this chapter on T test is a very simple one. We will determine whether a sample group mean is likely to come from a population for which we have no knowledge of the parameters.

Experiment

If I were to proceed with a random selection of subjects to whom I administered some treatment or manipulated an independent variable, this design would be an experimental design. For example, I might have taken a (randomly selected) group of middle/high teachers and trained them to use certain technology. Then, I would compare their knowledge of technology to the general knowledge in the population to determine if my teaching technique was effective.

Campbell and Stanley (1963) refer to this "experimental" design as the "one-shot case study" because the design does not have the requisite parts of a true experimental design. Figure 10.1 shows the diagram for this kind of design. There is only one group, the teachers being taught technology; there is no comparison or control group. There is no "pre-test" of the technology test to compare to the outcome test ("post-test").

A "true" experimental design would use a randomly chosen sample, but many times a sample is not chosen randomly. You can see that the potential problems

	Research Treatment Variable: Technology Training	Outcome Variable
Random sample??	Experimental Group: MH teachers	Technology knowledge test scores
	— (No control group)	—

FIGURE 10.1 The one-shot case study.

with this experimental design would lead to conclusions that are questionable. The research treatment might be effective, but it would be difficult to disentangle the experimental effects from the effects of the uncontrolled influences. No causal conclusions could be made on the basis of data from this design.

Post Facto Comparative Design

Because the data I will use are existing data, and I did not manipulate an independent variable, the design is a *post facto* comparative design. I have a sample group drawn from a larger population of data that I wish to compare to an (estimated) population mean to determine whether the sample is likely to come from this population. I cannot make causal conclusions with this design because I did not meet the conditions of a true experimental design (i.e., randomness, control group, control of extraneous variables). However, this is one context in which my conclusions would potentially be equal to those of the one shot case study. Some researchers would argue that an experimental design this "weak" would be no better than a comparative *post facto* design.

This design is not the strongest for many reasons. However, researchers often must use it because of the limitations of the study, the available data, and so on. The design can still provide meaningful data despite the inability of the researcher to make causal conclusions.

PARAMETER ESTIMATION

In research designs like the one I discussed above, the researcher is faced with making as strong a conclusion about the research question as possible with the available data. The first task, however, is to understand what data are available and how they can be used to make a strong conclusion.

Figure 10.2 shows the population, sample, and sampling distribution drawings that I presented in earlier chapters. As you can see, a research study will be able to provide sample information and perhaps have available an estimate of the population mean, but otherwise there will be no additional information for the researcher.

In this event, the researcher must *estimate the key parameters* in order to carry out the study as we did with the Z test. The most important parameter to estimate is the population standard deviation (σ_X). Without a notion of the variability of the population, we cannot make an accurate estimate of the overall likelihood of the sample mean representing the population mean.

Estimating the Population Standard Deviation

The method for estimating σ_X is fairly straightforward. In fact, it follows a commonsense logic. If you want to know what the population variance looks like, what is the best guess? The best guess is probably the variance of the sample. Obviously, it may not be a perfect representation, but *the sample variance is the best estimate of the population variance*.

PARAMETER ESTIMATION

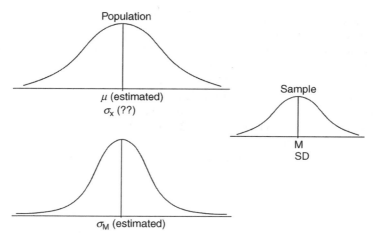

FIGURE 10.2 Using the sampling distribution with estimated population values.

The method that statisticians have devised to create a more likely estimate than the sample standard deviation is to "adjust" the sample value. The principle is that if you divide by a smaller number, the resultant calculation will be larger.

Estimating the population standard deviation from the sample follows the same principle. If we *adjust* the standard deviation formula for the sample by subtracting a value of "1" in the denominator, it will provide a *larger* estimate of the population standard deviation. A larger estimate is more likely to "capture" the true population standard deviation, and we can have greater certainty that is representative.

Figure 10.3 shows this process. The actual SD becomes s_x when corrected by the formula. When corrected, it becomes a more accurate estimate of the population standard deviation, σ_X.

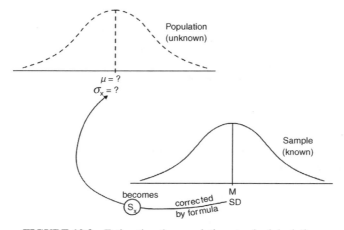

FIGURE 10.3 Estimating the population standard deviation.

Look at the following formula (SD) that I discussed in Chapter 6. This is the formula for calculating the standard deviation for a group of scores that constitute its own population. I talked about this formula in Chapter 6 in the context of identifying a difference in the way both Excel® and SPSS® calculate standard deviation. In fact, I labeled this the "population SD" or "actual SD." We learned that we use this calculation when we are simply trying to describe a set of data without making reference to a greater population from which the set of scores may have been derived. We simply consider the set of scores its own population.

Now that we are considering inferential statistics, we can modify this formula a bit to be a better estimate of the population standard deviation from a sample value.

$$SD = \sqrt{\frac{\sum X^2 - \frac{(\sum X)^2}{N}}{N}}, \qquad s_x = \sqrt{\frac{\sum X^2 - \frac{(\sum X)^2}{N}}{N-1}}$$

If you recall, I distinguished this actual SD with the "inferential SD" that is reported by Excel® and SPSS®. Although I did not introduce the formula, I wanted to ensure that you understood the difference between the actual SD and the Inferential SD. Review the discussion in Chapter 6 to make sure you understand the difference.

A New Symbol: s_x

The second formula above for the value of "s_x" is the "adjusted" formula for estimating a population standard deviation from a sample standard deviation. Many statisticians and researchers use s_x or simply "s" as the symbol for the *estimated standard deviation of the population*. You can see that the only difference between the formulas is the denominator: N in the SD formula and $N-1$ in the s_x formula.

We are estimating one parameter, the population standard deviation. As a way to be more likely to capture the true population standard deviation, we subtract a value of 1 from the divisor and therefore show a larger result. Review the student–teacher ratios in Chapter 6. You will see the following:

$$SD_{(Population)} \text{ or actual SD} = 4.74, \qquad SD = \sqrt{\frac{10055.02 - \frac{(605.13)^2}{40}}{40}}$$

Now, compare this value to the value for s_x which is obtained with the same values above except for the denominator (i.e., 39 instead of 40). This is the value reported by both Excel® and SPSS®.

$$\text{Calculating } s_x = 4.80, \qquad s_x = \sqrt{\frac{10055.02 - \frac{(605.13)^2}{40}}{39}}$$

Biased Versus Unbiased Estimates

Statisticians refer to the actual SD as a "biased estimator" because it does not necessarily provide a true picture of the population standard deviation. Since the sample mean is likely different from the population mean, and since the sample mean is used in the calculation of the actual SD, it will not provide a true representation of s_x. The estimated standard deviation of the population s_x is therefore an unbiased estimate when it is adjusted as we discussed.

A Research Example

Returning to the research question we posed at the beginning of this chapter, "Does place and income affect technology skills?" We can assess this question using the T test that I will describe below. First, however, we can get some practice with calculating s_x, the unbiased estimate of the population standard deviation.

The hypothesis for the research question above is: "Does the average technology skill index of our sample of MH teachers from wealthy urban schools represent the population of MH teachers?" Thus, we are asking whether our (known) mean is representative of an (unknown) population mean. Do teachers from wealthy urban schools have better technology skills than those from the overall population of schools?

The outcome variable is the MH teacher technology skills index that I described above from my factor analysis.[1] The sample group consists of 28 wealthy urban schools from the United States. Wealthy is defined as a low FR percentage (between 1% and 5%), and urban is a category defined by the national database we used in our TAGLIT study. Table 10.1 shows the data (in two columns) for the sample of schools.

Review the process in Chapter 6 for calculating the actual SD and s_x using Excel® and SPSS®. Remember, Excel® and SPSS® both provide s_x as the default calculation. Figure 10.4 shows the descriptive summary for the mhskills variable; the reported standard deviation (s_x) is 0.264. Figure 10.5, the SPSS® output, yields the same values as the Excel® values.

The actual SD from Excel® is 0.259 (use the "STDEVP" function). The discrepancy between the actual SD and s_x is not large; however, the larger s_x is a better estimate of the population standard deviation. Note that in both cases the sample group appears to be normally distributed as indicated by skewness and kurtosis figures.

---

[1] The range of the MH teacher skill index is small because the possible values range from 1 to 4. The overall conclusions of the study should take this into account in deciding the meaningfulness of the findings.

TABLE 10.1 The Outcome Data mhskills for the Research Study

mhskills	
3.70	3.09
3.69	3.07
3.63	3.06
3.53	3.05
3.50	3.04
3.50	3.03
3.41	3.03
3.37	3.02
3.37	3.02
3.33	3.02
3.23	2.99
3.22	2.93
3.16	2.87
3.10	2.65

T TEST FOR A SINGLE MEAN

Now that we have an estimate of the population standard deviation, we can use this information to test our hypothesis that our sample mean is likely to come from a specific population. We will use the same procedure that we used with the Z test, except now we must use the estimated parameter s_x.

Figure 10.6 shows the information we have in graphic form so we can visualize the process we will use. As you can see, we have estimated the population standard deviation, but we need the estimated population mean. In these types of research problems, the researcher must have at least an estimate of the population mean

mhskills	
Mean	3.200
Standard error	0.050
Median	3.093
Mode	#N/A
Standard deviation	0.264
Sample variance	0.070
Kurtosis	−0.455
Skewness	0.334
Range	1.053
Minimum	2.647
Maximum	3.700
Sum	89.603
Count	28.000

FIGURE 10.4 Excel® descriptive statistics for mhskills.

T TEST FOR A SINGLE MEAN 183

Descriptive Statistics

	N	Mean		Standard deviation	Skewness		Kurtosis	
	Statistic	Statistic	Standard error	Statistic	Statistic	Standard error	Statistic	Standard error
mhskills	28	3.2001	0.04983	0.26369	0.334	0.441	−0.455	0.858
Valid N (listwise)	28							

FIGURE 10.5 SPSS® descriptive statistics for mhskills.

from which the sample mean can be compared. Typically, there is an informed estimate based on previous research or practitioner observation, among other possibilities. In this case, we can use $\mu = 3.10$ as the estimated population technology skill level of the MH teachers.

As in the Z test, we need to make use of the sampling distribution since it is a more perfect comparison distribution for our sample mean. Because it is theoretically a standard normal distribution, we can transform our sample mean into the distribution to see how our sample mean compares to the population mean.

We calculate the (estimated) standard error of the mean using the same procedure as in the Z test. The following formula applies; note that because we use s_x, the estimated population standard deviation, we will be calculating the estimated standard error of the mean, which has a new symbol to identify it specifically: s_m.

$$s_m = \frac{s_x}{\sqrt{N}}$$

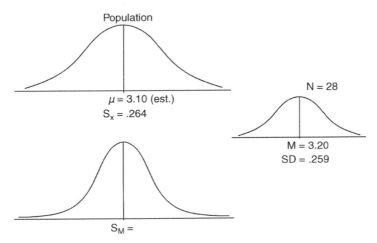

FIGURE 10.6 Visualizing the T-test process with the sampling distribution.

TABLE 10.2 Population and Sample Symbols for Inferential Statistics

M	The mean of the sample.
SD	The standard deviation of the sample (assumes the sample is its own population).
μ	The Greek letter "mu" is the symbol for the population mean.
M_M	This is the symbol for the mean of the sampling distribution. You can see how this works by observing that it is a mean (M), but a mean of the sampling distribution indicated by the subscript. Thus, it is the *mean of the distribution of means*, or the "mean of the means." Because it is (theoretically) created by all possible samples, it is a parameter.
σ_X	"Sigma X" is the standard deviation of all the population raw scores. This differs from SD in that it does not refer to a sample, but to the entire population of scores.
σ_M	The standard deviation of the sampling distribution of means; also called the *standard error of the mean*.
s_x	The estimated standard deviation of the population. The subscript x identifies this estimated value as belonging to the population of all scores.
s_m	The estimated standard error of the mean. The subscript m identifies this estimated value as belonging to the sampling distribution of means.

This new symbol indicates that it is estimated (by using the lowercase s instead of σ) and that it belongs to the sampling distribution (since the subscript is an m rather than an x). Thus, we add two additional symbols to our earlier list in Table 9.3. Table 10.2 reproduces the earlier symbols and adds the two new symbols.

Example Calculations

In order to perform a *T* test, we use a formula very similar to the *Z* test.

$$t = \frac{M - \mu}{s_m}, \qquad \text{where} \quad s_m = \frac{s_x}{\sqrt{N}}$$

As you can see, the *T*-test formula parallels the *Z*-test formula:

$$t = \frac{M - \mu}{s_m}, \qquad Z = \frac{M - \mu}{\sigma_M}$$

The numerator, the difference between sample mean and population mean, is the same except that with the *T* test we are using an estimated μ.

Using the formula with our current problem, we can calculate t. We must first calculate s_m because it will be the denominator in the *T*-test formula:

$$s_m = \frac{s_x}{\sqrt{N}}, \qquad s_m = \frac{0.264}{5.29}, \qquad s_m = 0.05$$

(Note that the s_m of 0.05 is reported in both Figures 10.4 and 10.5, the Excel® and SPSS® output files.)

T TEST FOR A SINGLE MEAN

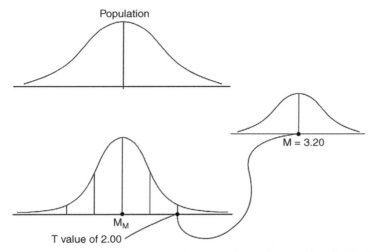

FIGURE 10.7 Transforming the sample mean value to the sampling distribution.

$$t = \frac{M - \mu}{s_m},$$

$$t = \frac{3.20 - 3.10}{0.05},$$

$$t = 2.00$$

Therefore, our sample mean of 3.20 is transformed into a t ratio of 2.00. This value indicates that the sample mean value lies 2 standard deviations above the population mean when transformed to a standard normal T distribution using the T-test formula. Figure 10.7 shows how to visualize this value.

It is interesting to note that had we used the Z test, we could have rejected the null hypothesis because the test value (2.00) lies above the Z-test rejection region (1.96). However, because of the necessity of using the T distribution (i.e., small samples, unknown population standard deviation), we will have a new 5% rejection region (2.052) that is based on the T distribution. I will discuss this new rejection region below; but for now, note that the new exclusion value of 2.052 is higher than our test value of 2.00, resulting in the inability to reject the null hypothesis.

Degrees of Freedom

Degrees of freedom (df) are important elements in statistics. Essentially, they represent *the restrictions on the values when we are estimating a population parameter*. Technically, df's identify how many values are free to vary when making the

mhskills	X − M	(X − M)²
3.70	0.50	0.25
3.69	0.49	0.24
3.63	0.43	0.18
3.53	0.33	0.11
3.50	0.30	0.09
3.50	0.30	0.09
3.41	0.21	0.05
3.37	0.17	0.03
3.37	0.17	0.03
3.33	0.13	0.02
3.23	0.03	0.00
3.22	0.02	0.00
3.16	−0.04	0.00
3.10	−0.10	0.01
3.09	−0.11	0.01
3.07	−0.13	0.02
3.06	−0.14	0.02
3.05	−0.15	0.02
3.04	−0.16	0.03
3.03	−0.17	0.03
3.03	−0.17	0.03
3.02	−0.18	0.03
3.02	−0.18	0.03
3.02	−0.18	0.03
2.99	−0.21	0.04
2.93	−0.27	0.07
2.87	−0.33	0.11
2.65	−0.55	0.31
3.20	0.00	1.88

FIGURE 10.8 Understanding the concept of degrees of freedom.

parameter estimate. We have run into this before, and we can discuss this important concept with the current research problem.

Figure 10.8 shows the sample schools' MH teacher skills indexes. As you can see, I created an Excel® table with two columns: $X - M$ represents the deviation scores in which the mean is subtracted from each index score. These are the "deviation amounts" that I discussed in Chapter 6 (see Figure 6.7). I also included the square of these values, $(X - M)^2$, as I would if I were to calculate the standard deviation.

Recall that the deviation scores, when added together, equal 0 for the reasons we discussed in Chapter 6. Here is the curious fact: When we use these values to estimate a parameter, we may be using a mean value that varies slightly from an unknown population mean. Therefore, the values in the $X - M$ column can change slightly, but one must be "fixed" to ensure that the deviations still add to 0. Look at the $X - M$ values; in order for the sum of the values to equal 0, the last value

T TEST FOR A SINGLE MEAN

("−0.55") must have this specific value. If the other values change, one value must take a specific value in order to ensure the outcome of 0.

This example is just to show that *estimating parameters places restrictions on our data*. In the case of the T test with one sample, only one value cannot vary. Thus we can calculate df to be $n - 1$. This represents the sample size minus one value. In our example of $n = 28$, the df $= 27$ ($28 - 1$).

Each statistical procedure that we encounter will have a different way of calculating degrees of freedom because each will be estimating different parameters. We will make note of df's as we discuss the hypotheses tests.

The T Distribution

Before we proceed to a hypothesis test, we need to discuss the T distribution. It is essentially the same as the Z distribution in terms of being a standard normal distribution. That is, it is normally distributed with known values between the mean and each standard deviation.

Because the T distribution is typically based on small sample sizes, however (along with estimated parameters), the T distribution varies by sample size. Figure 10.9 shows how different sample sizes affect the T distribution. As you can see, the larger the sample size, the more the T distribution looks like the Z distribution with the large "hump" in the middle with smaller tails on both

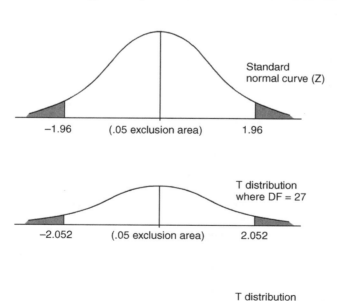

FIGURE 10.9 The nature of the T distribution.

sides. The smaller the T distribution, the fewer the comparison values that make it up, with the result that it appears flatter and flatter as the sample sizes decrease.

The result of these dynamics of size is that the T distribution is often thought of as a *series of distributions that are linked to the size of the sample to be compared.* Since the shape of the curve alters drastically with smaller sample sizes, the exclusion/rejection region for an hypothesis test changes as well. *It takes a higher calculated T ratio to reject the null hypothesis the smaller the sample size.* This is because the 5% exclusion area "moves" away from the mean with smaller sample sizes due to the changing shape of the distribution.

In our example, $n = 28$; therefore, df $= 27$ (since df $= n - 1$). In order for us to reject the null hypothesis at the 5% exclusion region, we would need the t ratio to be at least 2.052 (positive or negative) as indicated in Figure 10.9. The rejection values for the 5% exclusion area get smaller and smaller as the sample sizes increase. With very large numbers in the sample, the T distribution is indistinguishable from the Z distribution (the top curve in Figure 10.9).

The Hypothesis Test

Recall that our hypothesis asks whether a wealthy, urban sample of schools will have similar MH teacher technology skill indexes as the population of schools. In order to assess this hypothesis, we can proceed to the hypothesis test as we did with the Z test; we always use the same five steps.

1. **The Null Hypothesis or (H_0): $\mu_1 = \mu$.** Does the MH teacher technology skill index from wealthy urban schools represent the teachers' indexes in the population of schools?
2. **The Alternative Hypothesis or (H_A): $\mu_1 \neq \mu$.** The MH teacher technology skill index from wealthy urban schools does not represent the teachers' indexes in the population of schools.
3. **The Critical Value: T-Table Values of $= (\pm)$ 2.052 (5% Exclusion Area).** If the calculated value exceeds this value (either positively or negatively), we can reject the null hypothesis. In such a case, our sample value being that high would be considered not likely to occur just by chance; rather some reason other than chance would create a finding this extreme.
4. **The Calculated Value (2.00).** This is the t-ratio value that we calculated above. It represents the results of the t-test formula that transformed the sample value into a standard score on the T distribution so that we can compare it to other possible sample outcomes.
5. **Statistical Decision: Reject Null Hypothesis?** We cannot reject the null hypothesis since our calculated value ($t = 2.00$) does not exceed the t-table value (df $= 27$) for the 5% exclusion area (either positively or negatively). In our study, the sample schools' showed a value representative of the overall population of schools.

TYPE I AND TYPE II ERRORS

6. **Interpretation.** MH teacher average technology skills index scores in the sample group ($N = 28$) of schools ($M = 3.20$) were representative of schools in the overall population with an estimated mean of 3.10. In an actual research study, a t ratio (2.00) this close to the rejection area (2.052) might be considered significant.

Depending on how rigidly you interpret the "statistical rules," you might conclude that your sample of MH teachers tends to show higher technology skill indexes than schools in the population. We will return to this example finding after we discuss additional concepts that may help us with this dilemma.

TYPE I AND TYPE II ERRORS

As I have continually noted, statistics deals with uncertainty. You can begin to get a better picture of this through our discussion of inferential statistics. The hypothesis test attempts to establish *the extent to which it is likely* that a certain sample mean reflects a population mean, not the absolute certainty.

This being the case, statisticians and researchers often can make mistakes in their conclusions from hypotheses tests. These errors can be presented in two groups.

Type I (Alpha) Errors (α)

The Type I error is mistakenly rejecting the null hypothesis when it should be accepted. That is, we might reject the null hypothesis, as we did in our Chapter 9 Z-test example of the effects of student–teacher ratios on math achievement. However, it is possible that our sample mean was one of the few in the exclusion area just by chance and not a mean that reflected a different population. So, even though we rejected the null hypothesis on the basis of the size of the Z ratio, it perhaps was one of the small likelihood samples that just happened to be located in the tails of the distribution (i.e., the 5% exclusion area). We rejected the null hypothesis, but perhaps there really is no effect of student–teacher ratios on math achievement. Perhaps our sample group of schools was just a chance finding.

The 5% exclusion area that statisticians and researchers use for hypothesis tests is somewhat arbitrary. No matter what the exclusion area set, however, there is the small possibility that the sample mean comes from beyond the limit of the exclusion area by chance alone and has nothing to do with the research question. Figure 10.10 shows this error using the general Z-test process for testing the null hypothesis.

As you can see in Figure 10.10, if the Z-test ratio (the calculated value of the sample mean translated into a Z value in the standard normal distribution) lies beyond the exclusion value, we would reject the null hypothesis. In the figure, our hypothetical Z ratio lies in the tail so we would reject the null hypothesis.

Note, however, that there is 5% of the distribution located in the two tails. In terms of probability, we establish these as an arbitrary limit for ease of decision making. One possibility for the Z ratio lying that far into the exclusion area may

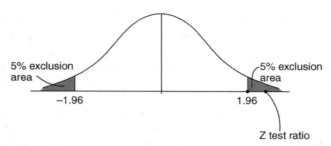

FIGURE 10.10 The Z-test exclusion areas.

simply be a matter of chance. We may have chosen a sample with a large mean that happened to differ greatly from the population mean. Therefore, our decision about the sample not representing the population may be in error. It may be more an issue of chance than the meaningfulness of the research treatment/variable.

When we use the designation $p < 0.05$ in announcing our statistical decision (from the hypothesis test), we are saying that there is less than 5 chances in 100 we would obtain a test value (Z ratio in this example) that large by chance alone. Therefore we reject the null hypothesis. However, note that our test result may be one of those chance findings that happen to be in the 0.05 area of the distribution.

The upshot of all this is that we must understand that our alpha (type I) error is 0.05. Whatever exclusion area we announce (5%, 1%, etc.) defines our alpha error.

Type II (Beta) Errors (β)

Beta errors are quite interesting for many reasons. We will examine several of these in the chapters ahead. For now, however, we will define the Type II (beta) error as *not rejecting the null hypothesis when it should be rejected.*

In some ways, this is the opposite of the alpha error. Just as there is the possibility that we could reject the null hypothesis due to chance, we could also *not* reject a null hypothesis when we really should have. Perhaps our sample was taken by chance from the left side of the curve and whatever research treatment we used "moved" it toward the right side, but not far enough to fall in the right-hand rejection region. Perhaps our distribution is skewed, resulting in the research treatment having to "work harder" to move the test value toward the rejection region.

There are many reasons why a research finding may not be able to reject the null hypothesis despite being a meaningful finding. This is the range of the type II error.

Figure 10.11 shows the type II error. The example is a general one, although it is similar to the results we observed in our *t*-test example of the hypothesis test of rich urban schools affecting MH teacher skill indexes in this chapter. As you can see in the figure, our sample mean was calculated to be very close to the rejection region of 1.96 (the 5% rejection region for the Z distribution), but not beyond the limit. (This is similar to the findings from our example above in which the *t* ratio of 2.00 was very close to the 5% exclusion area of 2.052 in the *T* distribution.)

EFFECT SIZE

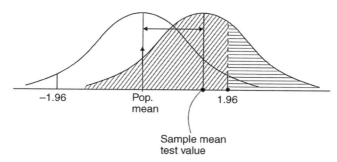

FIGURE 10.11 Visualizing beta and non-beta areas.

Perhaps there were other factors resulting in the inability to reject the null hypothesis (of teacher skills) rather than the effect of schools being rich and urban. Perhaps the sample was not randomly chosen, perhaps the variable was skewed, and so on. For whatever reason, we did not reject the null hypothesis. There is a chance, however, that we possibly should have rejected the null hypothesis.

Look in Figure 10.11 at the distance between the population mean and the sample mean (shown by a two-sided arrow). This shows that the research effect "moves" the sample mean away from the population mean some distance. The horizontally hatched region of the sample distribution is the area that lies above the rejection region in the comparison distribution. Any sample mean test value that would fall in this region would result in the null hypothesis being rejected and therefore would be a "correct" decision.

The possibility of the beta error would lie in the diagonally hatched region of the test distribution. This is the region that would not result in a rejection of the null hypothesis because this is the area that falls below the rejection region. As you can see, our sample test value fell just below the rejection region and therefore in the beta error section, representing a possible type II error.

The beta error can be measured like the alpha error. The diagonally hatched region would represent the proportion of beta (β), while the horizontally hatched section of the test distribution represents the non-beta area ($1 - \beta$).

EFFECT SIZE

Effect size is a very important concept in statistics as I have mentioned in past chapters. It refers to the *strength or impact of a finding*. (In Chapter 9, I discussed effect size as practical significance versus statistical significance.) Although related to statistical significance, it refers to the distance a research variable or treatment "moves" a sample test mean away from a population mean.

Figure 10.11 shows the effect size by the double-sided arrow separating the population and sample means. In our example above, the MH teacher average skill index was much higher than the population average due to place (urban) and wealth (low FR), even though it did not result in rejecting the null hypothesis.

Here are some things to note regarding effect size:

1. It measures the impact of the research treatment/variable to move a sample test mean away from a population mean.
2. It is a separate consideration from statistical significance. Even though these are related concepts, effect size addresses the extent of the sample mean being moved away from the population while statistical significance refers to whether this is a chance finding.
3. Statistical tests have effect size measurements even though the null hypothesis is not rejected.

Each statistical test procedure we will discuss in this book has an effect size measurement. For the single-sample t test, the effect size formula is

$$d = \frac{M - \mu}{s_x}$$

This formula is derived from Cohen's d, the classic formula for transforming the distance between two means (in the single sample t test, this is the distance of the sample mean from the population mean) into standard deviation units by dividing the $M - \mu$ distance by the standard deviation (s_x). Substituting our values from the example above, we obtain

$$d = \frac{M - \mu}{s_x}, \qquad d = \frac{3.20 - 3.10}{0.26}, \qquad d = 0.385$$

Therefore, the effect size of our research study is 0.385. This represents the difference in the population and sample mean in standard deviation units. Several values are suggested for interpreting the magnitude of effects size, like the following from Cohen (1988):

0.20 small effect
0.50 medium effect
0.80 large effect

Despite not being able to reject the null hypothesis, the research variable (wealthy, urban schools) resulted in higher MH teacher average technology skill indexes than those found in the population of schools. The population and sample means were over 1/3 standard deviations (0.385) apart. There is thus a low-to-medium effect size evern though we did not reject the null hypothesis.

Another Measurement of the (Cohen's *d*) Effect Size

You may notice that the effect size calculation d is similar to the t ratio. However, the t ratio transforms the $M - \mu$ distance into standard deviation units of the sampling distribution (s_m) instead of the population distribution (s_x). Because these are similar measures, we can use another calculation to achieve similar results [see Cohen (1988)].

$$d = \frac{t}{\sqrt{N}}, \qquad d = \frac{2.00}{\sqrt{28}}, \qquad d = .38$$

The chief difference between these methods is that the second formula is more sensitive to sample size.

POWER, EFFECT SIZE, AND BETA

There are relationships between Type II error, effect size, and power. Power represents *the ability of a statistical analysis to detect a "true" finding*. The larger the power, the greater the probability of rejecting the null hypothesis when it should be rejected. As we discussed above, Type II (beta) error is the probability of not rejecting the null hypothesis when it should have been rejected. Power and beta are therefore complementary: Power is therefore defined as $(1 - \beta)$.

Look again at Figure 10.11, which shows all these measures.

- Beta (β) is the area of the sample distribution diagonally hatched and represents sample test mean values that do not reject the null hypothesis.
- Power $(1 - \beta)$ is represented by the area horizontally hatched. Sample test means that fall in this section will reject the null hypothesis correctly; this represents the power of the relationship among the test variables.
- Effect size is represented by the double-sided arrow, the distance the sample test value moves away from the population mean and measured in standard deviation units.

Power tables exist in which the researcher can determine the sample size and magnitude of Cohen's *d* to achieve certain "levels" of power (Cohen, 1988).

ONE- AND TWO-TAILED TESTS

Up to now, we have assumed that when we create a rejection region, or exclusion area of the comparison distribution for the hypothesis test, we will split the exclusion area into both tails. Figure 10.12 shows the two tailed test in which the rejection region is split into the two tails of the distribution. The figure shows the Z distribution since it is easier to understand as an example. The top figure shows the 5% exclusion area. As you can see, the 5% must be split in half so that both tails will have half the rejection region. Thus, $2.50\% + 2.50\% = 5\%$.

Figure 10.12 also shows the 1% exclusion area in which the 1% is split into both tails of the distribution (0.5% in each tail).

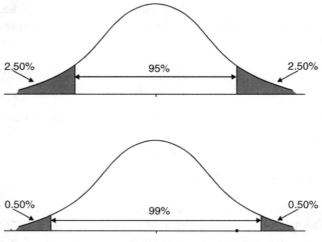

FIGURE 10.12 The two-tailed test.

Two-Tailed Tests

The "default" hypothesis test uses the two-tailed test because it allows for the possibility that a research finding may be changed in either direction, positive or negative. If you recall the example of the Z test we discussed in Chapter 9, we had a z ratio of -3.31. Because this was the Z test, we used the exclusion area of 5%, which translated to z values of ± 1.96. If we had used the 1% exclusion area with two tails, the exclusion values would have been ± 2.58. The result of the hypothesis test was that we could reject the null hypothesis because our test value (-3.31) exceeded the left exclusion value of -1.96.

As we discussed above, the exclusion values change when using the T distribution because the sample values are often smaller, and so on. In the example we discussed in this chapter, our t ratio of 2.00 was not large enough to reject the null hypothesis because the exclusion value was ± 2.052. If we had wanted to use the more stringent 1% exclusion area for the T distribution (with df $= 27$), the exclusion values would have been ± 2.771. The exclusion values for the T distribution must be evaluated differently for each statistical test because the sample sizes can vary.

One-Tailed Tests

Researchers can also establish one-tailed tests by "stacking" the entire rejection region in one tail or the other, but not both. Figure 10.13 compares the one- and two-tailed 5% exclusion areas for the Z distribution. The top figure shows the two-tailed test with the 5% split into both tails; the exclusion value is ± 1.96. The bottom figure shows a one-tailed test with all 5% loaded into one tail. In this case, the exclusion value is 1.65. (If we had loaded the 5% in the left side of the distribution, the exclusion value would have been -1.65.)

ONE- AND TWO-TAILED TESTS

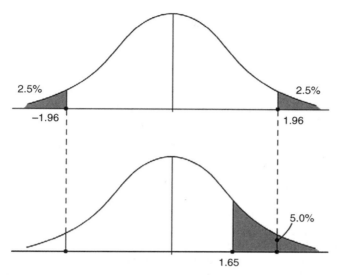

FIGURE 10.13 One- and two-tailed exclusion values.

Table 10.3 shows the exclusion values for the Z distribution for both the one- and two-tailed tests. As you can see, it matters a great deal which you choose for your research problem.

To give you a sense of how the exclusion values differ for the T distribution, Table 10.4 shows the values for $df = 27$, the sample size we used in our example above. Recall that our test value (t ratio) was 2.00. We did not reject the null hypothesis since I used the two-tailed test.

Consider what would have happened if I had chosen a one-tailed test at the 5% exclusion area. I would have rejected the null hypothesis! My t ratio of 2.00 would more than exceed the exclusion value of 1.703.

TABLE 10.3 Exclusion Values for the Z Distribution

	One-Tailed Test	Two-Tailed Test
5% Exclusion area	1.65 (either positive *or* negative)	± 1.96
1% Exclusion area	2.33 (either positive *or* negative)	± 2.58

TABLE 10.4 Exclusion Values for T Distribution (df = 27)

	One-Tailed Test	Two-Tailed Test
5% Exclusion area	1.703 (either positive *or* negative)	± 2.052
1% Exclusion area	2.473 (either positive *or* negative)	± 2.771

Choosing a One- or Two-Tailed Test

It is up to you as a researcher to choose which type of exclusion value you include in the hypothesis test. There is no absolute rule about which you choose. However, you should be ready to defend your choice. The 5% exclusion area of a two-tailed test is considered a standard for researchers. However, researchers exploring potential relationships among variables may choose a much lower exclusion region.

One potentially thorny issue in using a one-tailed test is which tail to identify as the exclusion area. The logical sequence of the hypothesis test requires researchers to choose, before they calculate the t ratio, whether they will use a one- or a two-tailed test. So, you should use some criteria for deciding which tail to use, if you decide on a one-tailed test.

One criterion is your expectation, based on the research literature, or other research findings, that your test result will tend toward one direction or the other. For example, if you have noted in the literature that student technology skills almost always eclipse teacher's skills, you might use the right tail as the exclusion area if you are comparing teachers and students on a technology learning task.

Another criterion is to think about the nature of your sample. If you are working with a sample group that is "extreme" on some measure (e.g., introducing a new method for learning math among special education students), you might anticipate that the students' scores would increase after the new instructional method. This is somewhat intuitive. If a group of students had really low achievement, for example, there would only be one way to go, namely, up! This is the principle of the "regression to the mean effect" in research. A group's scores will generally increase upon retesting if they are low to begin with. On the other hand, if they are high, retesting runs the risk of lowering the scores!

The example problem in Chapter 9 shows how a researcher may encounter difficulties by using a one-tailed test. As you recall, we tested schools with lower student–teacher ratios to see if they would have better achievement than the general population of schools that tend to have larger ratios. We rejected the null hypothesis, *but only because we used a two-tailed test*. Had we decided to load the rejection region into only the positive side of the distribution, under the general expectation that lower student–teacher ratios are better for learning, we could not have rejected the null hypothesis even though our z-test ratio was extremely high at -3.31.

So, in some sense, it is a gamble as to whether you use a one- or two-tailed test. It is easier to reject the null hypothesis with a one-tailed test, but it is also riskier if there is a reasonable possibility that the direction of your findings will tend to the opposite side from what you expect.

A NOTE ABOUT POWER

We discussed power above, but I wanted to note here that there are ways a researcher can increase the power of a finding. The following are some of the factors I have discussed or hinted at thus far that might increase power:

1. Using a one-tailed test, depending on the nature of the research question.
2. Using a lower exclusion value (5% versus 1%).
3. Increasing the sample size because this will generally lower the size of the standard error with the result of increasing the test ratio.
4. Making sure the sample is not skewed.

We will add to this list as we proceed. For now, remember that you as the researcher are in the driver's seat of your research. Use appropriate methods, pay attention to power and effect size, and be systematic in obtaining the most accurate results with the appropriate statistical procedures.

POINT AND INTERVAL ESTIMATES

In conducting the analysis of our t test of a single mean, we transformed our sample mean to see where it would fall on the sampling distribution of means. This allowed us to make a decision about whether the sample mean likely came from the population. To researchers, the t-test procedure is called a "point estimate." The object of the procedure is to create one point in the sampling distribution of means so that we can compare it to the population mean. In order to do this, we had to have some idea of the population mean. Ordinarily, the population mean is not known, so the researcher must posit a value, based on past research or other criterion. This value is more than a guess; however, it still is likely not exactly equal to an unknown population value.

Another way to proceed with a research study is to estimate the population mean within a certain range of values. This is known as a "confidence interval." *This interval is the range of values that will likely contain the true population mean within a certain percentage of certainty.* We often use the same benchmark for probability as we do for a point estimate to create the range of values: $p = 0.05$. Thus, we speak of the 0.95 confidence interval to mean the range of values within which the true population mean will fall with a 95% certainty. We might also choose a more "certain" range of values by using a benchmark of $p = 0.01$. I will show how this works.

Calculating the Interval Estimate of the Population Mean

What would this interval of values look like? How can it be created? The short answer is that we will use the "inclusion area" to create the range of values, rather than the exclusion area that we used in the t test above.

Look at Figure 10.12 again. I used this figure to discuss the two-tailed test. However, the figure also illustrates the inclusion area that will "capture" the true population mean. The top figure shows a 0.95 inclusion area (indicated by the double-sided arrow) and the bottom shows a 0.99 inclusion area. These areas conform to the 0.95 confidence interval ($CI_{0.95}$) and the 0.99 confidence interval ($CI_{0.99}$), respectively.

A statistical formula is used to identify the limits of this interval—that is, the specific values on the sampling distribution of means that bracket the inclusion area. These values represent the confidence interval. The formula is not new to you. If you recall, we learned in Chapter 8 to transform a z score into a raw score as follows:

$$X = Z(SD) + M$$

We can use this formula to create the confidence intervals by adapting it to the T distribution, since we are dealing with unknown population parameters. The adapted formula is

$$\text{Confidence interval} = \pm t(s_M) + M$$

As you can see, the Z value is replaced by $\pm t$ and SD is replaced by s_m in the formula. Beyond these changes in symbols that represent the difference between the Z distribution and the T distribution, the formula is the same. Examining each element of the formula, we observe the following:

$\pm t$ is the value of t from a statistical table that recognizes the appropriate df value. For example, in our research question above, the t value at df $= 27$ was ± 2.052.

s_m is the estimated standard error of the mean. In our example, $s_m = 0.05$.

M, the sample mean in our example, was 3.2.

Substituting these values into the formula will yield the CI. In this example, we can use $CI_{.95}$ because that is the assumption used to create the appropriate T value from the table (i.e., the 5% exclusion—95% inclusion area).

$$CI_{0.95} = \pm 2.052(0.05) + 3.2$$

You will notice that the t value from the T table is both a positive and a negative number because we are using it in a two-tailed procedure; we want to identify values on the left and right of the sample mean that will bracket the population mean estimate. Therefore, we need to calculate the confidence interval twice: once with t as a negative value and once as a positive value.

$$CI_{0.95} = -2.052(0.05) + 3.2$$
$$CI_{0.95}(\text{left bracket value}) = 3.097$$
$$CI_{0.95} = +2.052(0.05) + 3.2$$
$$CI_{0.95}(\text{right bracket value}) = 3.303$$

Figure 10.14 shows these values. The T-table values are shown capturing the 95% inclusion area. Below these brackets are the calculated values that define the $CI_{0.95}$ interval. We can summarize by stating that with a 95% certainty, the true population mean will fall between 3.097 and 3.303.

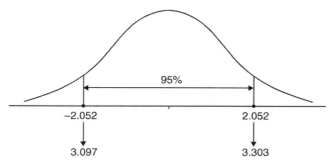

FIGURE 10.14 Confidence intervals values for the example.

This is quite a different process than the point estimate, but the procedures are created to answer different questions:

- *Point Estimate.* Is the sample mean of 3.20 likely to be from a population with an assumed mean of 3.10 (from our example)?
- *Confidence Interval.* Based on the sample values, what is the population mean likely to be? Between what range of values is the population likely to fall with a 95% certainty?

THE VALUE OF CONFIDENCE INTERVALS

As you will see, confidence intervals are quite important in statistics. We have learned how to create brackets that will contain a population parameter. That is, we have estimated a population parameter within a certain probability level.

Confidence limits can be placed around any parameter estimate. We will learn to create CIs with each procedure we discuss.

You may notice that you can create a 95% or a 99% CI. Both CIs are created in the same fashion; the difference being the tabled t value that defines the inclusion area. The value of the $CI_{0.99}$ is that it is *more likely to contain the true population estimate*, since it will create a wider interval of values. However, in doing so it will be *less precise*. Look at our two examples:

$CI_{0.95}$ Interval values: 3.097–3.303
$CI_{0.99}$ Interval values: 3.061–3.339

Figure 10.15 compares the $CI_{0.95}$ and $CI_{0.99}$ for our example. As you can see, the $CI_{0.99}$ is wider, and therefore more likely to contain the actual population mean. However, the width of the interval makes it a *less precise* estimate than the $CI_{0.95}$.

Our research findings provide an interesting example. If you recall, the population mean was assumed to be 3.10. The $CI_{0.95}$ lower boundary of 3.097 directly fell

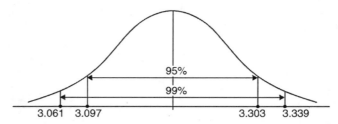

FIGURE 10.15 Comparing $CI_{0.95}$ and $CI_{0.99}$.

on this assumed value. It would thus barely be included in the $CI_{0.95}$. However, the assumed mean of 3.10 would easily fit in the $CI_{0.99}$. The researcher must decide which is the more important aspect of the estimate, the extent of confidence or precision.

USING EXCEL® AND SPSS® WITH THE SINGLE-SAMPLE T TEST

We can use both Excel® and SPSS® to evaluate a single-sample T test, although SPSS® is the more straightforward. Both provide the same results.

SPSS® and the Single-Sample T Test

SPSS® provides a very easy way to perform this test with a drop-down menu. As you can see in Figure 10.16, you can access the "One-Sample T Test . . . " menu through the Analyze menu on the main screen and then choose "Compare Means."

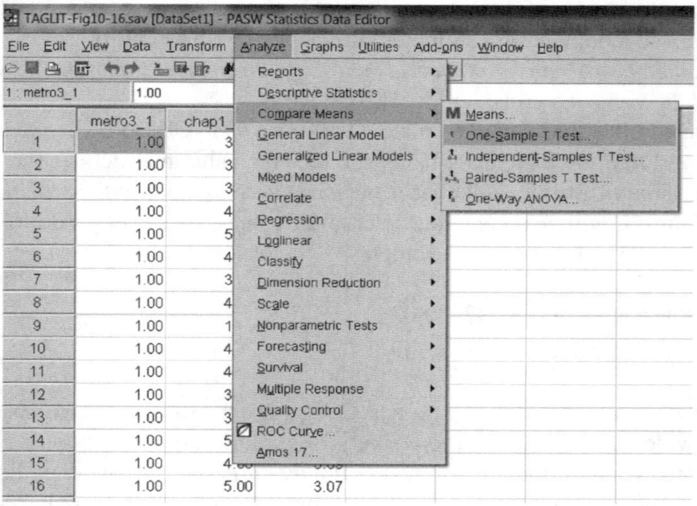

FIGURE 10.16 Using SPSS® to obtain the single-sample T test.

USING EXCEL® AND SPSS® WITH THE SINGLE-SAMPLE *T* TEST

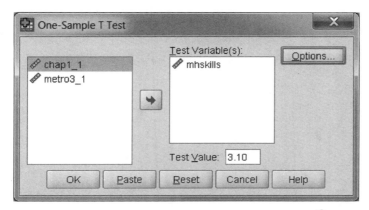

FIGURE 10.17 Using the single-sample *T* test in SPSS®.

When you choose the one-sample *T* test, you are presented with the menu in Figure 10.17.

This menu allows you to specify which dependent (outcome) variable you are testing by specifying the appropriate variable in the "Test Variable(s):" window using the arrow button. Include the assumed mean value in the "Test Value" window near the bottom of the menu. As you can see in Figure 10.17, I included 3.10 because this was the value I assumed the population mean to be in our research question.

Figure 10.18 shows the SPSS® output tables including descriptive statistics and the *T*-test results.

Figure 10.18 shows several important statistical calculations. The top panel provides the descriptive statistics on MH teacher technology skills, including mean

	One-Sample Statistics			
	N	Mean	Standard deviation	Standard error of the mean
mhskills	28	3.2001	0.26369	0.04983

	One-Sample Test					
	Test value = 3.10					
					95% Confidence Interval of the difference	
	t	df	Significance two-tailed	Mean difference	Lower	Upper
mhskills	2.009	27	0.055	0.10011	–0.0021	0.2024

FIGURE 10.18 The SPSS® output tables with *T*-rest results.

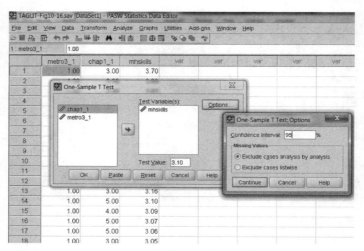

FIGURE 10.19 The SPSS® output tables with *T*-test results.

(3.2001) and standard deviation (0.26369) values. It also provides the s_m value (0.04983).

The bottom panel shows the calculated *t* ratio (2.009) and the df (27) values. The significance reported (0.055) is the *actual probability value* of values occurring "beyond" this result. Up to now, we have referred to a table of values to create numerical values that exclude a certain portion of the tails for our hypothesis test (i.e., 0.05, 0.01). SPSS® provides the exclusion percentage in the tail *from the point of the calculated t ratio* (2.009 in this case). If this actual probability value is less than (i.e., smaller) the tabled exclusion value of 0.05, the *T* test is considered a significant finding because the test value would fall into the exclusion area.

In this example, however, we cannot reject the null hypothesis because our test value (2.009) fell at the probability point that excluded 0.055 of the distribution in the tail. We needed this number to be smaller than 0.05. Therefore, as we found in our earlier analysis by hand, we cannot reject the null hypothesis. We can point out other findings for the CI through SPSS® in Figure 10.19.

When I specified the *T* test for a single mean in SPSS® (see Figure 10.18), I had the option of requesting CI values. Figure 10.19 shows this request. Note that you can call for CI values other than 0.95 or 0.99 by entering them in the top window. As you can see, I called for $CI_{0.95}$ values.

If you refer back to Figure 10.18, the SPSS® output file already includes the confidence intervals because it is a default procedure. The results look a bit different from those we discussed since SPSS® shows confidence interval values around the *assumed population mean*. This is the way to view the output values:

Assumed population mean ("Test Value") = 3.10
$$-0.0021 = 3.0979 \quad \text{Lower limit}$$
$$+0.2024 = 3.3024 \quad \text{Upper limit}$$

USING EXCEL® AND SPSS® WITH THE SINGLE-SAMPLE T TEST

These values correspond to the values we calculated by hand above. You can check the $CI_{0.99}$ values in the same way as above. Simply change the default 0.95 in the "One-Sample T Test: Options" window to 0.99 and SPSS® returns the following values for the Lower and Upper interval: Lower = −0.0380 and Upper = 0.2382. Using the same process as above, you can confirm the $CI_{0.99}$ values. The values are the same as our hand calculated CI values, accounting for slight differences due to rounding.

Assumed population mean ("Test Value") = 3.10

$$-0.0380 = 3.062 \quad \text{(Lower limit)}$$
$$+0.2382 = 3.338 \quad \text{(Upper limit)}$$

Excel® and the Single Sample T Test

The Excel® program makes you work a bit harder than SPSS®, but you can use it effectively to conduct a single-sample t test. The Data sub-menus do not include the single-sample t test. However, the descriptive analysis that you performed on the MH Teacher skills (see Figure 10.4) provides all the information you need to conduct the analysis.

If you recall, here is the relevant information taken from the descriptive analysis:

Mean = 3.20
Standard deviation $(s_x) = 0.264$
Estimated standard error $(s_m) = 0.05$
Count $(N) = 28$

You can simply substitute the values above in the t-test formula. The only thing you need to remember is the assumed population mean (3.10):

$$t = \frac{M - \mu}{s_m}, \quad t = \frac{3.20 - 3.10}{0.05}, \quad t = 2.00$$

As you can see, your calculation of t using Excel® descriptive summary is the same as our own calculation and that derived from SPSS®.

The next step in the hypothesis test is to determine whether the calculated t ratio (2.00) is "extreme enough" to reject the null hypothesis. You could proceed as you did in the hand calculation by comparing the t ratio to the tabled value of t (2.052). This comparison would lead you to be unable to reject the null hypothesis because 2.00 does not exceed the exclusion value of 2.052.

You may access the T table in Excel® through the use of the "TDIST" function. This function is available by choosing the statistical functions from the Formulas menu on the main page. (Review the material in Chapter 2

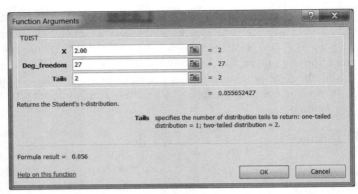

FIGURE 10.20   The TDIST function in Excel®.

dealing with the Excel® functions.) When you choose TDIST, the window in Figure 10.20 appears.

As you can see, I entered the relevant information in the TDIST function windows. The "X" window is the place to enter the calculated t ratio. Excel® compares this value to T-table values using df (27) and tails (2) information.

When this information is entered, you can select "OK" to paste the resulting value to your spreadsheet. However, note that the result (0.055652427) is returned in the middle of the function screen. Like SPSS®, this value is the actual probability of values greater than the sample mean occurring on the comparison distribution. If the value is beyond (smaller than) the 0.05 level (the standard exclusion area used by researchers), then you would reject the null hypothesis. However, in this example, 0.05565 is larger than 0.05, so you cannot reject the null hypothesis.

You must calculate the CI values by hand as we did above because Excel® does not provide this information for the single-sample t test. You need to have access to a T distribution table for this calculation because you need the 0.05 (or 0.01) exclusion values to calculate the estimated population mean interval. Because Excel® (TDIST) returns the actual probability of a result, you cannot use this for the CI. (The Excel® function "Confidence" calculates confidence intervals for the Z test, but not for the T test.)

TERMS AND CONCEPTS

Biased Estimator   A sample measure that does not provide an accurate measure of the population characteristic.

Confidence Interval   Creating a range of values that will likely contain the true population value within a certain percentage of certainty.

Degrees of Freedom   The restrictions on the sample values when estimating a population parameter.

Effect Size The strength or impact of a finding, typically the amount of distance the test value is "pushed" away from the population value.

Factor Analysis A statistical procedure that "reduces" a series of items to related groupings of items.

One-Tailed Tests Locating the exclusion area in one tail of the comparison distribution.

Point Estimate Transforming a sample value (e.g., sample mean) to a comparison distribution to determine whether it is statistically different from a population value (e.g., population mean).

Power The ability of a statistical analysis to detect a "true" finding.

Two-Tailed Tests Locating the exclusion area in both tails of the comparison distribution.

Type I (Alpha) Error The Type I error is mistakenly rejecting the null hypothesis when it should be accepted.

Type II (Beta) Error Not rejecting the null hypothesis when it should be rejected.

REAL-WORLD LAB VI: SINGLE-SAMPLE T TEST

This lab will use the STAR Classroom Observation Protocol^{TM} data provided by The BERC Group, Inc. I included a general description of this database in Chapter 3 as a process for measuring the extent to which Powerful Teaching and Learning^{TM} is present during a classroom observation. The BERC Group, Inc. has performed thousands of classroom observations of all grade levels and subject areas.

The sample group of observations in Table 10.5 consists of observations of math classrooms ($N = 18$) in grades K–8 across several schools. The study variable (i.e., outcome variable) is an overall measure of the extent to which Powerful Teaching and Learning^{TM} was present during a classroom observation in year four of a study.

TABLE 10.5 The STAR Classroom Observation Protocol^{TM} Data

Overall	Overall
4	3
3	3
3	3
2	4
2	2
3	2
4	4
2	2
3	3

The study variable "Overall" scored observations of teachers in the following categories:

Score	Category
1	Not at all
2	Very little
3	Somewhat
4	Very

The purpose of this lab is to respond to the following research question: "Do K–8 math classrooms demonstrate different levels of Powerful Teaching and Learning^{TM} than the population of all classrooms observed?" The population of classrooms of all grades and subjects is assumed to have an Overall average of 2.45.

1. Calculate the single sample t test by hand and perform the hypothesis test.
2. Calculate the effect size and CI$_{.95}$.
3. Perform the single sample t test through Excel® and SPSS®.
4. Provide a summary of your findings.

REAL-WORLD LAB VI: SOLUTIONS

1. Calculate the single-sample T test by hand and perform the hypothesis test.

$$s_x = \sqrt{\frac{\sum X^2 - \frac{(\sum X)^2}{N}}{N-1}}, \quad s_x = \sqrt{\frac{160 - \frac{2704}{18}}{17}}, \quad s_x = 0.758$$

$$s_m = \frac{s_x}{\sqrt{N}}, \quad s_m = 0.179$$

$$t = \frac{M - \mu}{s_m}, \quad t = \frac{2.89 - 2.45}{0.179}, \quad t = 2.458$$

- $H_0: \mu_1 = \mu$ (there is no difference between the Overall mean of the sample of K–8 math classrooms and the classrooms in the study population).
- $H_A: \mu_1 \neq \mu$ (the sample group Overall mean is not the same as all classrooms in the study population).
- The Critical Value: The two-tailed T-table value $(t_{0.05, 17df}) = 2.110$.
- Calculated $t = 2.458$.
- Statistical Decision: Reject the null hypothesis since the calculated value of the t ratio (2.458) exceeds the exclusion value on the distribution (2.110).
- Interpretation: The single-sample T test revealed that the sample group of K–8 math classrooms showed higher average observed Powerful Teaching

REAL-WORLD LAB VI: SOLUTIONS

and Learning™ scores than the population of classrooms observed at all grade levels and in all subject areas.

2. Calculate the effect size and $CI_{0.95}$.

$$d = \frac{t}{\sqrt{N}}, \qquad d = \frac{2.458}{4.24}, \qquad d = .58 \text{ (Medium effect)}$$

Confidence interval $= \pm t(s_M) + M, \qquad (t_{0.05, 17df}) = 2.110$
Lower interval value $= -2.110 \,(0.179) + 2.89 = 2.512$
Upper interval value $= +2.110 \,(0.179) + 2.89 = 3.268$

$CI_{0.95}$ consists of an interval of 2.512 to 3.268 for the population *from which this sample came*. Note that this interval does not contain the assumed population mean (2.45). Because we rejected the null hypothesis, we concluded that our *sample mean was so different that it must have come from a population mean much higher than the general population with an assumed Overall mean of 2.45*.

3. Perform the single-sample T test through Excel® and SPSS®.

Figure 10.21 shows the Excel® descriptive statistics output for Overall. The calculated values needed for the single-sample T test are shaded.

Using the shaded mean and s_m values from Figure 10.21, the calculated t ratio is the same as our hand calculation under #1 above.

$$t = \frac{2.89 - 2.45}{0.179}, \qquad t = 2.458$$

Using the Excel® TDIST function, the calculated value is $p = 0.025$. Because this value is lower (i.e., more extreme) than the 5% exclusion value, we can reject the null hypothesis. The hypothesis test results in #1 above are verified.

	Overall
Mean	2.889
Standard error	0.179
Median	3.000
Mode	3.000
Standard deviation	0.758
Sample variance	0.575
Kurtosis	−1.118
Skewness	0.195
Range	2.000
Minimum	2.000
Maximum	4.000
Sum	52.000
Count	18.000

FIGURE 10.21 The Excel® descriptive data for the sample group overall scores.

One-Sample Statistics

	N	Mean	Standard deviation	Standard error of the mean
Overall	18	2.89	0.758	0.179

One-Sample Test

Test value = 2.45

	t	df	Significance (two-tailed)	Mean difference	95% Confidence interval of the difference	
					Lower	Upper
Overall	2.455	17	0.025	0.439	0.06	0.82

FIGURE 10.22 The SPSS® results of the single-sample t-test.

The SPSS® single-sample t test results are shown in Figure 10.22. As you can see, the shaded t ratio (2.455) is the same as those calculated by hand and through Excel® (slight differences due to rounding).

Other relevant findings from the SPSS® output in Figure 10.22 are the following:

- The significance (0.025) like that reported by Excel® is the actual probability of a finding at least this extreme; that is, the t ratio (2.455) falls into the exclusion area at the point where only 0.025 probability of a sample value are more extreme. Thus, since this value of 0.025 is smaller (more extreme) than the 0.05 exclusion area, we can reject the null hypothesis.
- The confidence interval values are the same as those we calculated by hand under #1 above:

$$\text{Assumed population mean (``Test Value'')} = 2.45$$
$$+0.06 = 2.51 \quad \text{(Lower limit)}$$
$$+0.82 = 3.27 \quad \text{(Upper limit)}$$

4. Provide a summary of your findings.

A single-sample T test of a randomly chosen group ($N = 18$) of K–8 math classrooms revealed significant differences in their overall Powerful Teaching and Learning™ scores from the population of observed classrooms in all grades and subjects with an assumed mean of 2.45 ($t = 2.46$, $p < 0.025$). The K–8 math classrooms demonstrated higher average Overall scores than the population. This study revealed a medium effect size ($d = 0.58$) indicating a meaningful sample mean score difference from the population of classrooms.

11

INDEPENDENT-SAMPLES *T* TEST

We have arrived at an important benchmark in the book. By now, you have gained an understanding of descriptive statistics, and you are working your way toward an understanding of sampling distributions and how they are helpful in making statistical decisions. We have discussed two inferential tests: the *Z* test, and the single-sample *T* test.

In the chapters ahead, we will build upon your understanding of inferential statistics. This chapter extends what we learned in Chapter 10 about the *T* test. As you will see, there is a logical process of extending the Single-Sample *T* test to the two sample (independent) *T* test.

A LOT OF "*T*'s"

In the previous chapters, we discussed several "*t*" statistics and measures:

- *T* score—a transformed *z* score.
- Single–sample *T* ratio
- Two–tailed *T* table of values
- One–tailed *T* table of values

To these we will add yet another *t*, the "two-sample *T* test for independent samples." This is a "workhorse" test in statistics because it is so versatile and straightforward. The reason it is so common is that it allows us to perform a very

---

Understanding Educational Statistics Using Microsoft Excel® and SPSS®. By Martin Lee Abbott.
© 2011 John Wiley & Sons, Inc. Published 2011 by John Wiley & Sons, Inc.

basic function in statistics and common practice, *compare*. The nature of statistics is comparison, as we have learned in the inferential process of comparing sample measures to population measures.

Perhaps you have heard an advertisement like the following: New Ox Vomit toothpaste is 25% better! The immediate question that should spring to mind is, Compared to what? Old Ox Vomit toothpaste? A wooden twig? Other brands of toothpaste? In order to find out whether this claim has any merit, we must compare it to something else to see if there is really a difference.

In Chapter 10, we used a sample mean to estimate a population mean. In essence, we were comparing a sample value to an (assumed) population value. We will extend this comparison to include a second sample group; the independent-samples T test is designed to do just that. This statistical procedure assesses whether two samples, chosen independently, are likely to be similar or sufficiently different from one another that we would conclude that they do not even belong to the same population.

RESEARCH DESIGN

As we encounter new statistical procedures, I want to draw us back to our examination of research design. It is important to understand *how a statistical procedure should be used* rather than simply how to calculate it.

If you recall, we distinguished experiment from post facto designs in Chapter 9. One of the key differences was whether the researcher manipulated the independent variable (experiment) or simply measured what data already exists (*post facto*). The independent T test can be used with either design.

Experimental Designs

Figure 11.1 shows the diagram for an experimental design in which there are two groups: experimental group and control group. As you can see, the treatment variable is manipulated by assigning one group (experimental) one level of a treatment and the other group (control) a different level (or no treatment at all). The dependent (outcome) variable is tested after the independent variable has been changed to see if there is a difference between the outcome measures of the two groups.

Research Treatment Variable (Independent or Predictor)			Dependent Variable (Outcome)
Random selection and assignment?	Pretest scores	Experimental group	Outcome test scores (post-test scores)
Random selection and assignment?	Pretest scores	Control group	Outcome Test Scores (post-test scores)

FIGURE 11.1 The research design with two groups.

If there is a difference, the researcher attributes the change to the presence or action of the independent variable (i.e., makes a causal attribution).

The researcher may only make causal attributions if there is randomization in which subjects are chosen and assigned to groups randomly. If randomization is present, then the assumption is that both groups (experimental and control) are equal at the outset. Then, if they are different on the dependent variable measure after the experiment is over, the differences must be because of the treatment that was introduced to change the experimental group and not the control group.

Under these conditions, the researcher compares the outcome measures of both groups when the experiment is over to see if there is a difference. The independent T test is the statistical procedure to use because we are comparing two sample groups. Formally stated, we are testing the difference between two sample groups to see if they belong to the same population after the experimental manipulation. It is assumed that they belong to the same population at the beginning (before the independent variable is manipulated).

Independent and Dependent Samples. In research designs like those shown in Figure 11.1, the researcher compares the (post-test) outcome measures of both groups. Note that this statistical test uses *independent samples*. This means that choosing subjects for one group has nothing to do with choosing subjects for the other group. Thus, if I randomly select Bob and assign him randomly to group 1 (high noise), it has nothing to do with the fact that I choose Sally and assign her randomly to group 2 (low noise). This is an important assumption because it assures the researcher that there are no "built-in linkages" between subjects. The power of randomization will result in the comparability of the two groups in this way.

Dependent samples would consist of groups of subjects that had some structured linkage, like using the same people twice in a study. For example, I might use pre-test scores from Bob and Sally and *compare them with their own post-test scores*. Using dependent samples affects the ability of the randomness process to create comparable samples; in such cases, the researcher is assessing *individual* change (before to after measures) in the context of the experiment that is assessing *group* change.

Figure 11.2 shows how dependent samples might be used in an experimental context. As an example, we might have a "pre–post" design where we take a pre-test measure on the subjects in a group, expose the group to some experimental condition, and then take a post-test of the same subjects. This would mean that the pre-test group would be very specifically related to the post-test group (since they would be the same people) and would therefore not be independent.

Research Treatment Variable (Independent or Predictor)			Dependent Variable (Outcome)
Random selection and assignment?	Pretest scores	Experimental group	Outcome test scores (post-test scores)

FIGURE 11.2 Using dependent sample measures in experimental designs.

Research Treatment Variable (Independent or Predictor)			Dependent Variable (Outcome)
Matching	Pretest scores	Experimental group	Outcome test scores (post-test scores)
Matching	Pretest scores	Control group	Outcome test scores (post-test scores)

FIGURE 11.3 Using matched groups in experimental designs.

Another dependent samples design is using *matched samples*, a situation in which we purposely choose people to be in two groups to be compared rather than choosing randomly. For example, we might be concerned about gender and purposely assign equal numbers of men and women to two groups. As you can see in Figure 11.3, the randomness criterion in the first column is replaced by "matching" to indicate that we are using two groups of different people (not the same people twice), but they are *structurally linked by our decision to purposely "build in" some group similarity*.

Between and Within Research Designs. Research designs differ as to whether they measure independent samples, dependent samples, or both. The design in Figure 11.1 illustrates all three processes. *Between-group designs are those in which the researcher seeks to ascertain whether the groups demonstrate unequal outcome measures.* That is, are there differences *between* the groups' post-test scores? In Figure 11.1, this would be comparing the post-test scores for the experimental group versus the control group. It is represented by the "vertical" distance in the "Outcome" measure column.

Within-group designs are those in which the researcher seeks to ascertain whether subjects in a group change over time. In Figure 11.1 the within-group design would be the change from a group of subjects' pre-test scores to their post-test scores. This would be represented by the "horizontal" difference of a group; that is, are the scores *within* a specific group different after the treatment than they were at the beginning?

Mixed designs are those in which both within- and between-group measures can be taken. The design in Figure 11.1 is one such design. The vertical comparison of post-test scores is between groups, and the horizontal comparison of pre-test to post-test difference is within subjects.

Using Different T Tests. Depending on the nature of the design, an experiment may call for different *T*-test procedures. With independent samples, the researcher would use an independent-samples *T* test by comparing the post-test measures between two treatment groups. This procedure is the focus of this chapter.

If the researcher uses dependent samples, they would need to use another kind of *T* test called a *dependent-samples T test*. Other names for this test are: repeated-measures *T* test, within-subjects *T* test, and paired-samples *T* test. Both Excel® and SPSS® refer to these as paired sample *T* tests.

Pretest or No Pretest

Experimental research designs differ in terms of whether they include a *pre-test* of the dependent variable measure. A pre-test is simply administering the dependent variable measure before the experiment begins to ensure that the two groups are in fact equal. Some experimental designs that include full randomization do not use a pre-test because the researchers assume that randomization results in equal groups; therefore, there is no need for a pre-test. In fact, under these conditions, eliminating a pre-test might eliminate potential problems because research subjects can often be affected by receiving a test of the outcome measure before the study. (This is known as *pre-test sensitivity*.)

Example of Experiment. When I was studying experimental psychology as an undergraduate student, I performed an experiment on the effects of noise on human learning. I randomly selected students and randomly assigned them to either a high or low noise condition (by using a white noise generator with different decibel levels). Then, I gave the students in both groups the same learning task and compared their performance. The learning task (the outcome measure) was simple word recognition. Figure 11.4 shows the research design specification for this experiment.

Note some features of the experiment shown in Figure 11.4:

- I randomly selected students and then randomly assigned them to the two treatment groups.
- I did not use a "control group" (the "absence of the treatment") but rather a second level or condition of the treatment variable to yield two treatment groups.
- I did not pre-test the subjects on word recognition before exposing them to different experimental treatment conditions.
- I administered the word recognition test (outcome measure) to both groups after exposing them to different experimental treatment conditions.

I randomly selected and assigned students, which allowed me to assume that they were equal on all important dimensions (to the experiment). I exposed the two groups to different conditions, which I hypothesized would have differential effects on their learning task. Thus, if I had observed that one group learned differently (either better

Research Treatment Variable (Noise)			Dependent Variable (Word Recognition)
Random selection and assignment	No pretest	Experimental group (high noise level)	Outcome test scores (number of recognized words)
Random selection and assignment	No pretest	Control Group (low noise level)	Outcome Test Scores (number of recognized words)

FIGURE 11.4 Example of experimental research design using *t* test with two groups.

or worse) than the other group, I could attribute this difference to the different conditions that I exposed them to (high or low noise). If their learning was quite different, I could conclude, statistically, that the groups were now so different that they could no longer be thought to be from the same population of students I started with. That is the process I used for testing the hypothesis of difference. Specifically, I used the *T* test with independent samples to detect difference in post-test scores.

I will discuss this experiment in subsequent chapters as well. The short answer as to whether or not I observed statistical differences between the high and low noise outcome measure is no. This didn't mean that noise doesn't affect learning; it just gave me a way to look at the problem differently. As we will see, this example shows several features of the theory testing process as well as the *T*-test procedure.

Post Facto Designs

In Chapter 9, I discussed the *post facto* design as one in which I compare group performance on an outcome measure after group differences have already taken place. These designs can be correlational or comparative, depending on how the researcher relates one set of scores to the other (i.e., using correlation or difference methods, respectively).

A *post facto* design compares conditions with one another. Thus, for example, rather than perform an experiment to detect the impact of noise on human learning, I might ask a sample of students to indicate how loud their music is when they study and their GPA. Then, I could separate the students into two groups (high and low noise studiers) and compare their GPA measures.

In this design, therefore, I would not manipulate the noise measure; I would simply create groups on the basis of already existing differences in noise conditions. If the outcome measure (GPAs) was different between the groups, I would conclude that noise would *possibly* be a contributing factor to GPA. I could not speak causally about noise because many other aspects of studying may have affected GPA (and most decidedly would!).

Figure 11.5 shows how the *post facto* design might appear using my noise research question. I would simply use an independent samples t test to compare the GPA measures for high and low noise studiers.

Studying Under High Noise	Studying Under Low Noise
GPA scores	GPA scores
GPA scores	GPA scores
—	—
—	—
Mean GPA$_{High}$	Mean GPA$_{Low}$

FIGURE 11.5 The *post facto* comparison for independent *t* test.

INDEPENDENT *T* TEST: THE PROCEDURE

This same design could use dependent samples if I deliberately "stack" the two samples to be the same on some issue. For example, I might equate the numbers of women and men students and equivalent numbers of freshmen, sophomores, juniors, and seniors in both noise groups. If I did this, I would be *matching* the groups and therefore creating dependent samples. Under these circumstances, I would need to use the dependent-samples *T* test.

INDEPENDENT *T* TEST: THE PROCEDURE

In the independent *T* test, the researcher takes a pair of samples to see whether these can be said to be from a single population. The experimental and *post facto* designs that we discussed above both would yield sample data for two samples. Figure 11.6 shows how the two sample process works.

The chief concern with this test is the difference in the means of the two samples. If the samples are chosen randomly, by chance the means will both be close to the actual population mean (the value of which is unknown). By chance alone, the difference between the means should be fairly small.

If we chose two sample groups (or, in an experiment, we randomly chose a group and randomly assigned them to two groups), we would therefore expect the group means to be similar. In research, we start with this assumption but observe whether the two sample means are still equal after an experimental treatment, or if the group means are different when we compare different conditions of the research variable.

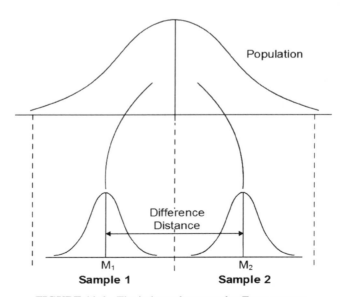

FIGURE 11.6 The independent samples *T*-test process.

Using my *post facto* example above, my reasoning would be as follows:

- I have two groups of students, some who study under high noise and some who study under low noise.
- I assume the groups of students were equivalent before they developed their habits of studying under different amounts of noise.
- My task is to determine whether, now that they have developed their habits of study, they are still equivalent or different on a word recognition task; if I reject the null hypothesis, it will indicate that they no longer belong to the same population of students.
- If they are different now, then I can say that the different noise may have affected their ability to recognize words. However, there were surely other influences that led them to develop their study habits, so I cannot say that the different word recognition ability is caused only by the noise.

But how large does this difference have to be before it could be said that a difference that large could not reasonably be explained by chance and therefore the two groups do not represent a single population? That is the nature of the T-test process that we examined with the single-sample T test in Chapter 10.

Creating the Sampling Distribution of Differences

In Chapter 10, we learned that in order to decide whether a (single) sample mean came from a (assumed) population, we had to use the sampling distribution of means as a standard of comparison. Because we will now ask a similar question with two sample groups, we need to think about a sampling distribution created not by repeated sampling of a single group, but by a *pair of groups*.

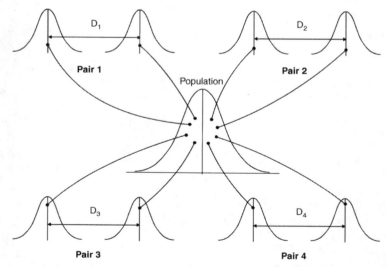

FIGURE 11.7   All possible pairs of samples.

INDEPENDENT T TEST: THE PROCEDURE

The *sampling distribution of differences* is the sampling distribution that we will use as a standard of comparison in the independent samples T test. It is called the sampling distribution of differences because it focuses on the *differences between the means of pairs of sample groups*. Figure 11.7 shows the process used to create the sampling distribution of differences (remember, this is not something the researcher does, it is simply to show the conceptual steps for creating the sampling distribution of differences).

Figure 11.7 shows pairs of samples being randomly selected from a population. The sampling distribution is made up of *all possible pairs of sample means*; Figure 11.7 simply shows four such pairs as an example of how it works. When all possible pairs of samples are taken from a population, a distribution can be created on the basis of the *differences between the means of the pairs* (designated in the figure by "D"). Figure 11.8 represents the process of creating the D values (differences between pairs of sample means) that form the sampling distribution of differences.

The Nature of the Sampling Distribution of Differences

In Figure 11.8, the sampling distribution of differences is created by differences between all possible pairs of sample means. This sampling distribution of differences will now serve as the standard of comparison to see whether a pair of sample means that a researcher randomly selects could be said to be representative of a research population. In effect, *how does the difference in the sample means that we select compare to all possible sample mean differences?*

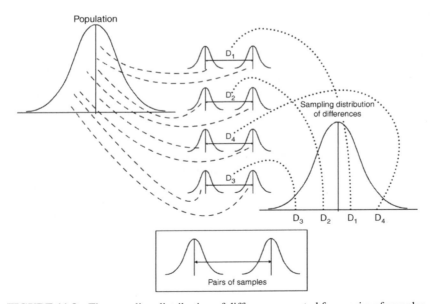

FIGURE 11.8 The sampling distribution of differences created from pairs of samples.

The Mean and Standard Deviation of the Sampling Distribution of Differences.
The mean of the sampling distribution of differences ($\mu_{m_1-m_2}$) should be equal to 0, because if we select means randomly, they will by chance come from both sides of the distribution; the positive and negative values of the means, when added, would cancel one another out resulting in a total of 0.

Because all pairs of samples are hypothetically taken to create the sampling distribution, *this distribution should be normally distributed.* Therefore, if a researcher observes that a pair of samples in their study results in a difference in means that is significantly removed from 0 on the sampling distribution of differences, the researcher can conclude that the research samples cannot be said to come from the same population.

In order to transform the difference in sample means from our research study to the sampling distribution of differences, we need to be able to specify the standard deviation of this sampling distribution of differences. The standard deviation of the sampling distribution of differences is known as the standard error of difference and is symbolized by s_D. The lowercase s identifies this as an estimated standard deviation, and the subscript D identifies it as belonging to the distribution of differences. The shorthand designation for s_D is the *estimated standard error of difference*.

Technically, s_D is the *estimate of a parameter*. If you can imagine it, the standard deviation of all possible mean differences that forms the sampling distribution of differences is symbolized by $\sigma_{m_1-m_2}$. Because we could never calculate this, we must *estimate* this population parameter, which is our estimated value of s_D.

We can now introduce these new symbols to our list we compiled in Table 10.2. Table 11.1 shows the entire list of symbols including the two new entries. Figure 11.9 shows the symbols relevant to the distribution of differences.

Calculating the Estimated Standard Error of Difference

Figure 11.9 shows the conceptual method for calculating s_D. As you can see, in order to create the estimated standard error of difference, we must *combine the information from two separate samples*. Each sample standard deviation is important to the overall estimate, so we must "pool" the sample standard deviations to create a single sampling distribution standard deviation. Technically, we *pool the variances* from the two research samples to obtain s_D.

The pooled variance is a way of weighting the sample variances (with df's) to ensure a better estimate. This is simply the *average of the variances from which we can derive s_D as long as the sample sizes are equal*. Because sample sizes (n_1 and n_2) are so critical in estimating variance, the formula for this process requires equal sample sizes. (If sample sizes are unequal, we would need to estimate s_D with a different formula because estimates of the population variance based on the samples, s^2, will not be equally weighted.)

Below is the formula for calculating s_D *when there are equal sample sizes*. Note the sequence of calculations necessary to calculate s_D. You must calculate the estimated standard deviation (s) from each sample; then you must use s to calculate

INDEPENDENT *T* TEST: THE PROCEDURE

TABLE 11.1 New Entries to the List of Population and Sample Symbols

M	The mean of the sample.
SD	The standard deviation of the sample (assumes the sample is its own population).
μ	The lowercase Greek letter "mu" is the symbol for the population mean.
M_M	This is the symbol for the mean of the sampling distribution. You can see how this works by observing that it is a mean (M), but a mean of the sampling distribution indicated by the subscript. Thus, it is the *mean of the distribution of means*, or the "mean of the means." Because it is (theoretically) created by all possible samples, it is a parameter.
σ_X	"Sigma X" is the standard deviation of all the population raw scores. This differs from SD in that it does not refer to a sample, but to the entire population of scores.
σ_M	The standard deviation of the sampling distribution of means; also called the *standard error of the mean*.
s_x	The estimated standard deviation of the population. The subscript x identifies this estimated value as belonging to the population of all scores.
s_m	The estimated standard error of the mean. The subscript m identifies this estimated value as belonging to the sampling distribution of (single) means.
$\mu_{m_1-m_2}$	The mean of the sampling distribution of differences; a parameter. This value is 0 because if all possible samples are taken, half will be negative and half will be positive, resulting in a 0 value when the means are subtracted.
$\sigma_{m_1-m_2}$.	The population standard deviation of all mean differences; a parameter.
s_D	The estimated standard deviation of the sampling distribution of differences. Also known as the estimated standard error of difference.

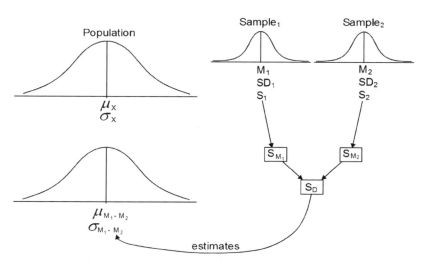

FIGURE 11.9 Symbols in the distribution of differences.

s_m for each sample. This is the same process we used in the single-sample T test. We are simply performing the calculation twice because there are two samples involved. Finally, the separate s_m values are pooled to create s_D. This is the point at which we combine the sample values to create the standard error of difference.[1]

$$s_D = \sqrt{s_{m_1}^2 + s_{m_2}^2}, \quad \text{where} \quad s_{m_1} = \frac{s_1}{\sqrt{n_1}} \quad \text{and} \quad s_{m_2} = \frac{s_2}{\sqrt{n_2}}$$

Using Unequal Sample Sizes

Unequal sample sizes present the researcher with the problem I mentioned above. Because we need to pool the estimated population variances, unequal sample sizes will affect the overall estimate according to the magnitude of the differences. An *average* places the estimated value directly in the middle of both sample estimates. However, when one sample is larger, the pooled "average" is influenced more greatly by the value with the bigger sample size.

To account for the different "weights" presented by different sample sizes, researchers use a different formula for s_D. This formula looks quite complex, but it makes conceptual sense if you look carefully. The formula uses elements that you have learned before. Here is the formula:

$$s_D = \sqrt{\left(\frac{(SS_1) + (SS_2)}{(df)}\right)\left(\frac{1}{n_1} + \frac{1}{n_2}\right)}$$

If you recall from Chapter 6, the sum of squares (SS) value is the way we calculated a global measure of variation for a set of values.

$$SS = \sum X^2 - \frac{(\sum X)^2}{N}$$

When the SS is divided by the sample size, it is a calculation for population variance (or the *actual* variance of a set of scores).

$$\text{Variance} = \frac{SS}{N} \quad \text{or} \quad \text{Variance} = \frac{\sum X^2 - \frac{(\sum X)^2}{N}}{N}$$

Under the square root sign, this became the (actual) standard deviation

$$SD = \sqrt{\frac{SS}{N}} \quad \text{or} \quad SD = \sqrt{\frac{\sum X^2 - \frac{(\sum X)^2}{N}}{N}}$$

---

[1] Note that the formula for s_D combines the estimated *variances* from both samples but produces the standard error of difference, which is a standard deviation measure.

INDEPENDENT T TEST: THE PROCEDURE

We used these calculations to estimate the population standard deviation (s_x) by dividing the values by the degrees of freedom. In the case of a single sample, the df $= n - 1$. (See Chapter 10 to review the formula and calculation for s_x.) Looking again at the formula above using the pooled estimate of variance for calculating s_D with unequal sample size, you will see these primary elements.

$$s_D = \sqrt{\left(\frac{(SS_1) + (SS_2)}{(df)}\right)\left(\frac{1}{n_1} + \frac{1}{n_2}\right)}$$

As you can see in the *left half* of the equation under the radical sign above, we are simply combining SS values from both samples and dividing by the degrees of freedom for both. *This is the pooled variance and is shown below as (Pooled)* σ_X^2. Taking the square root of the pooled variance yields the *pooled standard deviation*, which is useful when we calculate effect size below. It is shown below as (Pooled) σ_X.

$$\text{(Pooled)} \; \sigma_X^2 = \frac{(SS_1 + SS_2)}{df}, \quad \text{(Pooled)} \; \sigma_X = \sqrt{\left(\frac{(SS_1) + (SS_2)}{(df)}\right)}$$

When you have two samples, you are combining the degrees of freedom as well as the variation measure. In the independent T test with two samples, the degrees of freedom is thus equal to

$$(df_1 = n_1 - 1) + (df_2 = n_2 - 1) = n_1 + n_2 - 2$$

The right half of the equation for s_D under the radical is tantamount to dividing the left half by the combined sizes of samples to yield s_D from the pooled variance.

Remember what the estimated standard error of difference (s_D) is. It is the *estimated standard deviation of the sampling distribution of difference* (or estimated standard error of difference). Whether s_D is calculated from samples that have equal n sizes or unequal n sizes, the s_D measure is valuable because it allows the researcher the ability to *transform* the difference in sample means ($m_1 - m_2$) to a point on the sampling distribution. This enables the researcher to compare this point with all possible sample differences to determine if the resultant point (difference between samples) is too large to be obtained by chance.

The Independent T Ratio

What I described above are the elements used to calculate the independent T test. It is an extension of the single-sample T test in that we are simply adding a second sample value. Once we have calculated s_D, we can calculate the overall T ratio.

$$t = \frac{(M_1 - M_2) - \mu_{M_1 - M_2}}{s_D}, \quad df = N_1 + N_2 - 2$$

Compare this formula with the single-sample T ratio we discussed in Chapter 10:

$$t = \frac{M - \mu}{s_m}, \qquad \text{df} = N_1 - 1$$

As you can see, the independent T test adds a second sample mean value and uses a calculated standard error of difference (s_D) that incorporates the second sample (according to whether the sample size is equal or unequal). The population of all paired sample differences ($\mu_{m_1-m_2}$) is subtracted from the difference in sample means ($M_1 - M_2$). The distance of the sample mean difference from the population value is transformed to standard deviation units on the sampling distribution when it is divided by s_D. This value (the independent t ratio) indicates that it is likely not a chance finding if it exceeds the tabled probability value for the exclusion.

INDEPENDENT *T*-TEST EXAMPLE

As an extended example of the independent T test, I will use the STAR Classroom Observation Protocol™ data I used in Real-World Lab VI: Single-Sample T Test. This time, the research question will focus on the impact of school level on overall observation of Powerful Teaching and Learning™.

Put in the form of a question, the research focus is, Do elementary/middle schools differ from high schools in average Overall measures of Powerful Teaching and Learning™ among math teachers? The data are a random sample ($N = 50$) from all math teacher observations in year four (2008) of a STAR Classroom Observation Protocol™ study.

Table 11.2 presents the data I will use for this example. As you can see, the study presents two groups: classroom observations from elementary/middle schools ($n_1 = 23$) constitute group 1, and classroom observations from high schools constitute group 2 ($n_2 = 27$). The data in each of these groups represent the classroom observation data for selected classrooms in the schools. The data are the "Overall" scores that represent the extent to which Powerful Teaching and Learning™ is present during a classroom observation. (See Chapter 10, Real-World Lab VI: Single-Sample T-Test for a description of how this variable is measured.)

The research question addresses whether the Overall scores are equal between these two groups. That is, is there a difference between elementary/middle and high school classrooms' STAR Powerful Teaching and Learning™ scores? Responding to this research question requires a hypothesis test of difference using the independent-samples T test.

The Null Hypothesis

The null hypothesis for the independent-samples T test is similar to that for the single-samples T test.

$$H_0: \quad \mu_1 = \mu_2$$

INDEPENDENT T-TEST EXAMPLE

TABLE 11.2 STAR™ "Overall" Data from Elementary/Middle and High Schools

Elementary/Middle	High School
3	1
2	3
2	2
3	1
2	2
2	2
4	1
2	1
2	1
1	1
1	3
1	2
3	3
4	1
3	1
4	2
4	2
3	4
1	1
3	3
2	2
2	2
4	2
	2
	3
	2
	2

Technically, the null hypothesis states that the population from which group 1 came is the same as the population from which group 2 came. This is a formal way of stating that the sample group means are equal.

The Alternative Hypothesis

The alternative hypothesis is that the populations from which the samples came are not equal.

$$H_A: \quad \mu_1 \neq \mu_2$$

The Critical Value of Comparison

We will use the T table as we did for the single-sample T test. Despite the fact that our overall sample is 50, and therefore considered large, our group sample sizes

(23 and 27) warrant the use of the T table. Furthermore, we have no knowledge of the parameter values.

Remember that the degrees of freedom must be identified so we can specify the exclusion value for the T table of values. For the independent-samples T test:

$$\mathrm{df} = n_1 + n_2 - 2,$$
$$\mathrm{df} = 23 + 27 - 2 = 48.$$

Note that the overall sample group size of 50 is indicated by a capital N, whereas each group size is indicated by the lowercase n. This is an important distinction to keep in mind as we proceed through this and subsequent statistical procedures.

Referring to the T table with df = 48, we find the comparison value is 2.0106. If the T table does not show your specific df, as it does not in this case, use the closest value. Remember that the value of 2.0106 is the t ratio value that identifies the 5% exclusion area on our comparison distribution. If our calculated t ratio exceeds this number, we would reject the null hypothesis and conclude that our two sample means are different from one another.

The comparison value is identified for a two-tailed test. Therefore, if the calculated t ratio exceeds this value either positively or negatively, we would reject the null hypothesis. The comparison value is identified:

$$t(0.05, 48) = 2.0106$$

The Calculated T Ratio

Recall our formula for the independent-samples t ratio, using the s_D formula for unequal cell sizes. Using the data from Table 11.3, calculate the t ratio:

$$t = \frac{(M_1 - M_2) - \mu_{M_1 - M_2}}{s_D}, \quad s_D = \sqrt{\left(\frac{(SS_1) + (SS_2)}{(df)}\right)\left(\frac{1}{n_1} + \frac{1}{n_2}\right)},$$

$$t = \frac{(2.52 - 1.93) - 0}{0.264}, \quad s_D = \sqrt{\left(\frac{(23.7391) + (17.8519)}{(48)}\right)\left(\frac{1}{23} + \frac{1}{27}\right)},$$

$$t = 2.23, \quad s_D = 0.264$$

The calculated t ratio for the independent samples T test (with df = 48) is 2.23. This value[2] represents the transformed difference in the sample-group means of a value in standard deviation units on the sampling distribution of differences. It shows where the two-sample-group difference in means falls on the comparison distribution so we can decide either if it is a chance difference or whether the difference is too large by chance to be considered a common population.

---

[2] The calculated t ratio is 2.26 if you carry out the operations with several decimals. Our calculated t ratio used only three decimal places, so the t ratio will differ a bit due to rounding.

INDEPENDENT T-TEST EXAMPLE

TABLE 11.3 STAR™ "Overall" Data Analyses

	Elementary/Middle		High School	
	X_1	X_1^2	X_2	X_2^2
	3	9	1	1
	2	4	3	9
	2	4	2	4
	3	9	1	1
	2	4	2	4
	2	4	2	4
	4	16	1	1
	2	4	1	1
	2	4	1	1
	1	1	1	1
	1	1	3	9
	1	1	2	4
	3	9	3	9
	4	16	1	1
	3	9	1	1
	4	16	2	4
	4	16	2	4
	3	9	4	16
	1	1	1	1
	3	9	3	9
	2	4	2	4
	2	4	2	4
	4	16	2	4
			2	4
			3	9
			2	4
			2	4
$\sum$	58	170	52	118
$\sum X^2$		170		118
$\left(\sum X\right)^2$	3364			2704
SS	23.7391			17.8519
Pooled variance		0.86648		
Pooled standard deviation		0.9308		

Statistical Decision

Because the calculated t ratio (2.23) exceeded the exclusion value (2.0106), we would reject the null hypothesis. The t ratio, representing the transformed difference in our study groups in standard deviation units of the sampling distribution of differences, thus is so extreme that it could not be assumed to occur by

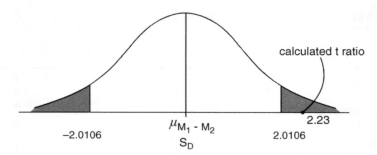

FIGURE 11.10 The statistical decision for the STAR™ group research question.

chance alone. Thus, we would conclude that the sample means came from different populations. Figure 11.10 shows how these values compare.

Interpretation

The independent-sample T test revealed that the sample group of elementary/middle classrooms showed higher average observed Powerful Teaching and Learning™ scores than the high-school sample of classrooms. The t ratio of 2.23 is statistically significant ($p < 0.05$), indicating that the difference in group means is not likely due to chance.

This is a "global" test of the effects of school level on the presence of Powerful Teaching and Learning™ among math classrooms. Because the sample ($N = 50$) was randomly drawn, we can reasonably generalize the conclusions to the overall population of math classroom observations among study schools.

BEFORE–AFTER CONVENTION WITH THE INDEPENDENT T TEST

Which group to consider group 1 or group 2 in an independent-samples T test is an important consideration. As you can see from the formula, the researcher can consider *either group* to be group 1 or group 2. *Depending on how they are entered into the formula, the sign of the t ratio will differ.*

If we had considered high-school classrooms to be group 1 and Elementary/MS classrooms to be group 2 in the previous example, the resultant t ratio would have been -2.23 instead of $+2.23$. This would not have altered the finding (rejection of the null hypothesis) unless we had specified a one-tail test in the positive exclusion region.

This example points out that you need to use caution in specifying group identification. This is as much an issue for experimental studies as it is for *post facto* studies because the researcher can assign the treatment group as either group 1 or group 2.

Let me suggest a convention for this dilemma. When you are dealing with a research design that uses "before" and "after," you can think of the "before" as what you are measuring the "after" against. By convention, *subtract the "before" score from the "after" so that you can show the change "left over" after the before score is subtracted out.* This convention would yield the following conceptual design (the general experimental pattern):

$$t = \frac{\text{After} - \text{Before}}{\text{StError}}$$

If you subtract out the before from the after scores, you will just have the increment left over that is due to your program, or whatever you designed to move the sample data "away" from the population data. Thus, for example, if you are dealing with a program that reduces stress, and you posit that a treatment program will result in lower stress scores after the program is completed, then using the above formula revision should show negative values if your research hypothesis is accurate. If your research hypothesis suggests that higher scores should result from training designed to increase management potential, then using the above should yield positive values if your research hypothesis is accurate.

This convention is perhaps less clear with *post facto* designs because the researcher may not be aware which direction the results may tend. The key issue is *being aware of the nature of the data and how the signs should fall if your alternate (research) hypothesis is true.* In the *post facto* example above, the researcher may have been aware from previous observation or research literature that high school math classrooms might show lower Powerful Teaching and Learning™ scores than elementary/middle-school classrooms. In this case, the researcher might place high school classrooms as group 2 so their scores would be subtracted out of elementary/middle-school classrooms to yield the *increment of difference as a positive value.*

There are exceptions to this process. It is up to you as the researcher to make sure that you understand which value you place first in the formula and how that decision will affect your conclusions. One obvious exception is if your research hypothesis is not true, or if in fact it is the opposite of what you expect!!

CONFIDENCE INTERVALS FOR THE INDEPENDENT T TEST

Remember that the T test above is a point estimate. We transformed the group mean difference to a point on the sampling distribution in order to see if it fell into the exclusion area (which it did). However, if we were interested in estimating the population value, we would use the CI procedure we learned in Chapter 10 with a couple of changes to the formula.

The formula below is for the single sample (refer to Chapter 10 for a complete analysis of the formula):

$$\text{CI}_{0.95} = \pm t(s_M) + M$$

The independent-samples T-test formula is almost identical, but include some changes:

$$CI_{0.95} = \text{(table value of } \pm t)(s_D) + (M_1 - M_2)$$

As you can see, the tabled value of t is the same. The changes are due to using a different sampling distribution with consequent changes in the standard error (s_D rather than s_M) and due to having two sample means rather than a single mean (thus, using $M_1 - M_2$ rather than M).

Using the data available from the analyses above, we obtain

$$CI_{0.95} = \pm 2.0106(0.264) + (2.52 - 1.93)$$
$$CI_{0.95} = -2.0106(0.264) + 0.596$$
$$CI_{0.95}(\text{left bracket value}) = 0.065$$
$$CI_{0.95} = 2.0106(0.264) + 0.596$$
$$CI_{0.95}(\text{right bracket value}) = 1.127$$

Therefore, we can say that the true population mean value from which our sample group differs is between 0.065 and 1.127. This interval does not include 0, the assumed population mean, thereby indicating that our sample group means are significantly different.

EFFECT SIZE

Effect size is very important, so we need to calculate it as we did for the single-sample T test. As you recall, we calculated *Cohen's d* according to the following formula:

$$d = \frac{t}{\sqrt{N}}$$

Because we added another sample group, we need to adjust the formula. Cohen (1988) lists the calculation as follows:

$$d = \frac{M_1 - M_2}{\sigma}$$

Cohen noted that the denominator (σ) is the population standard deviation. He further specified that for the formula, the researcher could use the standard deviation for *either group because they are assumed to be equal*. (Another approach is to use σ for the *entire set of scores* as the denominator.) Because we have learned to create the pooled variance and the pooled standard deviation as the combined estimate of the population standard deviation, we could use this as the denominator.

$$d = \frac{M_1 - M_2}{(\text{Pooled})\sigma_X}$$

Using this adjusted formula, we find the following:

$$d = \frac{2.52 - 1.93}{0.931}, \quad \text{thus } d = 0.6337$$

We can use the same criteria for judging the magnitude of the effect size (d) that we did in Chapter 10: 0.20, Small; 0.50, medium; 0.80, large. In this example our effect size is judged to have a medium effect.

An alternate method (Cohen, 1988) that does not include the pooled standard deviation is

$$d = t\sqrt{\frac{n_1 + n_2}{(n_1)(n_2)}}, \quad d = 2.23\sqrt{\frac{23 + 27}{621}}$$

This alternate formula provides essentially the same value of $d = 0.6328$. In both cases, the effect size is judged to be medium. In terms of the research problem, this indicates that the independent variable "level of school" had an impact on the Overall scores such that the difference between elementary/middle- and high-school groups "moved" the average Powerful Teaching and Learning™ about 0.63 standard deviation units apart (based on the population standard deviation).

EQUAL AND UNEQUAL SAMPLE SIZES

The example I used above included unequal sample sizes. While it is a bit more complex, I wanted to use it to explain features not present in the equal samples case. The unequal samples example demonstrates several important features of the pooled variance that are important to effect size and will also help to illustrate features of other statistical procedures we cover in subsequent chapters.

I also wanted to use an example with unequal sample size because that is typically what researchers find with real data. And, as we will see in the next sections, Excel® and SPSS® provide s_D despite sample size differences.

THE ASSUMPTIONS FOR THE INDEPENDENT-SAMPLES T TEST

All statistical tests require that the researcher first assess whether the conditions are appropriate for using a specific procedure. *Using the correct statistical procedure for a given research problem ensures a greater likelihood of not committing Type I and Type II errors.* This is a general statement that applies to all statistical procedures, not just the independent T test. Many statistical tests, including the T test, are called "robust" because they can be relied upon to deliver valid results even if some of the assumptions are not perfectly met, but the researcher should always assess the assumptions prior to using a statistical procedure.

Here are the requirements for the independent-samples T test:

- The samples are independent of one another.
- Dependent variable is interval level.
- Sample populations are normally distributed.
- Both populations have equal variance (this is also known as the "test of homogeneity" because we are assessing "sameness").

The researcher can assess the first two of these assumptions easily. Whether or not the samples are independent of one another is connected to the method and purpose of the research. As we discussed earlier, the cases for both of our sample groups must not be connected to one another. That is, the membership of one group must not rely on the membership of another. We will discuss the "dependent samples T test" in a later chapter. The researcher likewise can assess whether or not the outcome variable is interval level.

The third and fourth assumptions require a bit more investigation. To determine whether the samples are normally distributed, the researcher must use the descriptive statistical procedures we discussed in Chapter 5. Are skewness and kurtosis values "in bounds"? Does the graphical "evidence" match the numerical assessment of skewness, kurtosis, and so on? These procedures should be assessed before the researcher proceeds to the independent T test.

The fourth assumption requires a separate statistical test. The descriptive statistical summary produced to check the third assumption will reveal the values for variance and standard deviation. However, the researcher cannot conclude that they are equal or unequal simply by looking at them!

The Excel® "F-Test Two Sample for Variances" Test

Excel® provides a way to assess the equality of variances in the two samples used in a research study. The "F-Test Two Sample for Variances" is available through the "Data–Data Analysis" series of menus. Figure 11.11 shows the sub-menu called out from the Data Analysis menu.

When you choose this option, you will need to specify the range of values for groups 1 and 2. The output with the results of the test appear in a separate sheet, as shown by Figure 11.12. As you can see, the output includes some descriptive analyses in addition to the F-test results.

We will discuss the F test in much greater detail in later chapters. But for now, think of the F test as a way to compare variances rather than means. We have compared sample means in Chapter 10 (single sample) and in the current chapter (two sample means) using the sampling distribution of means. The F distribution is a *sampling distribution of variances*. So, when we are interested in whether the variances of two sample groups are equal, we compare their differences to the sampling

THE ASSUMPTIONS FOR THE INDEPENDENT-SAMPLES T TEST

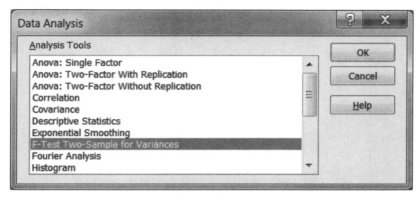

FIGURE 11.11 The Excel® equal variance test menu.

distribution of variances. The following is the formula used to create the F ratio to test whether it exceeds chance expectation:

$$F = \frac{s^2_{\text{Group 1}}}{s^2_{\text{Group 2}}}, \qquad F = \frac{1.08}{0.69}, \qquad F = 1.57$$

You can compare the results of our analysis ($F = 1.57$) with the value in the Excel® output result in Figure 11.12. If you look at the formula, it simply compares the variances to see if one is substantially greater or lesser than the other. If both are relatively equal (or comparable), the result would be $F = 1$. The further the F ratio departs from 1, the less likely the group comparisons are equal. But how different do the variances have to be before we judge them not equal? We use the same logic as we did with the T tests. If our test value (the transformed score applied to the sampling distribution) exceeds the exclusion value, we conclude that it is too extreme to be a chance finding. (Thus, we are performing a sort of "mini hypothesis test"!)

The F distribution looks different than the sampling distribution of means because we are sampling all possible variances. Figure 11.13 shows how this sampling distribution of variances appears.

F-Test Two-Sample for Variances

	Elementary/Middle	High School
Mean	2.52	1.93
Variance	1.08	0.69
Observations	23.00	27.00
df	22.00	26.00
F	1.57	
P(F≤f) one-tail	0.13	
F Critical one-tail	1.97	

FIGURE 11.12 The Excel® output for the two-sample variance test.

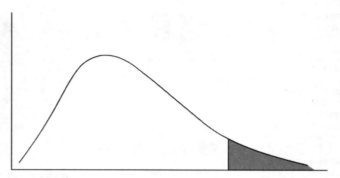

FIGURE 11.13 The F distribution and exclusion area.

The F distribution appears to be a normal distribution that has skewed to the right. This is because variances are never negative values. As you can see, all values tend to the right (positive) with the majority of observations near the left (smallest) side of the distribution. When we perform a hypothesis test for equal variance, the calculated F-ratio values can fall to the left or the right side of the distribution, depending on which variance (i.e., for group 1 or group 2) has the greatest value.

You could place either group in the numerator or denominator in the formula; it depends on which group you identify as group 1. However, please note that "left"-side values for this F test are difficult to determine, so the best rule of thumb is *always place the largest variance in the numerator and the smallest variance in the denominator* because we are only trying to determine the proportion of the two variances. Excel® will calculate both exclusion regions, however.

If group 1 has the greater value, the resulting proportion will likely tend toward the right side, in which case the right exclusion value is checked for an excessive value. If group 1 is smallest, then the left exclusion area is the one that will determine excessive calculated F ratios.

In our case, group 1 variance (2.52) is largest, so we would expect the F ratios to tend toward the right exclusion area. We do not need a separate F table of values to identify the exclusion area because Excel® reports the exact probability of a finding. F tables of values are common in statistics, so you may check statistical authorities [like Cohen et al. (2003), for example].

Figure 11.12 also reports two additional values. These are the values we use to assess equality of variances. In Figure 11.12, the following outcome indicates the likelihood of a chance finding:

$$P(F \leq f) \text{ one-tail} \qquad (0.13)$$

We can interpret this result as a 0.13 probability that a calculated F ratio that high (1.57) could be considered a chance finding, using a one-tailed table of probability. The next statement identifies the exclusion value for a one-tailed table of probability that our calculated *F* would need to exceed before it would fall into the exclusion range:

$$F \text{ Critical one-tail} \quad (1.97)$$

Taking both of these findings into account, we would conclude that our two group variances, while different, are nevertheless in the range of chance expectation. Because our calculated *F* ratio does not exceed the critical value that defines the exclusion area (1.97), we can conclude that our variances are statistically equivalent.

The SPSS® "Explore" Procedure for Testing the Equality of Variances

SPSS® also has a procedure for testing the equality of variances. It is known as "Levene's Test" and is reported through the "Explore" menu. Figure 11.14 shows how to access this procedure.

When you choose Explore, two further dialog boxes appear that allow you to specify the procedure. Figure 11.15 shows the box in which you can specify the variables of interest. As you can see, I entered the STAROverall variable as the dependent variable because that is the variable I want to check for equal variances *according to groups of "elemmidhs."* The procedure will therefore assess whether the elementary/middle/high school classroom values for STAROverall have an equivalent variance to the values for high school. I also chose "Both" so that the output would include visual (plots) as well as numerical (statistics) output.

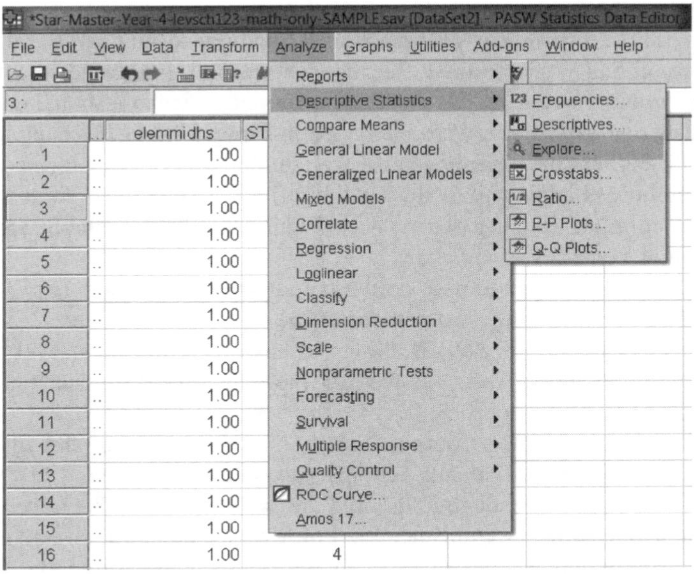

FIGURE 11.14 Accessing the SPSS® Explore procedure.

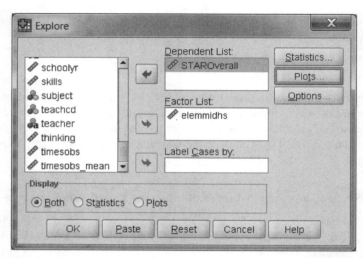

FIGURE 11.15 Specifying the variables of interest for equality of variances.

Figure 11.16 shows the further specification of the Explore procedure that I accessed by choosing the "Plots..." button on the preceding Explore menu (see Figure 11.15). Choosing Plots results in a further specification of the Explore procedure in which I can ask for several additional criteria to be applied. I will not explain all of these here, but I would only point out the choices for "Spread vs Level with Levene Test" in the bottom of the panel.

As you can see, there are several choices for how to manage the data to be analyzed for equality of variance. Because we are only interested in a "global" measure of equality, we can choose "Untransformed," which tests the data values "as is" or as raw data. If we were interested in "transforming" group values that may include outliers or extreme scores, we could choose "Transformed" in this panel to see how changing the values might affect the outcome. As it is, we are only interested in the equality of variances. Figure 11.17 shows the SPSS® output for assessing the equality of variance.

There are several things to note on the output shown in Figure 11.17, although our primary interest is in the assessment of the equality of variance. As you can see, the Levene Statistic (3.469) is shown to have a significance level of 0.069. Technically, the Levene's Test tests the hypothesis that the difference between the variances of the two groups is 0.

Essentially, the SPSS® output states that the elemmidhs groups of STAROutcome have statistically comparable variances since the Levene's value (3.469) is not large enough to fall into the exclusion area. The Levene's value would have had to show a significance level of less than 0.069 to fall into the exclusion area (usually defined by $p = 0.05$). Therefore, we can assume that the group variances are equivalent.

THE ASSUMPTIONS FOR THE INDEPENDENT-SAMPLES T TEST

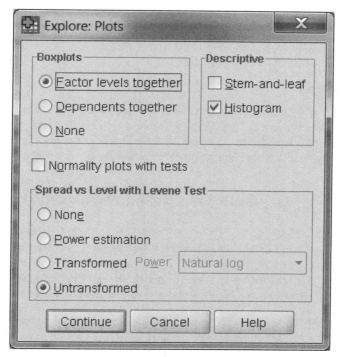

FIGURE 11.16 Specifying the Levine's Test for equality of variances.

Test of Homogeneity of Variance

		Levene Statistic	df1	df2	Significance
Overall score revised	Based on mean	3.469	1	48	0.069
	Based on median	2.165	1	48	0.148
	Based on median and with adjusted df	2.165	1	44.559	0.148
	Based on trimmed mean	3.190	1	48	0.080

FIGURE 11.17 The SPSS® output assessing equality of variance.

The Homogeneity of Variances Assumption for the Independent *T* Test

Despite using different statistical procedures, both Excel® and SPSS® concluded that the group variances were equal. Therefore, we can consider the fourth assumption to be met. Had the equality of variances test (*F* test) or Levene's Test indicated that the group variances were not equal, we would need to discuss further whether we could proceed with the *T* test. As I mentioned, the *T* test is robust and will provide valid results even with slightly different variances. However, with large differences, we might need to transform the values or use a different statistical procedure.

Note here, however, that Levene's test is very sensitive and sometimes overly conservative. You might use both the Excel® and SPSS® procedures to confirm your decision to proceed with the T test.

A Rule of Thumb

A simple rule of thumb might be helpful as well as these more formal procedures. If the F ratio we calculate according to the formula above is 2 or less, we can generally consider the variances to be equal for the purposes of using the independent T test. We only need a global indication of homogeneity of variance to proceed.

USING EXCEL® AND SPSS® WITH THE INDEPENDENT-SAMPLES T TEST

When we are assured that our data meet the assumptions for the independent T test, we can proceed using Excel® and SPSS®, which are both very straightforward procedures. Each provides the calculation for the t ratio and the pertinent information for hypothesis tests, and the output for both preclude the necessity for using the T table of values.

Both Excel® and SPSS® provide separate T tests, depending on whether the group variances are equal. Therefore, even if the homogeneity of variance test or Levene's Test show unequal variance, the researcher can rely on the separate formulas in Excel® and SPSS® to provide a meaningful T-test result.

Using Excel® with the Independent T Test

The T-test procedure is available in Excel® from the "Data–Data Analysis" menus. Figure 11.18 shows the menu of procedures with the "t-Test: Two-Sample

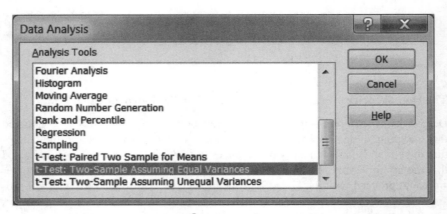

FIGURE 11.18 The Excel® specifications for the independent T test.

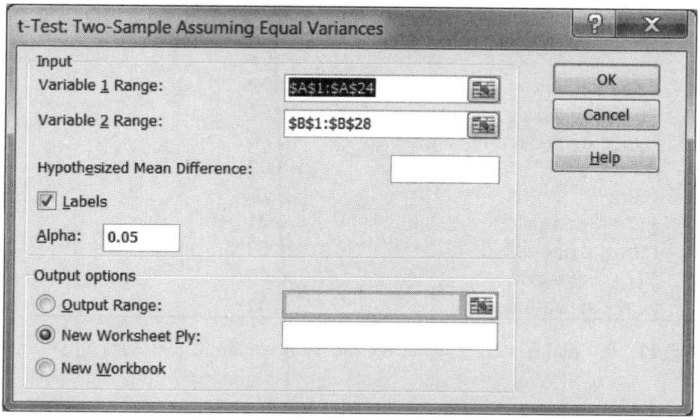

FIGURE 11.19 The Excel® call-out window for locating data.

Assuming Equal Variances" option highlighted. Note that another test option, directly below the highlighted selection, allows the researcher to choose the same T test, but one in which the variances are not assumed to be equal. Excel® uses a special formula to account for the unequal variances. It is best to check that the equal variance assumption is met prior to using the T test (as I discussed in the previous section of this chapter); however, you can make use of this special procedure. In the current example, we determined that we had met the homogeneity of variance assumption, so we can choose the first option.

In Excel®, you need to create the file structure in such a way that the program will recognize the groups appropriately. If you look at Table 11.2, you will see that the data are presented in two columns. The first column are the raw Group 1 values (Elementary/Middle) and the second column consists of the raw Group 2 values (High School). When you specify the location of the data for the T test, be sure that you specify the columns correctly.

When you choose this option, the call-out window in Figure 11.19 appears. In this window, the researcher must specify the location of both group values (in the "Input" windows), whether labels are included in the locations ("Labels" box), and which exclusion value you wish to use (in the "Alpha" window showing "0.05" in the current example). You can choose to have Excel® print the output in different places ("Output options"). The default location is a separate sheet within the data file.

Figure 11.20 shows the Excel® output for the T test of the two groups. You can compare the values with those we computed by hand above. All the values in the output match those we calculated. The "new" aspects of the output table related to the last four values. These show the probability of an extreme t ratio. Essentially,

	Elementary/Middle	High School
Mean	2.52	1.93
Variance	1.08	0.69
Observations	23.00	27.00
Pooled variance	0.87	
Hypothesized mean difference	0.00	
df	48.00	
t Stat	2.26	
$P(T \leq t)$ one-tailed	0.01	
t Critical one-tailed	1.68	
$P(T \leq t)$ two-tailed	0.03	
t Critical two-tailed	2.01	

FIGURE 11.20 The Excel® output for the independent T test with equal variances.

with these values, we do not need to make reference to a T table of values. Here is how to interpret the results:

- $P(T \leq t)$ one-tailed: 0.01. This value (0.01) represents the probability of our finding ($t = 2.26$). Essentially, this t-ratio value is the transformed mean difference score placed in the sampling distribution of differences. The exclusion value of the one-tailed T test, according to a Table of Values, is 1.68. Because our calculated value of 2.26 far exceeded this value, we can reject the null hypothesis because the probability of a t-ratio value that high by chance alone is $p = 0.01$.
- t Critical one-tailed: 1.68. This is a "companion" value to the immediately preceding value. It represents the exclusion value for a 0.05 one-tailed test.
- $P(T \leq t)$ two-tailed: 0.03. This probability is the probability of our finding in a two-tailed test. In a two-tailed (0.05) test, our calculated t ratio (2.26) exceeded the critical tabled value of T and thus could reject the null hypothesis. The actual probability of a t-ratio value that high by chance alone is $p = 0.03$ on the two-tailed comparison chart. The value below (2.01) is the exclusion value for the 0.05 two-tailed test.
- t Critical two-tailed: 2.01.

Taken together, the results indicate that we can reject the null hypothesis that Elementary/Middle and High School Powerful Teaching and Learning™ Overall scores for math classrooms are equal. We can conclude that the scores for math classrooms in elementary and middle schools are *statistically significantly higher* than those of math classrooms in high schools.

As I noted earlier, CI and effect size calculations are not available in Excel®. These can be easily calculated by hand, however. Use the formulas and procedures we discussed in the sections above.

Using SPSS® with the Independent *T* Test

Like Excel®, SPSS® provides a very straightforward way of conducting the independent samples *T* test. You can use the "Analyze" menu to access several *T*-test options that provide the output you need to conduct a research analysis. SPSS® provides CI output routinely, so this is an advantage over using the Excel® procedure. SPSS® is also limited, however, in that it does not provide results for Cohen's *d* effect size measure. However, this is a simple hand calculation, so the researcher can provide this critical information.

Figure 11.21 shows the Analyze menus for accessing the independent-samples *T* test. As you can see, the "Compare Means" sub-menu provides the option for the independent *T* test.

In Chapter 10 we discussed the single-sample *T* test, which is the choice immediately preceding the two-sample *T* test. When you choose the "independent–samples *T* test" from the menu, the call-out window appears that allows you to specify the variables you wish to use in the analysis, as shown in Figure 11.22.

As you can see in Figure 11.22, the dependent variable "overall" is placed in the "Test Variable(s):" window. The "Options" button at the top right of the window allows the researcher to specify values for the CI. The default value is 95%, but you may change this depending on how you wish to balance the "confidence–precision" question I discussed in Chapter 10.

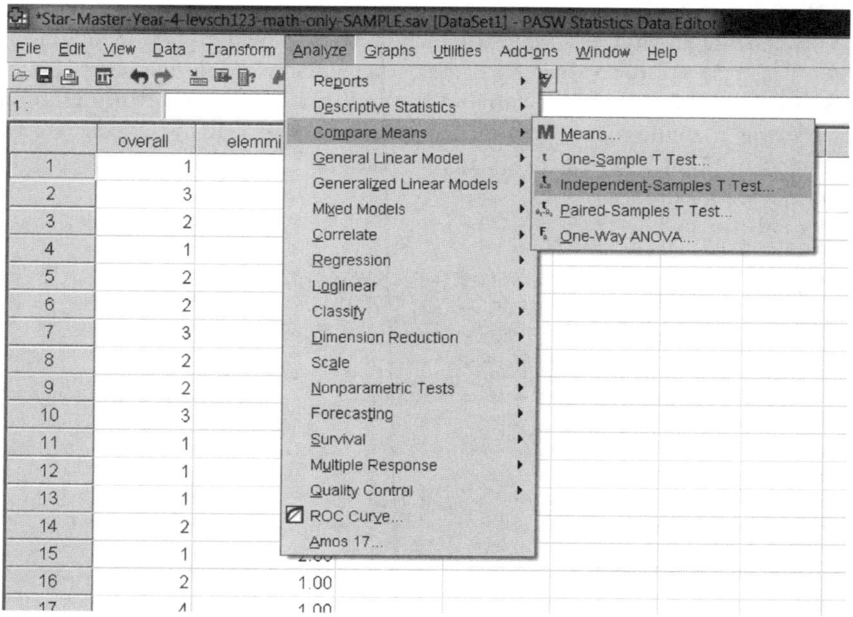

FIGURE 11.21 The SPSS® menus for the independent-samples *T* test.

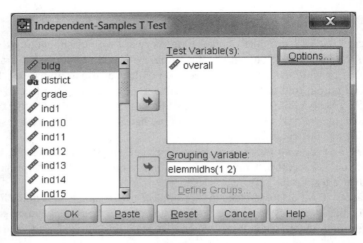

FIGURE 11.22 The Independent-Samples *T*-Test call-out window for specifying the analysis.

The independent variable is placed in the "Grouping Variable:" window. This allows the researcher to specify the values of elemmidhs in the database that will serve as group 1 and group 2. When you add the name of the variable to the window by using the arrow, SPSS® prompts you to "Define Groups" as shown in Figure 11.23.

You can see in Figure 11.23 that I entered "1" for Group 1 and "2" for Group 2. These are the values I assigned Elementary/Middle and High School classrooms for the elemmidhs variable in the SPSS® database. If my grouping variable was a continuous variable (like percentages of F/R by school), I could use the "Cut point:" button to specify a value that would divide my grouping variable into two groups.

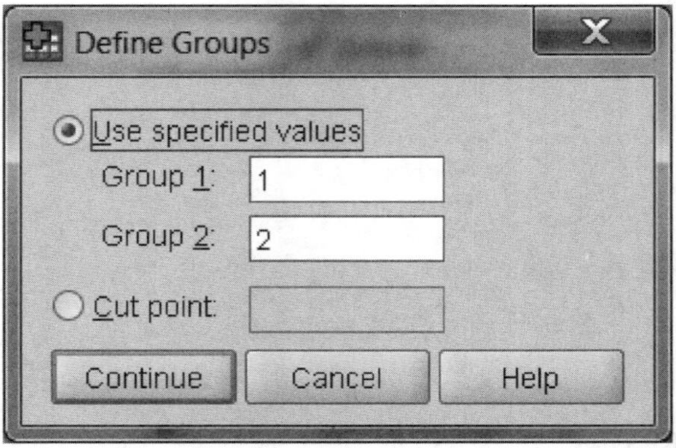

FIGURE 11.23 Specifying group values for the independent *T* test.

Group Statistics

	elemmidhs	N	Mean	Standard deviation	Standard error of mean
Overall score revised	1.00	23	2.52	1.039	0.217
	2.00	27	1.93	0.829	0.159

Independent-Samples Test

		Levene's Test for Equality of Variances		t-test for equality of means						
		F	Significance	t	df	Significance (two-tailed)	Mean difference	Standard error of difference	95% Confidence interval of the difference	
									Lower	Upper
Overall score revised	Equal variances assumed	3.469	0.069	2.256	48	0.029	0.596	0.264	0.065	1.127
	Equal variances not assumed			2.215	41.898	0.032	0.596	0.269	0.053	1.139

FIGURE 11.24 The SPSS® output for the independent T test.

When you create the analysis by choosing "OK" as shown in Figure 11.22, SPSS® creates the output showing the results of the analysis. Figure 11.24 shows the two panels of data produced in the output file. The top panel is the descriptive statistics shown for each group. You can compare these values to our hand calculations above, as well as with the Excel® output.

The second panel of Figure 11.24 provides the specific results for the independent-samples T test. Here are some important "parts" of the analysis:

- I shaded the left half of the panel to show that this is a separate part of the analysis. It is the equality of variance test, Levene's Test, that assesses whether we meet the assumption of equal group variances. As we discussed, if the significance is smaller than 0.05, this indicates that the equality of variance falls in the exclusion area, and we would have to conclude that the variances were not equal. However, as you can see in the panel, the significance level of 0.069 is larger than 0.05, so we may assume that the variances are statistically equal.
- The second column of the panel shows two "groups" of analyses: "Equal variances assumed" and "Equal variances not assumed." These are the two possibilities for Levene's Test. Since, in our current analysis, our group variances were considered equal, we can use the first row of results. If Levene's Test would have shown a significance level smaller than 0.05, we could use the second row of results because the group variances would be considered not equal.
- Considering the first row of data, the first statistic is the calculated t ratio (2.256). This value is the same as that produced by Excel®. Our hand calculation was a bit different due to rounding.

- The "Significance (Two-Tailed)" column indicates a significance level of 0.029. Remember, SPSS® provides the exact calculated probability of a certain finding. Therefore, this finding shows that it is well into the 0.05 exclusion area of a two-tailed test. Therefore, the t ratio would result in a rejection of the null hypothesis. This is a finding so extreme that, it could not be considered a chance finding.
- The standard error of difference (s_D) is provided in the next column and is the same value we calculated by hand. Recall that Excel® does not produce the s_D but shows the pooled variance. I calculated both of these by hand so you can compare the values.
- The $CI_{0.95}$ values are provided in the final two columns. With the independent samples T test, the confidence interval values shown are the actual interval values (lower and upper). Remember, the population of mean differences is 0. The interval of the population mean estimate of 0.065 to 1.127 does not include this value. We would not expect it to contain 0 because we rejected the null hypothesis, indicating that our groups were so different that they did not belong to the same population (i.e., one in which the population of mean differences was 0). These values match our hand calculations above.

As I mentioned above, neither Excel® nor SPSS® provide Cohen's d as a measure of effect size. However, because both provide the t ratio, you can calculate the effect size by hand using the "alternate" formula:

$$d = t\sqrt{\frac{n_1 + n_2}{(n_1)(n_2)}}, \qquad d = 2.26\sqrt{\frac{23 + 27}{621}}$$

This calculation finds $d = 0.64$, which differs from my hand calculation only slightly due to rounding. The conclusion remains that the differences in Powerful Teaching and Learning™ scores (as shown by the overall variable) attributed to grouping math teachers by level of school (Elementary/Middle or High School) shows a medium effect; it is a meaningful "separation" of scores.

PARTING COMMENTS

Remember that statistical tests need to be "fitted" carefully to the nature of the data and the research situation. If it is not carefully fitted, the power of the test will be diminished. Some very important things to remember in terms of the independent T test, in this regard, is that there are specific formulas to be used when the groups have *different sample sizes* and when the *sample groups are dependent, rather than independently formed*. In the former case, we learned to use a special formula for s_D that accommodated unequal sample sizes. In the latter case, you need to understand how to use the "repeated measures" T test that uses dependent samples. I will discuss this procedure in a later chapter.

You should congratulate yourself at this point. You learned about one of the most common and practically useful statistical tests available. The T test is common in all research literature, and now you will be able to understand how the results are interpreted. You also have the knowledge to assess whether test results, published or unpublished, were properly reported and used.

NONPARAMETRIC STATISTICS

To this point, we have studied *parametric statistics*, which are procedures that make reference to parameters (mean and standard deviation) of populations in helping us to make statistical decisions. If you recall, we have made use of sampling distributions to compare sample values (i.e., means and standard deviations) to estimated population parameters. These are procedures that use interval data.

There is another "class" of statistics that do not make these assumptions. These *nonparametric* statistical procedures directly calculate test values rather than estimating parameters or referring to sampling distributions. For these reasons, they are also known as *distribution-free tests*. They typically make decisions with ordinal or nominal data.

Nonparametric procedures are also helpful for statistical decisions that involve (interval level) sample data that do not meet certain assumptions. As we have seen, for example, the assumption of normally distributed variables is often crucial to an analysis—so much so that extremely skewed data might prevent a researcher from proceeding with a parametric procedure. With extreme violations of normally distributed data the researcher might resort to *transforming* the skewed data (i.e., changing the skewed data so that it becomes normally distributed). A second option for violations for normal distributions is to use nonparametric statistical tests.

There are several nonparametric statistical tests comparable to the independent-samples T test. Possibly the most well known is the Mann–Whitney U Test. Excel® does not provide nonparametric analyses, but SPSS® allows the researcher to use a variety of these tests.

As a brief (hypothetical) example, consider the data in Table 11.3X. These data show the teachers' rankings of (male and female) students on fear of math. Since the data are ranks (ordinal data), we cannot use the independent T test because that test requires at least interval data. The Mann–Whitney U test will help us determine whether the two sets of ranks are likely to be from the same population.

TABLE 11.3X Mann–Whitney U-Test Data

Fear Test	Gender
8	1
9	1
2	1
5	1
4	1
10	1
25	2
11	2
28	2
21	2
14	2
16	2

I will not detail the hand calculations for this test but rather show how to use SPSS® to obtain the results. If you would like to explore this and other nonparametric tests further, you might consult Siegel and Castellan's (1988) seminal work.

As you can see from Table 11.3X, the rankings are presented by the student's gender (1 = male, 2 = female). The higher ranks indicate greater math fear.

Using SPSS® is straightforward. Figure 11.24a shows the specification for the test. Note that there are two routes to conduct the test. When you choose "Nonparametric Tests" you have the option of selecting "Independent Samples," which allows you to use a template for this and related tests. However, using the second route, "Legacy Dialogs—Two Independent Samples," affords you more control over the test specification.

Using the second route produces the menu window shown in Figure 11.24b. As you can see, I specified the Fear rankings as the test variable and gender as the grouping variable. At this point, I can choose the Mann–Whitney U as well as a number of related tests. In the "Options" button (upper right part of the window), I chose to create descriptive statistics.

Running this test ("OK") results in the output shown in Figure 11.24c. These are the primary output to determine whether the sample groups are from different populations.

The output in Figure 11.24c shows the average of the ranks for both groups in the top panel. The bottom panel shows the results of the Mann–Whitney U test. As you can see, the test resulted in a significant difference between males and females in terms of fear of math ($p = 0.004$). Female ranks were much higher on average (9.5) than those of males (3.50).

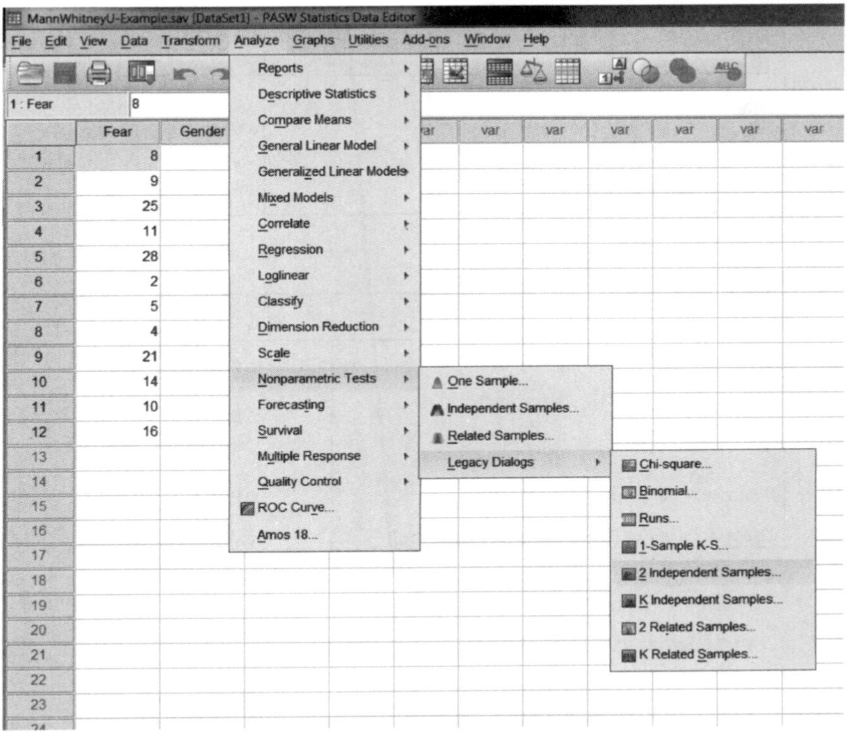

NONPARAMETRIC STATISTICS

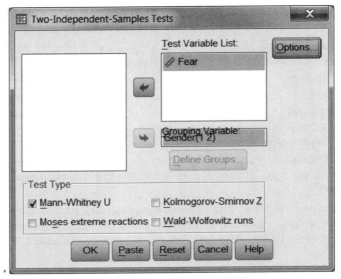

FIGURE 11.24a The SPSS® options for the Mann–Whitney U test.

FIGURE 11.24b The SPSS® specification for the Mann–Whitney U test.

Ranks

	Gender	N	Mean rank	Sum of ranks
Fear	1	6	3.50	21.00
	2	6	9.50	57.00
	Total	12		

Test Statistics[a]

	Fear
Mann–Whitney U	0.000
Wilcoxon W	21.000
Z	−2.882
Asymptotic significance (two-tailed)	0.004
Exact significance [2*(one-tailed significance)]	0.002[b]

[a]Grouping variable: Gender.
[b]Not corrected for ties.

FIGURE 11.24c The SPSS® output for the Mann–Whitney U test.

Note that the Mann–Whitney U-test statistic (0.000) is very low! In this test, the lower the result, the greater the significance.

TERMS AND CONCEPTS

Between Group Designs Those studies in which the researcher seeks to ascertain whether *entire groups* demonstrate unequal outcome measures.

Dependent Samples Groups that have some structural linkage that affects the choice of group membership. Examples are using the same subjects (pre–post) for an experiment, or comparing the results of a test using "matched groups." Tests for dependent samples are also known as "repeated measures," "within subjects," and "paired samples" tests.

Independent Samples For statistical testing purposes, samples are independent if choosing the elements of one group has no connection to choosing elements of the other(s).

Matched Samples Samples that have been intentionally created to be equivalent on one or several characteristics. Such a process affects the independence assumption of the groups. Matched groups are therefore considered "dependent" samples.

Mixed Designs Those studies in which there are both between and within components.

Within Group Designs Studies in which the researcher seeks to ascertain whether *subjects* in a group or matched groups change over time.

REAL-WORLD LAB VII: INDEPENDENT *T* TEST

In this lab, you will conduct an independent T test with our school database to see if *class size* affects *achievement*. Education researchers are just beginning to take a closer look at this research question because it is such a widely held opinion and it has resulted in funding opportunities for schools that have low student achievement.

The t test is particularly useful in this situation because it can be used with *post facto* or experimental designs, as we have discussed. The analytical procedures are the same. There are two schools of thought on this question that we can assess through our t-test analysis.

- It is a popular assumption among the public, education practitioners, and even education researchers that the smaller the class size, the greater the likelihood of increased achievement. The primary determinants are thought to be things like more time for responding to individual students, more time to devote to students of different learning styles, more physical space for students to utilize, and similar factors.
- A different school of thought indicates that class size, by itself, will not necessarily lead to higher achievement levels. Rather, it is the *opportunities to change the way learning occurs occasioned by smaller classes* that is the key ingredient in achievement gains. Just having smaller classes may mean that the teacher teaches the same way they always have, but with the luxury of fewer students unless they seize the opportunity to change their instructional approach in ways they could not with larger classes. My research colleagues and I discuss these issues as "first- and second-order changes" (Abbott et al., 2010).

We can use the school data to assess these views. This lab allows you to test the assumptions with the t test, which focuses on the difference between schools with "small" class size and schools with "large" class size in terms of school math achievement.

Procedures

In this lab, we will analyze the relationship between class size and math achievement by creating two groups of schools' math achievement according to a split of their class size. In order to have confidence in our t test, we want to first make sure that our assumptions are met.

Assumptions. As we noted, it is important to examine the test variables to make sure we can use the statistical test we are planning. For the independent T test, we will focus on the following:

- The samples are independent of one another.
- Dependent variable is interval level.
- Sample populations are normally distributed.
- Equal variance.

The Database. Table 11.4 shows the data for this Lab. The database is sorted by student–teacher ratio to illustrate the sorting procedures that follow.

Real-World Lab VII Questions

1. Discuss whether the assumptions are met for the independent T test.
2. Calculate the independent T test by hand and perform the hypothesis test.
3. Calculate the effect size and $CI_{0.95}$.
4. Perform the independent T test with Excel® and SPSS®.
5. Provide a summary of your findings.

REAL-WORLD LAB VII: SOLUTIONS

The solutions to this Lab are presented below according to the Lab questions.

1. Are Assumptions Met for the Independent T Test?

The assumptions for the independent T test are listed below. The first two can be assumed to be met by the research variables. Both variables are school level percentages and therefore interval level, and groups based on student to teacher ratios are not dependent on one another to form groups.

1. The samples are independent of one another.
2. Dependent variable is interval level.
3. Sample populations are normally distributed.
4. Equal variance.

Assessing the next two assumptions requires that we create our two study groups from the database in Table 11.4. As you can see, the student–teacher ratio data are listed as a continuous variable. In order to compare schools that differ on this variable, we need to create two groups based on the magnitude of student–teacher ratio.

TABLE 11.4 The Math Achievement and Student–Teacher Ratio Database for Lab VII

MathPercent-MetStandard	StudentsPer-ClassroomTeacher	MathPercent-MetStandard	StudentsPer-ClassroomTeacher
73	9	79	16
50	10	10	16
38	12	62	17
36	12	25	17
73	12	66	17
40	12	34	17
56	12	89	17
27	13	56	17
46	14	74	18
40	14	53	18
50	14	58	18
51	15	42	18
50	15	50	19
77	15	41	20
24	15	63	20
49	15	30	21
28	15	59	21
50	15	72	21
37	16	69	24
63	16	35	27

Assessing Normal Distribution. We can assess this assumption by creating descriptive information on both groups of schools that differ on the size of the student to teacher ratios. You can do this in a number of ways. One way is to use the "median split," which splits the database into two groups on the basis of the median value of the independent variable, the student–teacher ratio. The median teacher ratio in this dataset is 16 (you can easily identify this by the methods we learned in past chapters). Using this value, we can create two groups of schools based on the size of their student–teacher ratio, a variable we can call "class size."

You can use Excel® to sort "StudentsPerClassroomTeacher" from low to high. Then:

- Create a new variable called "Class Size".
- Assign a group of *Small sizes* ("1") on the new variable (Class size) if the values on the original variable ("StudentsPerClassroomTeacher") are 15 or less.
- Assign a group of *Large sizes* ("2") on Class Size if the values are 16 or greater on StudentsPerClassroomTeacher.
- Create two sets of Math Achievement averages according to the Class Size (Small or Large).

TABLE 11.5 The Study Database with Math Achievement in Small and Large Class Sizes

Class Size Small—Group 1 Math Achievement	Class Size Large—Group 2 Math Achievement
73	37
50	63
38	79
36	10
73	62
40	25
56	66
27	34
46	89
40	56
50	74
51	53
50	58
77	42
24	50
49	41
28	63
50	30
	59
	72
	69
	35

The data file we have created is shown in Table 11.5.

As you can see, using the median split of 16 resulted in two groups of schools: Those with "small" class sizes (<16 student–teacher ratio) and "large" class sizes (≥16 student–teacher ratio). We can now compare the math achievement values in both groups.

Assessing whether the two sample groups are normally distributed can be checked by skewness and kurtosis values, as well as graphical methods. Using Excel® for the descriptives yields the values in Figure 11.25.

According to the data in Figure 11.25, both groups of schools show approximately normally distributed groups since skewness and kurtosis figures are small. Confirm this with SPSS®, which shows similar results (i.e., skewness and kurtosis figures that are "in range" when the figures are divided by their standard errors). The graphical "evidence" follows in Figures 11.26 and 11.27, which were obtained through Excel® "Descriptive" and "Histogram" menus, as we discussed in earlier chapters. SPSS® shows similar graphs obtained through the "Graphs" menu.

Although both figures show histograms that do not appear perfectly normal, the profiles tend toward normal when the sample size is taken into account. Combined

REAL-WORLD LAB VII: SOLUTIONS

	Math Achievement by Class Size	
	Small Classes	Large Classes
Mean	47.66	53.01
Standard error	3.62	4.13
Median	49.25	57.15
Mode	50.00	#N/A
Standard deviation	15.36	19.38
Sample variance	235.94	375.74
Kurtosis	−0.15	−0.24
Skewness	0.48	−0.33
Range	53.10	78.80
Minimum	24.30	9.70
Maximum	77.40	88.50
Sum	857.90	1166.30
Count	18.00	22.00

FIGURE 11.25 Descriptive analyses for real-world lab VII.

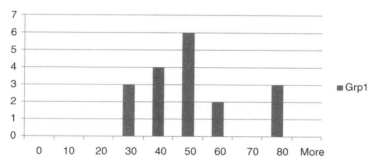

FIGURE 11.26 Math achievement in schools with small class size.

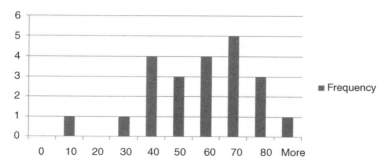

FIGURE 11.27 Math achievement in schools with large class size.

F-Test Two-Sample for Variances

	Variable 1	Variable 2
Mean	53.01	47.66
Variance	375.74	235.94
Observations	22.00	18.00
df	21.00	17.00
F		1.59
$P(F \leq f)$ one-tailed		0.17
F Critical one-tailed		2.22

FIGURE 11.28 The Excel® output for the F test of equal variance.

with the numerical data, we can assume that the normal distribution assumption is met for this procedure.

Assessing Equal Variance. Assessing this assumption requires that we use the Excel® and SPSS® procedures we used above in this chapter. The Excel® 'F test for equal variances' results is shown in Figure 11.28. As you can see, the one tail $p = 0.17$ which shows that the F ratio does not fall in the exclusion area ($p < 0.05$). Therefore, the two variances are statistically equal.

Note: I placed the group with the largest variance as group 1 and the group with the smaller variance as group 2. In this way, I would receive a result on the right side of the comparison distribution. I could have reversed this procedure, but "left tests" often are a bit trickier to decipher with actual F tables. Of course, the Excel® works with both processes.

The SPSS® "Explore" procedure output is shown in Figure 11.29. According to the analyses, Levene's Statistic is not significant. In every case, the "Significance" value (ranging from 0.199 to 0.253) does not fall in the exclusion area (of $p < 0.05$). Thus, by these criteria, the sample variances are equal.

Meeting the Assumptions for the Independent T Test. According to the foregoing analyses, the four assumptions for the Independent T test are met with this sample of data. We can now proceed to the remaining questions for Real-World Lab VII.

Test of Homogeneity of Variance

		Levene Statistic	df1	df2	Significance
MathPercentMetStandard	Based on mean	1.707	1	38	0.199
	Based on median	1.348	1	38	0.253
	Based on median and with adjusted df	1.348	1	37.042	0.253
	Based on trimmed mean	1.642	1	38	0.208

FIGURE 11.29 The SPSS® "Explore" output.

2. Calculate the Independent T Test by Hand and Perform the Hypothesis Test

$$t = \frac{(M_1 - M_2) - \mu_{M_1-M_2}}{s_D}, \quad s_D = \sqrt{\left(\frac{(SS_1) + (SS_2)}{(df)}\right)\left(\frac{1}{n_1} + \frac{1}{n_2}\right)},$$

$$t = \frac{(47.66 - 53.01) - 0}{5.62}, \quad 5.62 = \sqrt{\left(\frac{(4011) + (7891)}{(38)}\right)\left(\frac{1}{18} + \frac{1}{22}\right)},$$

$$t = -0.952,$$

H_0: $\mu_1 = \mu_2$ There is no difference in the populations from which the sample groups came.

H_A: $\mu_1 \neq \mu_2$ There is a difference in the populations from which the sample groups came.

$t_{(0.05, 38)} = 2.024$
$t_{(\text{calculated})} = -0.952$

Decision: Do not reject the null hypothesis; the two groups are not different.

Interpretation: Schools with large and small student to teacher ratios do not differ in terms of their average math achievement scores (the percentage of students who pass the math assessment).

3. Calculate the Effect Size and $CI_{0.95}$

$$d = \frac{M_1 - M_2}{(\text{Pooled})\sigma_X}, \quad d = \frac{47.66 - 53.01}{17.70}, \quad d = 0.30$$

$$d = t\sqrt{\frac{n_1 + n_2}{(n_1)(n_2)}}, \quad d = .952\sqrt{\frac{18 + 22}{(18)(22)}}, \quad d = 0.30$$

$CI_{0.95} = (\text{table value of } \pm t)(s_D) + (M_1 - M_2)$

$CI_{0.95} = \pm 2.024(5.62) + (47.66 - 53.01)$

Lower Limit: -16.72

Upper Limit: 6.02

4. Perform the Independent T Test with Excel® and SPSS®

Excel® Results. Figure 11.30 shows the Excel® results of the t test. The shaded t ratio (-0.95) agrees with our hand calculation. The actual probability of the result

t-Test: Two-Sample Assuming Equal Variances

	Small Class Size	Large Class Size
Mean	47.66	53.01
Variance	235.94	375.74
Observations	18.00	22.00
Pooled Variance	313.20	
Hypothesized Mean Difference	0.00	
df	38.00	
t Stat	−0.95	
$P(T \le t)$ one-tail	0.17	
t Critical one-tail	1.69	
$P(T \le t)$ two-tail	0.35	
t Critical two-tail	2.02	

FIGURE 11.30 Excel® results of independent t test.

($p = 0.35$) is also shaded and indicates that the t ratio does not extend into the exclusion region. Therefore, we do not reject the null hypothesis.

SPSS® Results. Figure 11.31 shows the SPSS® results of the t test. I shaded Levene's Test portion to point out that we have equal variances, as we discussed above. All other values agree with our hand calculations and the Excel® output. (The CI values in Figure 11.31 are slightly different from our hand calculations due to rounding.)

5. Provide a Summary of your Findings

The findings of an independent T-test analysis revealed no differences between low and high class sizes on school-based math achievement. Two groups of schools ($n_1 = 18$ and $n_2 = 22$) showed no statistically significant differences in

Independent Samples Test

		Levene's Test for Equality of Variances		t test for Equality of Means						
		F	Significance	t	df	Significance (Two-Tailed)	Mean Difference	Standard error Difference	95% Confidence Interval of the Difference	
									Lower	Upper
MathPercentMetStandard	Equal variances assumed	1.707	0.199	−0.952	38	0.347	−5.35	5.62	−16.74	6.03
	Equal variances not assumed			−0.974	37.973	0.336	−5.35	5.49	−16.48	5.77

FIGURE 11.31 SPSS® results of independent t test.

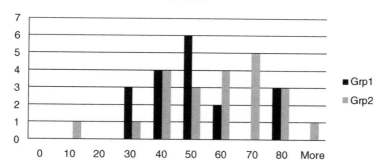

FIGURE 11.32 Graph comparing schools with low (Grp1) and high (Grp2) class size with regard to math achievement.

math achievement scores ($t = -0.952$, $p < 0.347$). However, math achievement scores tended to be higher in schools with larger student–teacher ratios, with a small effect size ($d = 0.30$). Figure 11.32 shows a graph comparing the groups' math achievement

12

ANALYSIS OF VARIANCE

We now come to a very popular test with educational researchers and educational psychologists. Analysis of variance (ANOVA) is popular because the researcher is able to compare several different groups—unlike the T test, which only compares two groups. ANOVA is really quite an ingenious test in the way in which it allows statistical decisions by analyzing *components of the variance of the groups* to be compared.

There are several variations of this test. We will focus primarily on the "One-Way" ANOVA in this chapter, but I will cover an extension of the test briefly at the end of the chapter. One way ANOVA refers to the number of independent variables. *In research, independent variables are known as "factors."* Therefore, if we have a research problem that has several levels of one independent variable, we can use One-Way ANOVA to detect any differences among the sample groups.

We will make brief reference to the "Factorial ANOVA" at the end of the chapter. This will introduce you to an important elaboration of the One-Way ANOVA. In Factorial ANOVA, we introduce a second (or more) independent variable to our research analysis. "Factorial" refers to the fact that we have more than one factor in the procedure. Thus, if we have two independent variables, our test would be a factorial ANOVA, or a 2XANOVA.

Understanding Educational Statistics Using Microsoft Excel® and SPSS®. By Martin Lee Abbott.
© 2011 John Wiley & Sons, Inc. Published 2011 by John Wiley & Sons, Inc.

A HYPOTHETICAL EXAMPLE OF ANOVA

In Chapter 11, I discussed an experiment I conducted on the effects of noise on human learning. Because there were only two groups to compare in that example (high versus low noise), I showed how the independent T test could be used to test the null hypothesis. When I conducted the actual experiment, I compared four groups on their rates of learning. Figure 12.1 shows the design of the experiment.

As you can see, there were four groups, three of which used different levels of noise and one control group in which no noise was present. (Recall that I used white noise in different magnitudes to distinguish the experimental groups.) The noise level in Group III, 30 decibels is comparable to quiet conversation; 60 decibels is equivalent to normal conversation; 90 decibels is equivalent to heavy truck traffic. (Sustained exposure to 90 decibel noise could result in hearing loss.)

There is (still) only one independent variable (noise) in the experiment, but now there are groups in four levels of the independent variable. A T test would be inappropriate because I would need to conduct six T tests to compare all the group results! This would involve the group comparisons in Figure 12.2.

Even if I had conducted all these T tests, I would have no idea of the "whole" test result. I needed the ANOVA because it conducts *all the comparisons within the same procedure at the same time*. I needed one "omnibus" answer to the question of whether noise affects human learning. If this result

Research Treatment Variable (Noise)			Dependent Variable (Word Recognition)
Random selection and assignment	No pretest	Experimental Group I (90 decibels)	Outcome test scores (number of recognized words)
Random selection and assignment	No pretest	Experimental Group II (60 decibels)	Outcome test scores (number of recognized words)
Random selection and assignment	No pretest	Experimental Group III (30 decibels)	Outcome test scores (number of recognized words)
Random selection and assignment	No pretest	Control Group (No noise level)	Outcome test scores (number of recognized words)

FIGURE 12.1 The four groups in the noise—learning experiment.

Group Comparisons
Experimental Group I versus Experimental Group II
Experimental Group I versus Experimental Group III
Experimental Group I versus Control Group
Experimental Group II versus Experimental Group III
Experimental Group II versus Control Group
Experimental Group III versus Control Group

FIGURE 12.2 The two group comparisons in the experiment.

rejected the null hypothesis, I could then proceed to dig further into the results to examine *which* of the groups were different from the others and therefore may have been responsible for the overall omnibus finding.

THE NATURE OF ANOVA

When a researcher conducts multiple (separate) tests on the same data in the research study, the alpha error "accumulates." This is known as *familywise error* since we consider each comparison to be part of the same "family" of tests. Thus, although it is not this simple, if we have six tests and the alpha error is 0.05 for each, we might have a 0.30 alpha error (6 * 0.05) in the entire family of comparisons! The more tests we conduct, the greater the likelihood that we will make alpha errors. Thus, the "overall" test is compromised.

The ANOVA "conducts" all these comparisons together by analyzing components of variance and therefore contains familywise error. We have seen how the T test used variance measures (i.e., sum of squares) to help detect differences between two sample groups. ANOVA identifies three sources of variance: the variation that exists *within* each sample group, *between* each sample group and the overall (grand mean), and the *total* variance from all sources.

Because the ANOVA process analyzes all these sources of variance simultaneously, we do not have to rely on multiple tests using the same data. The null hypothesis in the hypothetical experiment would thus be

$$H_0: \quad \mu_1 = \mu_2 = \mu_3 = \mu_4$$

This null hypothesis states that all four group means come from the same population; there is no difference between the group means. As you can see, there are a variety of ways in which the null hypothesis could be rejected. The following represent two of several possibilities: All the Groups could be from different populations (the top example); or, Group 1 could be from a different population than Group 2, and Groups 3 and 4 could differ (the bottom example). There are several other possibilities!

$$H_A: \quad \mu_1 \neq \mu_2 \neq \mu_3 \neq \mu_4,$$
$$H_A: \quad \mu_1 \neq \mu_2 = \mu_3 \neq \mu_4$$

If the means of the groups are markedly dissimilar to one another (in a statistically significant sense), we could reject the null hypothesis and conclude that at least one of the group means differs from the others. If we are able to reject the null hypothesis, ANOVA will not indicate which mean (or means) is different from the others. In that event, we would perform a separate test, called a *"post hoc* comparison," in order to identify which of the means were significantly different from one another. We will discuss this process after we discuss the overall or Omnibus ANOVA test.

THE COMPONENTS OF VARIANCE

If you recall the earlier chapters, you will remember that variance is a measure of the dispersion of scores in a distribution. We used variance measures in Chapter 11 when we discussed the T test, and we can use these measures for ANOVA. Variance is a more "global" measure of variance, so we will use it instead of the standard deviation (another measure of dispersion) as a way of determining how sample groups differ from one another. Look at Figure 12.3 to see how our experimental groups might be represented, using the noise experiment I introduced above.

Figure 12.3 shows the sample groups arrayed from left to right on an X axis of noise groups that show different "number of learning errors," which is the dependent variable for this experiment (represented on the Y axis). Remember that my experimental question was whether noise affected human learning. I operationally defined learning as the number of simple words recalled during a memorization task. The experiment sought to see if different amounts of noise would decrease the number of words recalled.

Figure 12.3 shows the three sources of variance in an ANOVA analysis. The four *sample groups have their own distributions, so all their scores "spread out" around their means. This is known as "Within Variance"* (V_{Within}) *because the variance is measured within each sample distribution.* Figure 12.3 also shows a "shadow" distribution that represents a total distribution if all the individual scores from all the sample groups were thrown in to one large distribution. If we were to measure *the variance of that large, composite distribution, this would constitute the "total*

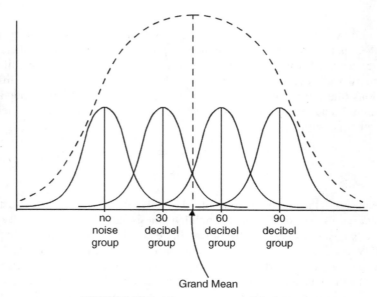

FIGURE 12.3   The components of variance.

THE PROCESS OF ANOVA

variance" (V_{Total}). As you can see, the composite distribution has a mean value, indicated by the dashed line. This total mean is known as the "grand mean." *Between variance ($V_{Between}$) is the variation between each sample group mean and the grand mean.*

- V_{Total} is the total variance of all individual scores in all sample groups.
- $V_{Between}$ is the variation between the sample means and the grand mean.
- V_{Within} is the variation of individual scores within their own sample groups.

THE PROCESS OF ANOVA

In effect, the ANOVA process determines *whether the sample means vary far enough from the grand mean that they could be said to be from the different populations*. What makes this question difficult to determine is that within each group, the scores vary around their own group mean. If the sample group means are close enough to one another, single scores could lie within the area where the groups overlap making group identification difficult for these individual scores.

Figure 12.4 shows two possible results that illustrate the process of ANOVA. The top panel shows the results of an experiment in which the groups are so "squished together" that the group variances are intermingled. In the bottom example, the groups are far enough apart that individual group variances do not have extensive overlap. That is, the variances *within* each sample group do not confuse the distance *between* the groups. If we have an actual result like the bottom example, we would be likely to determine that the groups are statistically different from one another. This might not be the case in the former.

ANOVA seeks to determine whether the $V_{Between}$ *would be large relative to the* V_{Within}, as appears to be the case with the bottom example. In the top example, the

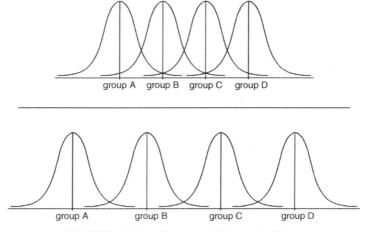

FIGURE 12.4 Different ANOVA possibilities.

V_{Within} "muddles the picture" of how the groups relate to one another. It is for this reason that within variance is often referred to in research (and in SPSS®) as "error variance." If we want to understand how far apart the group means are from the grand mean, the error variance (V_{Within}) "gets in the way." When there is less error variance, then the distance between the group means and the grand mean (i.e., the V_{Between}) becomes easier to distinguish.

CALCULATING ANOVA

The calculation of ANOVA compares V_{Between} to V_{Within}. When V_{Between} is large relative to V_{Within}, there is a greater likelihood that the researcher can reject the null hypothesis. We can suggest a "conceptual equation" that expresses this relationship:

$$F = V_{\text{Between}} / V_{\text{Within}}$$

ANOVA uses the "F distribution" as a comparison distribution for the relationship of the between-to-within variance. We introduced the F test in Chapter 11 when we discussed how to test whether two sample variances were statistically equal. This same distribution (see Figure 11.13) can be used to see where a calculated ANOVA falls on the sampling distribution of all possible samples. After all, we are still comparing variances. But in the ANOVA test we are comparing between to within variance, whereas in the Equal Variances Test we are comparing two sample variances.

The equation above thus expresses the relationship of between to within variance as a point on the F distribution of all possible samples to determine whether the sample group means are far enough from the grand mean that the calculated F ratio would fall in the exclusion area of the comparison distribution. If a researcher obtained an actual result like the bottom example in Figure 12.4, there would be a greater likelihood that the F ratio (calculated F) would fall in the exclusion area than if the top example was obtained.

Calculating the Variance: Using the Sum of Squares (SS)

ANOVA helps the researcher identify and compare all the components of variance present in the research study. Therefore, each of the variance components must be calculated in order to make a statistical decision. The sum of squares (SS) is used to measure the components. As I explained in past chapters, SS is a large number when the scores of a distribution are spread out but a small number when the scores are close together.

In Chapter 11, I specified the formula to calculate SS:

$$SS = \sum X^2 - \frac{(\sum X)^2}{N}$$

CALCULATING ANOVA

TABLE 12.1 Hypothetical Experiment Data

Control	30 Decibels	60 Decibels	90 Decibels
5	7	12	15
4	5	11	14
3	6	10	18
6	8	12	12

We can use variations of this formula to calculate each of the three sources of variance. Moving from conceptual to computational definitions, we can identify these sources as follows:

SS_T is the sum of squares of the total distribution of all individual scores (V_{Total})

SS_B is the sum of squares between the group means and the grand mean ($V_{Between}$)

SS_W is the sum of squares within each sample group (V_{Within})

Calculating the three SS values is somewhat easier when you consider that the total variation (V_{Total}) is comprised of the between ($V_{Between}$) and within (V_{Within}) variances. Therefore,

$$SS_T = SS_B + SS_W \quad \text{and} \quad SS_W = SS_T - SS_B$$

I will introduce a simple (hypothetical) set of findings for my noise–learning experiment to demonstrate how to calculate these components. Then we can explore an example with real data. Table 12.1 shows hypothetical data for the experiment. The dependent variable is number of learning errors.

Table 12.2 shows the data with appropriate squares to be used to calculate the sums of squares. The general SS formula follows the data; then the calculations are shown for each variance component.

TABLE 12.2 Hypothetical Experiment Data with Squares

	Control		30 Decibels		60 Decibels		90 Decibels	
	X_1	X_1^2	X_2	X_2^2	X_3	X_3^2	X_4	X_4^2
	5	25	7	49	12	144	15	225
	4	16	5	25	11	121	14	196
	3	9	6	36	10	100	18	324
	6	36	8	64	12	144	12	144
$\sum$	18	86	26	174	45	509	59	889

$$SS = \sum X^2 - \frac{(\sum X)^2}{N}$$

Calculating SS$_T$

$$SS = \sum X^2 - \frac{(\sum X)^2}{N}$$

$$SS_T = (86 + 174 + 509 + 889) - \frac{(18 + 26 + 45 + 59)^2}{16}$$

$$SS_T = 1658 - 1369$$

$$SS_T = 289$$

Calculating SS$_B$. Calculating SS$_B$ uses a variation of the SS formula, but the focus is on the *deviation of each group mean from the grand mean*. Therefore, we will calculate "group means" in variance terms and subtract the variance measure of the grand mean. Here is the revised formula:

$$SS_B = \frac{(\sum X_1)^2}{n_1} + \frac{(\sum X_2)^2}{n_2} + \frac{(\sum X_3)^2}{n_3} + \frac{(\sum X_4)^2}{n_4} (-) \frac{(\sum X)^2}{N}$$

This equation looks intimidating, but if you look at it, the main "pieces" make sense. As you can see, a *global measure* of each group's mean $((\sum X)^2/n)$ is divided by the group's sample size. Then, the grand mean measure $((\sum X)^2/N)$ is subtracted from the total of the separate group mean measures. I shaded this portion of the equation so you can see the components clearly. There is an important thing to note in this equation:

$$n \neq N$$

If you look carefully at the equation, you will note that the separate group mean measures have the "group sizes" (n) as the divisor. The grand mean measure at the end of the equation has the total individual score size (N) in the divisor. This is because the grand mean is calculated for all individual scores ($N = 16$) whereas the individual group mean measures are divided only by the scores that make up that group (all four sample groups have $n = 4$).

$$SS_B = \frac{(\Sigma X_1)^2}{n_1} + \frac{(\Sigma X_2)^2}{n_2} + \frac{(\Sigma X_3)^2}{n_3} + \frac{(\Sigma X_4)^2}{n_4} (-) \frac{(\Sigma X)^2}{N}$$

$$SS_B = \frac{324}{4} + \frac{676}{4} + \frac{2025}{4} + \frac{3481}{4} (-) \frac{(18 + 26 + 45 + 59)^2}{16}$$

$$SS_B = 81 + 169 + 506.25 + 870.25 (-) 1369$$

$$SS_B = 257.5$$

As you can see, I left the shaded portion in the equations to show that this portion is the same in the SS$_T$ and SS$_B$ formulas. It "represents" the grand mean, so its value will not change, just the way it is used. In the SS$_T$ formula, all the individual scores (X) deviate away from the (grand) mean. Thus, the shaded portion of the

CALCULATING ANOVA

formula shows that this is the value from which all the scores deviate. In the SS_B formula, the shaded portion represents the grand mean as well. Here, all of the sample group means are added together and then subtracted from the grand mean. It is as if each of the group means is treated as a single score, and we are calculating the variance of the group means around the grand mean (which, of course, we are actually doing!).

Calculating SS_W. From the above equation, we know that if we have calculated SST and SSB, we can derive SSW without additional calculations:

$$SS_W = SS_T - SS_B$$
$$SS_W = 289 - 257.5$$
$$SS_W = 31.5$$

Creating a Data Table. Now that we have calculated the key SS values, we can use a table to display the results. As you will see in the steps ahead, this is a very helpful visual step, and it is one that both Excel® and SPSS® use to display the ANOVA results. You should therefore learn to use the table so you can keep all the calculated values in order and so that you can understand how ANOVA results are reported by most all statistical software.

Table 12.3 shows the template we will use with our results thus far. As you can see, we have calculated the sums of squares for each "source" of variance. Remember that the SS is a measure of variance based on the raw scores we have in our data file.

Using Mean Squares (MS)

As you recall, inferential statistics addresses the issue of how sample values represent (or do not represent) population values. In the Z and T tests, we learned to transform sample values to points on a sampling distribution so that we could show how our calculated sample ratio compares to all possible sample values.

We need to treat ANOVA in the same way. *It is also an inferential process, so we will use sample values to estimate population values.* We calculated the sample variances using SS values, so we need to use these values to estimate population variances. This is akin to our using sample values (SD) to create an estimate (s_x) of

TABLE 12.3 The ANOVA Results Table

Source of Variance	SS	df	MS	F Ratio
Between	257.5			
Within	31.5			
Total	289			

the population standard deviation (σ_X) in the T test. In order to estimate the population standard deviation, we had to use the sample standard deviation and "adjust" it by dividing it by its degrees of freedom (in the single-sample T test, the df $= n - 1$) so that it would be a better estimate of the population parameter.

With ANOVA, we will use degrees of freedom to make each of our SS values better estimates of population values. These estimates are known as "mean squares" because they are created by dividing the SS by degrees of freedom, much like we create a mean by adding together the values in a sample and divide by the number in the sample.

Degrees of Freedom in ANOVA

We can use the degrees of freedom to help estimate population values because the size of the samples greatly affect the variance estimates. By using degrees of freedom in order to obtain better estimates, we will create mean squares that average out the variance regardless of sample size. Here are the degrees of freedom for each SS value:

- df_t (Degrees of freedom for total SS) $= N - 1$ ($N =$ the total number of individual scores in all groups combined).
- df_b (Degrees of freedom for between SS) $= k - 1$ (where $k =$ number of sample groups)
- df_w (Degrees of freedom for within SS) $= N - k$

When we make specific population estimates from sample values, we use ("lose") degrees of freedom. As you see, estimating the total variance uses 1 df because we are using the entire set of individual scores. Estimating the between variance uses another 1 df because we are using the set of sample means. Estimating the within variance uses several more (depending on how many sample groups there are) degrees of freedom, since the sample groups are separate "sets" of scores that deviate around their respective group means.

Calculating Mean Squares (MS)

Calculating MS values is simple when you look at our ANOVA table of values. Table 12.4 shows the MS values that are derived from dividing the SS values by their respective degrees of freedom.

TABLE 12.4 The ANOVA Results Table with Calculated MS Values

Source of Variance	SS	df	MS	F Ratio
Between	257.5	3	85.83	
Within	31.5	12	2.625	
Total	289	15	—	

CALCULATING ANOVA

You can derive the MS values directly from the table by dividing the SS values by their df values as follows:

$$MS_B = \frac{SS_b}{df_b}, \quad MS_B = \frac{257.5}{3}, \quad MS_B = 85.83,$$

$$MS_w = \frac{SS_w}{df_w} \quad MS_w = \frac{31.5}{12} \quad MS_w = 2.625$$

You can note other features from Table 12.4:

- The between and within SS values sum to the total SS value
- The between and within df values sum to the total df value

The *F* Ratio

You can see from Table 12.4 that I did not include the MS_T value. That is because we do not need it for our ANOVA analysis! In an earlier section of this chapter, I noted that the ANOVA process determines *whether the sample means vary far enough from the grand mean that they could be said to be from the different populations*. What we need in order to determine this is a comparison of the variation of the means from the grand mean (MS_B), to the variance within each of the sample groups that make the earlier estimate difficult to determine (MS_W).

Here is the formula for calculating the *F* ratio. Using the values from Table 12.4, I showed the actual, calculated *F* ratio from our example problem.

$$F = \frac{MS_B}{MS_W}, \quad F = \frac{85.83}{2.625}, \quad F = 32.70$$

As you can see, we are simply comparing the variation of the sample means around the grand mean (MS_B) to the variation within each sample group (MS_W). I mentioned before that the within variance measure is known as "error" just for this reason. If the sample group values did not vary at all, it would be easy to see how far away the sample means vary from the grand mean! Consider the two panels in Figure 12.5.

The top panel of Figure 12.5 shows a hypothetical example of four sample groups with almost no within variance. In this case, it is easy to see how distinct the group means are and how clearly they differ from the grand mean. The bottom panel shows just the opposite. There is so much within variance that it is almost impossible to see how distinct the group means are and how they vary around the grand mean. The within variance is essentially "noise" and is therefore called error.

It is for these reasons that I said that the *F* ratio is a measure of the between variation *relative to the within variance*. That is why the *F* ratio is simply

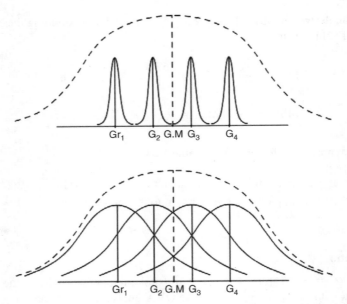

FIGURE 12.5 ANOVA possibilities of groups with different within variances.

comparing the between to the within variation. When you divide the two, you would anticipate a much higher ratio in the top panel than in the bottom panel because the divisor in the F ratio would be much smaller in the top panel.

In our example, the between variance measure was 85.83 relative to the within measure of 2.625. These represent quite differing variance measures! It is fairly easy to assume that the resulting F ratio (32.70) represents a situation where the sample group means are quite spread out around the grand mean. The question is, How far from the grand mean do the sample means have to vary before they could be said to represent different populations? This is the essence of the ANOVA hypothesis test.

Table 12.5 shows the final ANOVA table with the F ratio included. The structure of the table allows you to "see" how the calculation for the F ratio was performed. The between MS (85.83) was divided by the within MS (2.625) to yield 32.70.

TABLE 12.5 The Final ANOVA Results Table

Source of Variance	SS	df	MS	F Ratio
Between	257.5	3	85.83	32.70
Within	31.5	12	2.625	
Total	289	15	—	

The F Distribution

If you refer to Figure 11.13, you will see that the F distribution does not appear to be similar to the Z or T distributions. Remember I mentioned in Chapter 11 that the F distribution is a sampling distribution of *variances*, not means. We saw in the F test for equal variance that most group variances will hover around 1 when the variances are equal (because if we divide similar variance measures the ratio will be close to 1.00).

In the ANOVA test, we are not comparing two group variances, but between to within variance. When we compare these measures, the resulting F ratio is often much larger than 1. Thus, *the F distribution is directional*. Together with the fact that variance measures cannot be negative (they can be very small, but not negative), the researcher needs to compare a calculated F ratio with a tabled value of F that represents all possible samples of variance.

The F distribution is therefore our sampling distribution of comparison. F tables have been created to identify the exclusion values at 0.05 and 0.01 probability, for example. The table is complex because it takes into account the degrees of freedom of both the between and the within measures since sample size is so critical to the analyses.

In our example, the tabled value of F at 0.05 and df's for between (3) and within (12) variance measures, the exclusion value is 3.49. Thus, 3.49 is the value on the F distribution that defines the exclusion area. If our calculated value exceeds this value, we can reject the null hypothesis. In our example, the calculated F equals 32.70, which far exceeds the tabled value of F. Therefore, we can conclude that our finding (32.70) falls in the exclusion area of $p < 0.05$ (3.49) and is therefore not likely a chance finding (i.e., it is a statistically significant finding).

If our example values represented the data in my experimental study, we could reject the null hypothesis. This would mean that, taken together, the means of the group noise levels vary so greatly around the grand mean that the groups represent different populations of people.[1]

EFFECT SIZE

Like our other statistical analyses, we can create an effect size that shows the "impact" of the independent (predictor) variable on the dependent (outcome) variable. In the case of our example, the effect size concern could be asked as, How much does the *grouping on the independent variable* affect the outcome measure?

---

[1] By the way, the actual outcome of my experiment was interesting! I found no significant differences between the groups. My speculation as to why was that I was using white noise, rather than "real" noise (like different kinds of music, street noise, people yelling, etc.), I didn't take into account what the subjects were used to (that is, they were all different coming in to the experiment; some were used to studying with music, others weren't, etc.), and they were all trying really hard to do well on the "test." The research literature on this issue is very interesting. My undergraduate experiment as yet has not won me any prizes for original research!

We are trying to determine how much noise impacts learning by measuring how much the specific noise groupings (0, 30, 60, and 90 decibels) push the outcome measure apart.

So far, we have discussed effect size in terms of Cohen's d measures where we analyzed the difference between two groups (independent T test) or between a group and a population (single-sample T test). The ANOVA determines the significance of group differences by examining variance, so we can use a variance measure to express effect size.

η^2 is the symbol that refers to effect size in ANOVA. It is the Greek lowercase eta. *It refers to the proportion of variance in the outcome measure explained by the grouping on the independent variable.* It is easy to calculate from the ANOVA table from the following formula:

$$\eta^2 = \frac{SS_{Between}}{SS_{Total}},$$

$$\eta^2 = \frac{257.5}{289},$$

$$\eta^2 = 0.89$$

Thus, 89% of the variance in learning errors is due to assigning subjects to our four different noise levels! The effect size is very large (only 100% is possible, obviously), but this is probably because we have only groups of size four, and this is a hypothetical example.

How large is large? That is, how large does η^2 have to be before you would say it is meaningful? That is a question for which there are many answers! The reason it has many answers is that it depends on the sample sizes, number of groups, and so on. I alluded to power analysis in earlier chapters. Statisticians have constructed tables that take into account factors that would have an impact on the size of the effect size value. For example, there are tables for the 0.05 level of significance, as well as for the group size as considerations in judging effect size (see Cohen, 1988).

There are therefore many standards offered by statisticians and researchers for judging effect size meaningfulness. I tend to take two approaches to this question. First, I can suggest the following as benchmark comparisons: 0.01 (small), 0.06 (medium), and 0.15 (large). These are ballpark figures because the next approach is the most crucial.

The second approach to the question is allowing the researcher to judge the meaningfulness of the effect size by the nature of the problem studied. It would therefore have a lot to do with the subject. If I am studying a hypothetical research question with very small sample sizes (as I did in the noise–learning example), I would not be very excited about a very large effect size (and in our study, 0.89 far exceeded the 0.15 guideline for "large effect size"). However, if I am studying a new drug that can lessen the death rate from AIDS, I would be ecstatic to find an effect size of 0.03 even though it would be judged "small" by the earlier benchmarks.

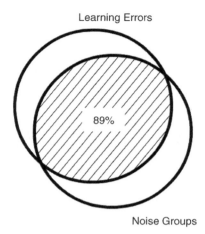

FIGURE 12.6 Venn diagram showing effect size.

One way to visually understand effect size as "explained variance" is to use Venn diagrams. Look at Figure 12.6. The top circle represents the outcome measure of learning errors. The bottom circle represents the noise groups. Effect size is represented in the portion of the top circle with diagonal lines. The 89% refers to our η^2 value of 0.89. If the top circle represents all the variation of learning errors, we can reduce that variance by 89% just by knowing the different noise conditions! There are many things that probably result in learning errors, but in our small study we have concluded that the different noise conditions reduced our lack of knowledge by 89%!

POST HOC ANALYSES

The question after we reject the null hypothesis is, What specific group mean variation(s) from the grand mean might be responsible for the overall size of the F ratio? Stated differently, are all the sample means similar in producing learning errors (obviously not since we rejected the null hypothesis), or are some group means much more responsible for affecting the F ratio? What configuration of group mean differences affected the overall F ratio?

Figure 12.7 shows two possibilities of group differences that might produce a significant F ratio. The top panel shows that the first three noise level groups are similar in producing learning errors, but the subjects in the fourth noise level are much more likely to produce learning errors than the other three. The bottom panel of Figure 12.7 shows that the first two groups are similar, but both are very different from the third and fourth groups (which are similar to one another). There are many other possible combinations of findings!

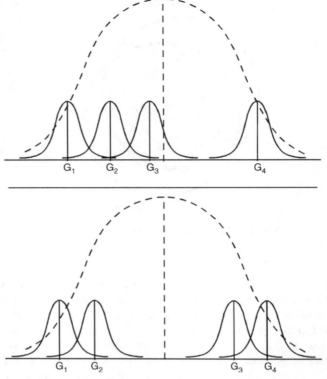

FIGURE 12.7 *Post hoc* test possibilities.

"Varieties" of *Post Hoc* Analyses

Post hoc analyses look at group comparisons within the overall ANOVA (significant) result to detect where sample group comparisons may be so large that they cannot be said to belong to the same population. The process is similar to individual T tests of pairs of samples within the overall ANOVA study. If you recall, I mentioned earlier in this chapter that our ANOVA would consist of six separate T tests if we wanted to conduct paired comparisons. Because we could not do this without creating familywise error, we performed the ANOVA test that compared all the sources of variance at once.

After the overall (omnibus) test of ANOVA is found to be significant, we must "return" to our initial question about which paired differences may be responsible for the ANOVA result. In order to avoid familywise error with the paired comparisons, statisticians have created processes for identifying paired sample differences that limits the overall error measures. Many are specially designed to fit specific research designs. Here are some examples of procedures used for "unplanned" comparisons (paired comparisons not identified before the study).

- *Scheffe Test*—A conservative *post hoc* analysis that works well with uneven group sizes and uses stricter decision criteria.
- *Tukey's HSD*—This Honestly Significant Difference procedure is used when the researcher wants to compare all possible pairs of samples.
- *Dunnett*—This test is used only with single comparisons, usually comparing a control group and experimental group(s).

There are several other *post hoc* analyses available for use, depending on the nature of the research issues. I will demonstrate the most common *post hoc* analysis for unplanned comparisons, the Tukey's HSD. Statistical programs such as SPSS® provide this and other *post hoc* analyses, so you can expand your range of understanding and expertise of *post hoc* analyses as you pursue further research.

The *Post Hoc* Analysis Process

The general process that researchers follow after concluding that the omnibus (overall) ANOVA test is significant is as follows:

1. Calculate the Tukey HSD critical value.
2. Create a comparison table that includes all the group means.
3. Compare each mean difference to the critical HSD value to determine which of the pairs are significantly different from one another.
4. Conclude the *post hoc* analysis with a general summary of results.

Tukey's HSD (Range) Test Calculation

This test is also called Tukey's Range Test because it requires the use of the "Studentized Range Table" to help identify a critical value of comparison. In essence, the HSD *is a critical value of exclusion* because it is based on a set of probabilities that define extreme values on a distribution that takes into account the degrees of freedom in the overall study and the number of groups in the analysis. Thus, *we use a formula to determine the value beyond which the paired group differences in the study would be considered extreme.*

The Tukey HSD formula is as follows:

$$\text{HSD} = q_{(\text{Range})} \sqrt{\frac{\text{MS}_w}{n}}$$

The "pieces" of the formula are as follows:

- HSD is the calculated point on the comparison distribution that identifies extreme values; any test value that is calculated larger than this value is considered extreme and would result in the rejection of the hypothesis that a pair of group means was equal.

- $q_{(Range)}$ is the value obtained in the Studentized Range Table that helps to establish the exclusion value. I will use the example of the 0.05 level of exclusion, but you can establish the value for 0.01 or other levels of exclusion. If I do not specify otherwise in the examples that follow, you can assume I am using the 0.05 level.
- MS_w is the mean square within value found in the ANOVA table. In our example above, it is equal to 2.625. Recall that I mentioned that the ANOVA within measures were known as "error." Thus, in the formula, the error measure is "adjusting" the critical value of exclusion. If the error is large, it will have a marked impact on the exclusion area, and so on.
- n is the group size used in the comparison(s). In our example, all the groups were $n = 4$. Thus, we can use this formula to establish one HSD critical value for each of our six paired comparisons. If the group size is unequal, there is a formula for Tukey's HSD that identifies the group sizes for each paired comparison. Therefore, if you have unequal group sizes, you would need to create more than one HSD value for the planned comparison. The SPSS® program adjusts for unequal group sizes.

Using the Range Table is straightforward if you remember that it is adjusted based on the ANOVA error measure (within groups MS). Table 12.6 shows a small part of a Range Table to give you an example of how to use it.

As you recall, we had four noise groups in the experiment, and the MS_W degrees of freedom was 12 (see Table 12.5). I shaded the relevant columns and rows in the example table in Table 12.6. The number of sample groups is the "4" column, and the MS_W df is the "12" row. Where these intersect is the $q_{(Range)}$ value we will use in the HSD calculation.

$$HSD = q_{(Range)} \sqrt{\frac{MS_w}{n}},$$

$$HSD = 4.20 \sqrt{\frac{2.625}{4}},$$

$$HSD = 3.40$$

TABLE 12.6 Example of Values from a Studentized Range Table

MS_W df	Number of Sample Groups						
	2	3	4	5	6	7	8
11	3.115	3.822	4.258	4.575	4.824	5.03	5.203
12	3.083	3.775	4.200	4.509	4.752	4.951	5.12
13	3.057	3.736	4.152	4.454	4.691	4.885	5.05

POST HOC ANALYSES

TABLE 12.7 The Group Means

Groups	Means
M_1	4.50
M_2	6.50
M_3	11.25
M_4	14.75

Means Comparison Table

Once the HSD critical value has been calculated, we can proceed to examine how the group means differ. If you refer to Figure 12.2, you will see the six group comparisons that are necessary when there are four sample groups. In order to see what the group differences are among the groups, it is a simple matter of subtracting the group means. Table 12.7 lists the means of each of the four sample groups.

We can calculate the six paired sample differences by subtracting each pair of means as follows:

$$M_1 - M_2 = -2.00$$
$$M_1 - M_3 = -6.75$$
$$\vdots$$
$$M_3 - M_4 = -3.50$$

Doing the comparisons this way would yield the appropriate paired mean differences. Note in the list, however, that if we subtract the higher group means (e.g., M_4) from the lower (e.g., M_1), the values are negative. In these paired comparison tests, we are interested only in the magnitude of the difference, not the sign. We can interpret the overall results by referring to the size of the means, but the paired tests do not use the sign. Therefore, you can subtract the smaller group means from the larger to get positive values.

Another method for doing this is to create a "matrix" of group means. A matrix is simply a table constructed of rows and columns. We will see arrays of data in matrix form later in the book, so it is a good idea to use them to display data. SPSS® uses matrix arrays for some output that we will examine (including the *post hoc* analysis).

Table 12.8 shows the group means in a matrix display. Above, we calculated the difference of M_1 and M_3 to be -6.75. I placed this value in the

TABLE 12.8 Matrix of Group Means

	M_1 (4.5)	M_2 (6.5)	M_3 (11.25)	M_4 (14.75)
M_1 (4.5)	—	2.00	6.75	10.25
M_2 (6.5)		—	4.75	8.25
M_3 (11.25)	−6.75		—	3.50
M_4 (14.75)				—

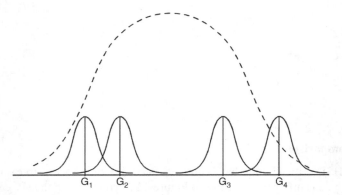

FIGURE 12.8 The *post hoc* summary for the example.

matrix under column M_1 and row M_3. Thus, when I subtract 11.25 from 4.5, I get −6.75.

However, I can obtain positive values by subtracting the lower mean values from the higher. I have shown these values in the table. For example, subtracting 4.5 from 11.25 results in a value of 6.75, which is located in the M_3 column and the M_1 row. Note that the matrix is symmetrical. The same values will appear in the lower "triangle" that appear in the upper 'triangle' (I shaded the upper triangle) in Table 12.8. I left the −6.75 in both the lower and upper triangles to show the symmetry of the table.

Compare Mean Difference Values from HSD

The next step in the post hoc analysis is to compare the group mean differences (from Table 12.8) to our HSD value of 3.40. If any of the group mean differences exceed the HSD value, we would consider those two means *significantly* different. As you can see from the table, all the mean pairs exceed HSD except the $M_1 - M_2$ difference (2.00).

Post Hoc Summary

Our analysis showed that all paired comparisons showed significant differences except the group 1 (control group)–group 2 (30 decibel) comparison. Figure 12.8 shows how the groups might be arrayed to yield this finding.

ASSUMPTIONS OF ANOVA

As with the other statistical tests we discussed, the researcher needs to assess whether the conditions of the data are appropriate for the ANOVA test. Here are the primary assumptions, although ANOVA is also somewhat robust with respect to slight variations of assumptions:

1. Population is normally distributed.
2. Population variances are equal. This assumption refers to the variance of the sample groups. You should make sure that the groups (in our case the noise conditions) have equal variance on the outcome measure (in our case, learning errors). (This is the homogeneity of variance assumption.)
3. Samples are independently chosen.
4. Interval data on the dependent variable.

ADDITIONAL CONSIDERATIONS WITH ANOVA

We will have more to say about these procedures, but here are some things to keep in mind as you think about the One-Way ANOVA.

1. You can still use ANOVA ("the F test") even if you only have two group means. In this case, remember that there is a relationship between F and T such that $F = T^2$.
2. The effect size is critical, as we have learned with the other tests we have covered. We learned to calculate η^2, which indicates the strength of the effects of F. Effect size is always calculated, even if the omnibus F test does not result in the reject of the null hypothesis. This is important to remember! Just because a study finding is or is not statistically significant does not indicate how much of a *practical impact* it has.
3. ANOVA, like the T test can be used in experimental contexts (where we consciously change the value of an independent variable to see what effects this has on a dependent variable) and for *post facto* situations (where we simply try to determine if there are existing differences among several sample groups without changing the value of the independent variable). In our example of noise and learning, I used ANOVA experimentally, since I directly "manipulated" or changed the value of the independent variable (noise level), to see if this affected the dependent variable (errors in learning performance).
4. Be sure to meet the assumptions of the test before you use it:
   - Populations are normally distributed.
   - Population variances are equal.
   - Independent selection.
   - Interval data on the dependent variable.

A REAL-WORLD EXAMPLE OF ANOVA

Shifting from a hypothetical to a real example of ANOVA, I will return to our state database showing schools with various percentages of students passing the math

TABLE 12.9 The Data for ANOVA Example

		Frequency	FR Percent	Valid Percent	Cumulative Percent
Valid	1—Low	27	45.8	45.8	45.8
	2—Medium	18	30.5	30.5	76.3
	3—High	14	23.7	23.7	100.0
	Total	59	100.0	100.0	

and reading assessments. In the extended example that follows, I will use Excel® and SPSS® to analyze the question, Does the percentage of low-income students in a school have an effect on school-based achievement levels? This is an important question because family income, while important, has not been the subject of a great deal of scrutiny as a potential factor in influencing achievement.

Our database cannot provide individual level student data, so we cannot analyze the individual impact of family income on students' achievement. However, we can study the question when it is viewed at the school level. In the latter case, we can use the database from the state of Washington, which shows the percentages of students who pass the math (and other subject) assessment and the percentage of students who qualify for free/reduced price meals (FR). Researchers use the FR variable as a way (often, the only way) to indicate low income.

I will use a dataset that is randomly chosen from all the schools in the state with fourth-grade classes ($N = 1232$). I called for a 5% sample within categories of low, medium, and high percentages of FR (using 33.33% and 66.67% to create the three groups). This method retains the proportion of the free reduced lunch groupings. The final sample ($N = 59$) only includes schools that provided achievement results for math and reading. Table 12.9 shows the data sample.[2]

ARE THE ASSUMPTIONS MET?

Before we proceed to the hand calculations for this example problem, we need to look at the assumptions for ANOVA. We need to determine whether the data are appropriate for the analysis we have planned.

1. Population is Normally Distributed?

As you recall from Chapter 5, I mentioned that skewness and kurtosis are good indicators to help assess normal distributions. For ANOVA, we need to make sure that the outcome data are normally distributed for each of the independent variable levels. In this case, we have three independent variable levels: low, medium, and high FR. Are the achievement data normally distributed? We can use the results in Figure 12.9 to test this assumption.

---

[2] I used a process similar to the one in Real-World Lab VII for creating FR groups. Refer to that discussion for creating the study datafile shown in Table 12.10.

ARE THE ASSUMPTIONS MET?

	Low	Medium	High
Mean	66.24	50.16	34.22
Standard error	3.40	3.93	3.87
Median	70.30	49.05	34.75
Mode	#N/A	#N/A	#N/A
Standard deviation	17.66	16.69	14.47
Sample variance	312.01	278.69	209.37
Kurtosis	2.98	−0.45	−0.53
Skewness	−1.47	0.19	−0.13
Range	80.10	59.80	48.60
Minimum	9.70	24.40	11.40
Maximum	89.80	84.20	60.00
Sum	1788.40	902.80	479.10
Count	27.00	18.00	14.00

FIGURE 12.9 The descriptive output (Excel®) to test the normal distribution assumption.

The Excel® output in Figure 12.9 indicate that all three groups are fairly normally distributed as determined by the similarity of mean and median values, and skewness and kurtosis figures. The Low group may possibly be a bit negatively skewed and leptokurtic judging by the skewness and kurtosis values. We can use SPSS® to investigate further. Figure 12.10 shows the descriptive values that include skewness and kurtosis standard errors, and Figure 12.11 shows the graphs for the groups.

As you can see, skewness for the Low group is generally within bounds when you divide the skewness value by its standard error ($-1.469/0.448 = 3.28$), although it does exceed the "2 to 3" benchmark I mentioned in Chapter 5. The Low group also appears to be leptokurtic by the same criteria ($2.981/.872 = 3.42$). The graphs in Figure 12.11 help to show these assessments.

If the researcher determined the Low group's skewness and kurtosis values were excessive, they could make a number of choices:

- Choose not to use ANOVA.
- Transform the outcome measure so it conforms to normal bounds.
- Examine and possibly eliminate the outliers that may change the distribution configuration.
- Proceed with the analysis and note the concerns in the conclusions.

MathPercentMetStandard

FR	Mean	N	Standard Deviation	Kurtosis	Standard Error of Kurtosis	Skewness	Standard Error of Skewness	Median
Low	66.24	27	17.664	2.981	0.872	−1.469	0.448	70.30
Med.	50.16	18	16.694	−0.450	1.038	0.190	0.536	49.05
High	34.22	14	14.470	−0.527	1.154	−0.128	0.597	34.75
Total	53.73	59	20.908	−0.742	0.613	−0.281	0.311	54.10

FIGURE 12.10 SPSS® descriptive output for normal distribution assumption: mathPercentMetStandard.

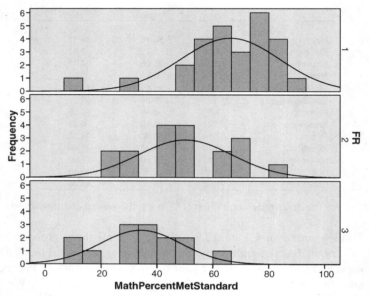

FIGURE 12.11   SPSS® graphs for FR groups.

Even though the skewness and kurtosis numbers are a bit high, I would proceed with the study. As it is, the coefficients are just at the highest bounds, but not overly excessive. At the end of the analysis section, I will include a comment that addresses the need for "diagnostic" analyses for study databases to help with these kinds of decisions. For illustration purposes, we will proceed with the analysis.

By the way, I used the "Analyze–Compare Means–Means" menus in SPSS® to generate the values in Figure 12.10. I could have use the "Analyze–Descriptive Statistics–Frequencies" menus as I have in past chapters, but I wanted to show this useful function. Figure 12.12 shows the menus and choices for this procedure. On the "Means" menu, you list the dependent variable and specify what groups of the dependent variable (FR in this example) to assess. As you can see, I specified means, kurtosis, skewness, standard errors, and so on, for the FR groups.

2. Are Variances Equal?

This second assumption is important for ANOVA because it relies on comparisons of variance measures. Excel® does not have a good way to assess equal variances for multiple groups. As we saw in Chapter 11 with the T test, Excel®'s "F-Test Two-Sample for Variances" analysis tool is very helpful for two sample groups, but not more than two. In Chapter 11, I showed the SPSS® Levene's Test output that is included with the T test. We can use the same process to test for equal

HAND CALCULATIONS

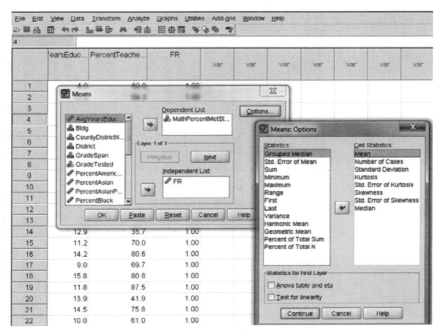

FIGURE 12.12 The SPSS® Means procedure.

variances in our sample groups. In the sections that follow, we will examine the Levene's Test results.

3. Samples are Independently Chosen?

This assumption is met by virtue of schools being randomly chosen and not placed in groups due to any structurally linking criteria.

4. Interval Data on the Dependent Variable?

This assumption is met. Percentages are interval data.

HAND CALCULATIONS

Before we proceed to Excel® and SPSS®, we can perform the calculations by hand for comparison. Because of the magnitude of the numbers (they are all in percentages), I prepared an Excel® spreadsheet showing the data and providing the key sums for the ANOVA calculations. Table 12.10 shows the data and the sums.

TABLE 12.10 The ANOVA Example Database with Key Values

	Low FR%		Medium FR%		High FR%	
	X_1	X_1^2	X_2	X_2^2	X_3	X_3^2
	9.70	94.09	24.40	595.36	11.40	129.96
	30.80	948.64	25.60	655.36	12.50	156.25
	50.00	2500.00	27.30	745.29	15.00	225
	50.80	2580.64	33.30	1108.89	27.90	778.41
	54.10	2926.81	41.70	1738.89	28.50	812.25
	54.40	2959.36	44.40	1971.36	33.30	1108.89
	55.80	3113.64	44.60	1989.16	34.10	1162.81
	55.90	3124.81	46.20	2134.44	35.40	1253.16
	63.20	3994.24	49.00	2401.00	35.60	1267.36
	64.70	4186.09	49.10	2410.81	42.30	1789.29
	64.80	4199.04	50.70	2570.49	46.30	2143.69
	65.00	4225.00	51.10	2611.21	46.80	2190.24
	65.80	4329.64	61.60	3794.56	50.00	2500
	70.30	4942.09	63.00	3969.00	60.00	3600
	70.60	4984.36	67.20	4515.84		
	72.10	5198.41	69.10	4774.81		
	73.90	5461.21	70.30	4942.09		
	76.70	5882.89	84.20	7089.64		
	78.30	6130.89				
	79.10	6256.81				
	79.30	6288.49				
	79.60	6336.16				
	82.40	6789.76				
	82.60	6822.76				
	83.70	7005.69				
	85.00	7225.00				
	89.80	8064.04				
Sums	1788.40	126,571	902.80	50,018	479.10	19,117
ΣX^2		126,571		50,018		19,117
$(\Sigma X)^2$	3198,375		815,048		229,537	

We can now use the formulas above to "populate" our ANOVA summary table. The calculations for SST, SSB, and SSW follow. Table 12.11 shows the values in the completed ANOVA summary table.

HAND CALCULATIONS

TABLE 12.11 The Completed ANOVA Summary Table for the Extended Example

Source of Variance	SS	df	MS	F Ratio
Between	9782	2	4891	17.59
Within	15,571	56	278	
Total	25,353	58	—	

Calculating SS_T ($SS_T = 25{,}353.49$)

$$SS = \sum X^2 - \frac{(\sum X)^2}{N}$$

$$SS_T = (126{,}571 + 50{,}018 + 19{,}117) - \frac{(1788.4 + 902.8 + 479.1)^2}{59}$$

$$SS_T = 195{,}706 - 170{,}352.6$$

Calculating SS_B ($SS_B = 9782$)

$$SS_B = \frac{(\sum X_1)^2}{n_1} + \frac{(\sum X_2)^2}{n_2} + \frac{(\sum X_3)^2}{n_3} (-) \frac{(\sum X)^2}{N}$$

$$SS_B = \frac{3{,}198{,}375}{27} + \frac{815{,}048}{18} + \frac{229{,}537}{14} (-) \frac{(1788.4 + 902.8 + 479.1)^2}{59}$$

$$SS_B = 118{,}458 + 45{,}280 + 16{,}395 (-) 170{,}352.6$$

Calculating SS_W ($SS_W = 15{,}571$)

$$SS_W = SS_T - SS_B$$
$$SS_W = 25{,}353.49 - 9782$$

The Hypothesis Test

- Null Hypothesis: $H_0 : \mu_1 = \mu_2 = \mu_3 = \mu_4$
- Alternate Hypothesis: $H_A : \mu_1 \neq \mu_2 \neq \mu_3 \neq \mu_4$
- Critical Value of Exclusion: $F_{(.05, 2, 56)} = 3.16$
- Calculated F: 17.59
- Decision: Reject the null hypothesis. (This test is significant at or beyond $p < 0.05$. (That is, the calculated F ratio was so large it surpassed the 0.05 exclusion value of 3.16 and therefore lies in the extreme portion of the comparison distribution.)
- Interpretation: The omnibus test indicates that the FR groups show different mean percentages of students who pass the math assessment. Schools with low FR percentages show higher percentages of students who pass the math assessment.

Effect Size ($\eta^2 = 0.39$)

The effect size is an easy calculation using the ANOVA summary table values. As the calculation indicates, 39% of the variance in math passing rates among the study schools is explained/affected by the FR groupings. This value far exceeds the 0.15 guideline for a high effect size, but the value of the effect size is contextualized by the research problem. In this case, I would think that one influence explaining almost 40% of schools' math passing rates is amazingly high. We will keep an eye on the FR variable in subsequent analyses.

$$\eta^2 = \frac{SS_{Between}}{SS_{Total}}$$

$$\eta^2 = \frac{9782}{25353}$$

Post Hoc Analysis

Using the Tukey Range Test, we can create the HSD critical comparison value to compare our mean differences. Using the Tukey Range table of values, we can identify $q_{(Range)} = 3.40$. We identified this value in the table by viewing the groups = 3 column and the MS_W df row = 60 because the table did not include the df = 56 row. If we wanted to be additionally conservative we could have used the df = 40 row (which identified the $q_{(Range)} = 3.44$) because this would make the exclusion value more stringent, but the actual df (56) was closest to the df of 60.

$$HSD = q_{(Range)} \sqrt{\frac{MS_w}{n}}$$

$$HSD = 3.40 \sqrt{\frac{278}{20}}$$

$$HSD = 12.68$$

Recall that this formula requires the sample sizes to be equal. Without introducing a formula that accommodates unequal sample size, you can take the average of the sample sizes ($(27 + 18 + 14)/3 = 20$).

Table 12.12 continues the *post hoc* analysis by presenting group mean differences to compare with the HSD value. As you can see, all group mean differences exceed the HSD value of 12.68. Therefore, the groups differ from one another statistically. Figure 12.11 shows how the group means might appear.

Figure 12.13 shows the groups graphed together. You can see that the percent passing rates are highest for the low FR group followed by the medium FR group and finally the high FR group.

USING EXCEL® AND SPSS® WITH ONE-WAY ANOVA

TABLE 12.12 The Group Mean Difference Matrix

	M_1 (66.24)	M_2 (50.16)	M_3 (34.22)
M_1 (66.24)	—	16.08	32.02
M_2 (50.16)		—	15.93
M_3 (34.22)			—

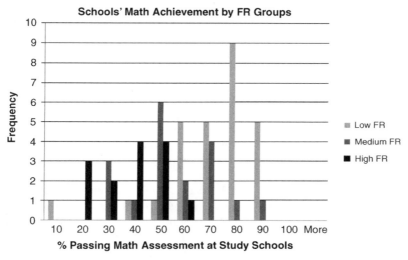

FIGURE 12.13 The three FR group distributions on percent passing math assessments.

USING EXCEL® AND SPSS® WITH ONE-WAY ANOVA

Now that we have used hand calculations to derive the ANOVA results, I will show how to use Excel® and SPSS® to derive the results quickly. We can compare all three sets of results.

Excel® Procedures with One-Way ANOVA

In order to use One-Way ANOVA in Excel®, the data need to be arrayed like the data table in Table 12.10. Show three columns, one for each of the FR groups, and populate the rows with the Math achievement data from the study schools. Simply select the entire set of data, including the labels in the first row, and use the "Data–Data Analysis" menus to bring you to the screen shown in Figure 12.14. As you can see, there are three choices for ANOVA. The One-Way ANOVA uses a single factor (one independent variable) so the top choice is the option we need for our analysis.

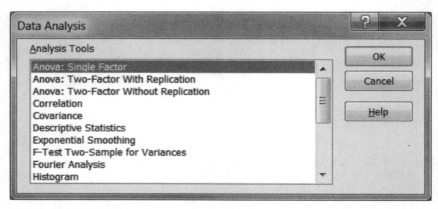

FIGURE 12.14 The Single Factor ANOVA menu option in Excel®.

When we run this operation, Excel® returns the output shown in Figure 12.15. The top panel is the descriptive statistics showing the mean and variance math achievement values for each FR group. The "Average" column will be helpful when we create the mean difference matrix for the *post hoc* analysis if we reject the null hypothesis in the omnibus F test.

The bottom panel of Figure 12.15 shows the ANOVA summary table. This is the same table we created by hand. It shows the sources and measures of variance and the calculated F ratio (17.59). Excel® also reports the exact probability of the F ratio in the "P-value" column. The reported value of "1.18E-06" does not look like our usual probability figures (e.g., $p < 0.05$) because it is written in scientific notation. The E-06 part of the value means to move the decimal place six places to the left of its current position. Thus, the probability value actually is $p = 0.00000118$. Our F ratio thus falls extremely far into the exclusion area and is highly unlikely to be a chance finding. Scientific notation is used as a shorthand mechanism to report numbers with many digits.

Anova: Single Factor

SUMMARY

Groups	Count	Sum	Average	Variance
Low FR	27	1788.4	66.24	312.01
Medium FR	18	902.8	50.16	278.69
High FR	14	479.1	34.22	209.37

ANOVA

Source of Variation	SS	df	MS	F	P value	F crit
Between groups	9781.66	2	4890.83	17.59	1.18E-06	3.161861
Within groups	15,571.83	56	278.07			
Total	25,353.49	58				

FIGURE 12.15 The Excel® single-factor ANOVA output.

The F crit (or, critical value for F) value in the summary table indicates the point value on the F distribution that defines the exclusion area. Our F ratio of 17.59 well exceeds this point, resulting in the extremely small p value.

Note that the Excel® output does not include information about effect size. However, you can calculate eta squared (η^2) easily from the summary table as we did in our hand calculations. Recall that $\eta^2 = 0.39$ calculated as follows:

$$\eta^2 = \frac{SS_{Between}}{SS_{Total}},$$

$$\eta^2 = \frac{9782}{25,353}$$

Because the omnibus test was significant (i.e., we rejected the null hypothesis), we can perform the *post hoc* analysis. However, no provision is made in Excel® to create the *post hoc* analysis. You can perform it as we did in our hand calculations if you are using Excel® as your only statistical software.

SPSS® Procedures with One-Way ANOVA

The one-way ANOVA procedure in SPSS® is easy to create and is thorough in what it reports. Figure 12.16 shows the menu choices that allow the one-way ANOVA procedure. As you can see, it is in the same menu group that includes the single-sample T test and the independent-samples T test that we discussed in previous chapters. This menu also includes the Means procedure we used above to create the descriptive data for the three FR groups.

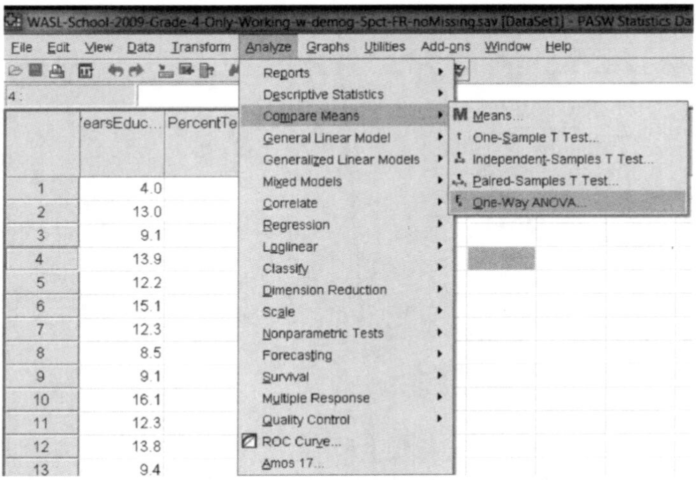

FIGURE 12.16 The SPSS® menu options for accessing the one way ANOVA.

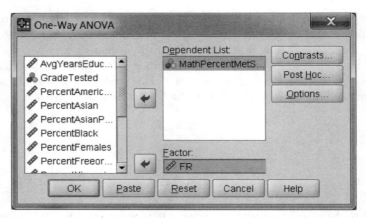

FIGURE 12.17 The One-Way ANOVA specification windows.

When you select the One-Way ANOVA procedure, the window shown in Figure 12.17 appears that allows the researcher to specify the analysis. As you can see, I specified the MathPercentMetStandard variable as the outcome measure, and I included FR as the grouping variable ("Factor"). The three buttons on the top right of the screen can be used for further specification.

Figure 12.18 shows the choices available from selecting the *"Post Hoc"* button from the One-Way ANOVA menu. As you can see, there are several *post hoc* procedures available for inclusion in the analysis. I selected the Tukey procedure to demonstrate the results alongside our hand calculations of the Tukey Range Test. Note that the *post hoc* procedures are "grouped" according to whether equal variances (of the FR sample groups) are assumed or not. There are a few *post hoc* procedures available if you find the variances to be unequal.

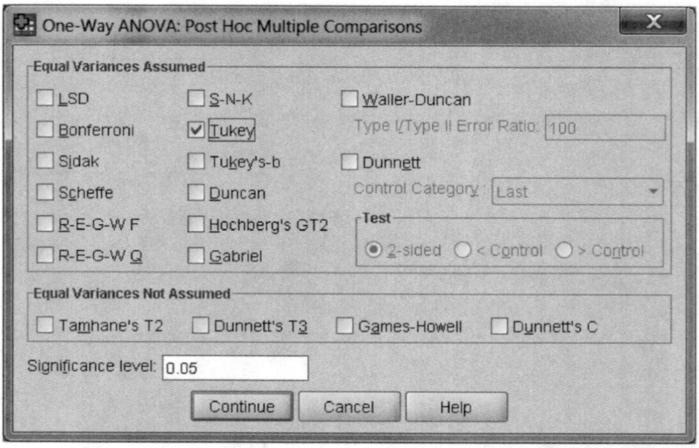

FIGURE 12.18 The *post hoc* choices from SPSS® One-Way ANOVA.

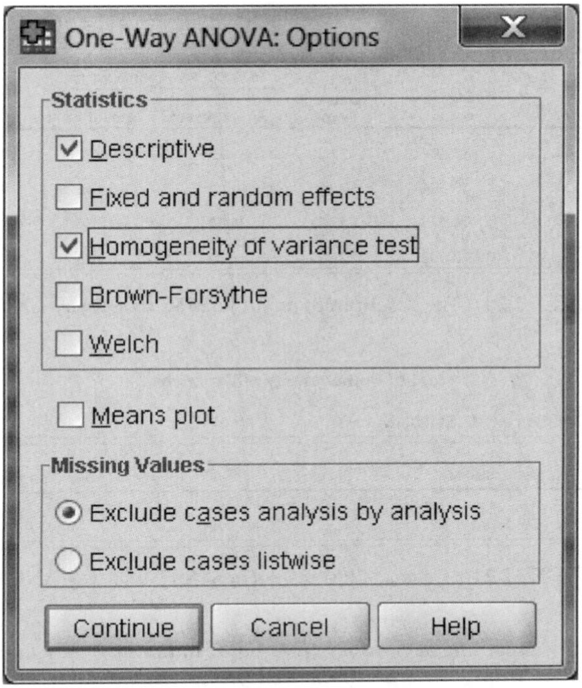

FIGURE 12.19 Options for SPSS® one-way ANOVA.

When you select the "Options" button from the One-Way ANOVA menu, you can select several important procedures that we can use in our analysis. Figure 12.19 shows these choices. As you can see, I chose "Descriptive" procedures to derive the group means and other data, along with the "Homogeneity of variance test" option that will produce Levene's Test for equality of variance. This will allow us to assess the equal variance assumption for one-way ANOVA that we discussed in the earlier section on the assumptions for ANOVA.

The output that results from these specifications is shown in Figures 12.20 through 12.23. The descriptive values are reported in Figure 12.20. You can compare this information with the Excel® report in Figure 12.15.

The next panel in the output is Levene's Test, shown in Figure 12.21.

As you see, Levene's Test was *not significant* because the "Significance" reported is $p = 0.827$. Therefore, the test statistic did not land in its exclusion region or it would have reported a much smaller chance probability (i.e., $p < 0.05$). Since Levene's Test was not significant, the FR groups can be considered to have equal variance in their school math achievement values. That is, because the test was not significant, the null hypothesis of equal variances among sample groups *was not rejected*. This is the finding that confirms that the equal variance assumption is met for one-way ANOVA and that allows the researcher to continue with the ANOVA analysis. *If Levene's Test would have shown a significant result, the*

Descriptives

MathPercentMetStandard

	N	Mean	Standard deviation	Standard error	95% Confidence Interval for Mean		Minimum	Maximum
					Lower Bound	Upper Bound		
1.00	27	66.24	17.664	3.399	59.25	73.22	10	90
2.00	18	50.16	16.694	3.935	41.85	58.46	24	84
3.00	14	34.22	14.470	3.867	25.87	42.58	11	60
Total	59	53.73	20.908	2.722	48.29	59.18	10	90

FIGURE 12.20 The descriptives report in SPSS® one-way ANOVA.

Test of Homogeneity of Variances

MathPercentMetStandard

Levene Statistic	df1	df2	Significance
0.191	2	56	0.827

FIGURE 12.21 Levene's test results in SPSS® one-way ANOVA.

"Significance" value would be smaller than 0.05, indicating that its calculated test value would have fallen in the exclusion area. Under this circumstance, the researcher would need to decide how to proceed because one of the ANOVA assumptions would not have been met.

The next panel of the results is the ANOVA summary table shown in Figure 12.22. This table is identical to the Excel® table (see Figure 12.15), but the "Significance" value is not reported in scientific notation. It is listed as 0.000, indicating that the p value for the F ratio is extremely small! If you change the decimal places for this value, it shows 0.00000118, which matches the Excel® value.

SPSS® does not routinely produce η^2, but you can confirm that the effect size is the same as our hand calculation ($\eta^2 = 0.39$) by dividing the appropriate values from the ANOVA summary table.

The final panel of the SPSS® output is the *post hoc* analysis that we specified as shown in Figure 12.18. The results are shown in Figure 12.23. I highlighted some

ANOVA

MathPercentMetStandard

	Sum of Squares	df	Mean Square	F	Significance
Between groups	9781.661	2	4890.831	17.589	0.000
Within groups	15571.831	56	278.068		
Total	25353.492	58			

FIGURE 12.22 SPSS® one-way ANOVA summary table.

of the values in the first row to explain the output. The first two columns indicate which two FR group means are being compared. According to the shaded values, the first comparison is between FR 1 (Low) and FR 2 (Medium). The difference in the group means (66.24 − 50.16 = 16.08) is shown in the third column ("Mean Difference $I - J$.") and the difference is considered significant as shown by the shaded value (0.007) in the Significance column. Thus, the difference in the group means (16.08) was large enough to fall in the exclusion region of the comparison distribution as determined by the Tukey HSD statistic (not shown). Therefore these two FR groups are significantly different in terms of schools' math achievement results. The remaining comparisons are interpreted in the same fashion.

The standard error of the HSD statistic (5.074 for the first comparison) is reported in the "Standard Error" column, but the calculated HSD value is not shown. Note that you could create confidence intervals around the mean differences as shown in the final two columns.

The Tukey report in Figure 12.23 is arranged a bit differently than our group mean matrix. However, you can see that both negative and positive values are reported, depending on which value is reported first. As with our matrix, the interpretation of the values depends on the context of the research question. In our study, the group means indicate percentages of students in the study schools who passed the math assessment. Therefore, positive values indicate stronger math performance. FR group 1 (Low FR) thus outperforms the other two groups, while FR group 2 outperforms FR group 3.

All the findings indicate that FR is a significant (because we rejected the null hypothesis) and meaningful (due to the magnitude of the effect size) influence on school math achievement. In fact, the results are dramatic. We may wish to quibble a

Multiple Comparisons

MathPercentMetStandard
Tukey HSD

(*I*) FR	(*J*) FR	Mean Difference (*I*−*J*)	Standard Error	Significance	95% Confidence Interval	
					Lower Bound	Upper Bound
1.00	2.00	16.081[a]	5.074	0.007	3.87	28.30
	3.00	32.016[a]	5.492	0.000	18.79	45.24
2.00	1.00	−16.081[a]	5.074	0.007	−28.30	−3.87
	3.00	15.934[a]	5.942	0.026	1.63	30.24
3.00	1.00	−32.016[a]	5.492	0.000	−45.24	−18.79
	2.00	−15.934[a]	5.942	0.026	−30.24	−1.63

[a]The mean difference is significant at the 0.05 level.

FIGURE 12.23 The Tukey *post hoc* results in the SPSS® one-way ANOVA procedure.

bit about the FR (Low) skewness and kurtosis, but the results are fairly clear. Family income (as measured by FR) appears to have a strong affect on school-based math achievement. We will extend our analysis with this variable in subsequent analyses.

THE NEED FOR DIAGNOSTICS

Every researcher knows that real data can be problematic. Not all sample group data conform to perfect distributions. Sometimes, even small discrepancies in the data or data entry errors can have a huge effect on the results of an analysis. Ordinarily, I include a section on diagnostics when I write about statistical procedures [e.g., see Abbott (2010)]. I will talk about these issues as we encounter them in this text, but I strongly encourage you to seek out further, advanced works when you begin to perform actual research with real data. Before you begin, pay attention to the nature of the data, and decide whether you are satisfied that there are no issues that may lead to invalid results. I encourage my students to always simply look at the data. You may be surprised what your eyes pick up that the analyses do not!

As an example of this point, consider our database shown in Table 12.10. The first value of the FR Low group is 9.7. You can see by looking at the other values that this value is quite low. In order to make this point, I eliminated it from the analysis. Simply dropping that single value resulted in the F ratio increasing to 24.98 (from 17.59) and the η^2 increasing to 0.48 (from 0.39)! In addition, the skewness and kurtosis values fell dramatically (both fell to below 1.00).

Eliminating one school from the study assured us that the assumptions for ANOVA were met and the power of the findings was increased substantially. This result shows the influence only one case can have in a study. Therefore, you need to be very careful about the assumptions and the meaningfulness of the data before you begin a study.

There can be a fine line between using diagnostic procedures to eliminate cases that should be eliminated and simply eliminating cases because it makes the results better! I intentionally left the extreme case in the data because it is a real school and these were real results. However, it is not a traditional school, it is an alternative school with nontraditional education for students whose needs cannot be met in a traditional school (e.g., students with behavioral problems).

Here is the key issue: Keep or eliminate cases based on the nature of the study and the research study guidelines. Because the school in my example above may not be typical of all the other schools in the study, I could have eliminated it as being noncomparable. Again, I chose to leave it in based on two things: (1) I wanted to discuss the general point of eliminating cases for diagnostic reasons, and (2) I wanted to point out that *declining to analyze a set of data because the data exceed skewness and kurtosis benchmark values ("from 2 to 3") may not be warranted.* In the current study, if I had eliminated the low case, the group mean would have increased to 68.41 (from 66.24). This would only have further separated the groups from one another. The slight skewness and kurtosis "violations" were not problematic in this study.

Taken together, I will underscore a theme I have stated previously. The researcher is ultimately in charge of the nature of the research. Statistical procedures can help to make sense out of data and help to make statistical decisions, but the researcher is ultimately the most important part of the process for conducting a study and making conclusions.

NONPARAMETRIC ANOVA TESTS

As I discussed in Chapter 11, there are nonparametric tests that correspond to the parametric tests that I cover in this book. One of the "parallel" nonparametric tests for the Independent-Samples T Test is the Mann–Whitney U that I discussed in Chapter 11. There are nonparametric tests that are similar to the ANOVA test as well.

The Kruskal–Wallis Test is one of the parallel nonparametric tests for one-way ANOVA. Like the Mann–Whitney U, this test uses ranks and determines whether groups of dependent variable ranks are different from one another.

The data in Table 12.12a are hypothetical data that compare the aggregate math achievement of urban (Setting = 1), rural (Setting = 2), and suburban (Setting = 3) groups of schools. Like the ANOVA test, the research question is whether these groups are equivalent or statistically different. The Kruskal–Wallis Test uses ranks to make the determination.

Using SPSS® to specify the Kruskal–Wallis Test is a similar process to the one we used for the Mann–Whitney U in Chapter 11. If you look at Figure 11.24a, you will see the process for specifying the Kruskal–Wallis Test. The only difference is that you would choose "k Independent Samples" rather than "2 Independent Samples" because Kruskal–Wallis tests the differences among more than two groups.

When you choose this option, you will see the window in Figure 12.23a. As you can see, I specified the MathAch variable as the test variable and entered the three

TABLE 12.12a Hypothetical Data for Kruskal–Wallis Test

Setting	MathAch
1	10
1	14
1	6
1	12
2	24
2	18
2	38
2	16
3	54
3	52
3	42
3	50

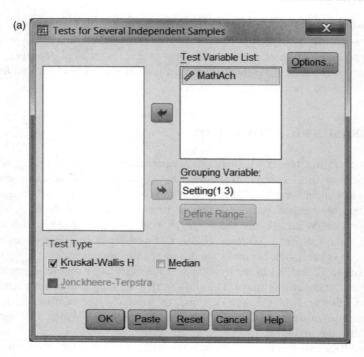

FIGURE 12.23 (a) The specification window for the Kruskal–Wallis test. (b) The output for the Kruskal–Wallis test.

setting groups in the "Grouping Variable" window. Note that you can choose other tests of this type in the "Test Type" box, but I will only show the results for the Kruskal–Wallis Test.

Figure 12.23b shows the results output for the Kruskal–Wallis Test. The top panel shows the descriptive results, which in this example are the mean ranks by setting groups. The bottom panel shows the test statistics that will help us determine whether there are significant differences between the setting groups. As you can see from the results, we can reject the null hypothesis at the $p < 0.007$ level. There are significant MathAch differences among the setting groups.

Like ANOVA, if there are significant differences among the different groups, we may wish to conduct post hoc analyses to see how the groups differ. One way of doing this is to conduct separate Mann–Whitney U Tests for all comparisons. In this case, there would be three paired comparisons (i.e., Setting groups 1 & 2, groups 1 & 3, and groups 2 & 3).

As I explained above with the post hoc process in ANOVA, conducting three separate Mann–Whitney U Tests may result in a problem because we would compound the overall alpha error. We can still use this process by using a "correction" to our alpha exclusion area. The Bonferroni method specifies that we can divide the overall alpha level (0.05) by the number of paired comparisons in order to create an appropriate target significance area for each. Because we have three comparisons, and we use the 0.05 level of significance, the Bonferroni technique would specify a new region of rejection at $p < 0.0167$ ($0.05/3 = 0.0167$).

This process creates a more conservative estimate of the significance of paired comparisons, but we can run into trouble if we have too many comparisons! In the current example, the overall Kruskal–Wallis results indicated a significant finding at $p < 0.007$. If we were to conduct all three comparisons, we would find the results shown in Table 12.12b. As you can see, each of the comparisons was significant at $p < 0.029$, which exceeds our new rejection region of 0.0167. If we were more selective in our comparisons, we could specify only one or possibly two comparisons, but beyond this we would not have confidence in our differences.

One further warning is in order for both the Mann–Whitney U Test and Kruskal–Wallis Test. If we have several tied ranks, the results may be somewhat compromised. My examples in this chapter and Chapter 11 were simple and did not have tied ranks. Actual data analyses would likely be larger and may involve several tied ranks.

TABLE 12.12b The Paired Comparison Results

	Mann Whitney U Significance
Group 1–2	0.029
Group 1–3	0.029
Group 2–3	0.029

TERMS AND CONCEPTS

ANOVA The Analysis of variance test that assesses the extent to which the variance between group means and the grand mean of a distribution is large relative to the variance of individual scores of different sample groups. This test is typically used with three or more sample groups.

***F* Test** The statistical test comparing the between and within variances in an ANOVA test. The calculated ratio of between to within variance is compared to the exclusion values of the F distribution.

Familywise Error The inflation of alpha error due to conducting multiple *post hoc* comparison tests.

Kruskal–Wallis Test The nonparametric test comparable to ANOVA but using ordinal data.

Levene's Test The statistical test assessing the assumption of homogeneity of variance.

***Post Hoc* Analyses** Individual comparison tests conducted among sample group values subsequent to a significant ANOVA finding.

Variance Between The variation of group means around the grand mean in an ANOVA analysis.

Variance Total The total variation (between and within) in an ANOVA analysis resulting from all individual scores varying around the grand mean.

Variance Within The variation of scores around their own group means in an ANOVA analysis. Also known as "error."

REAL-WORLD LAB VIII: ANOVA

This lab is an extension of our real-world example above. This time, however, the outcome measure is schools' reading achievement rather than math achievement. We will use the same predictor variable, FR. The overall research question is the same: Does family income affect schools' student achievement results?

We will use the same data sample (59 schools) with schools grouped according to low, medium, and high FR lunch (as an operational definition of family income). You can use Excel® to create the database the same way I did above for ease of calculation.

1. Discuss whether the data meet the assumptions for ANOVA.
2. (Temporarily) Eliminate the school in the low FR category with an extreme result and compare the results with your findings for #1 above.
3. Calculate the one-way ANOVA test by hand showing results (including the extreme case.)
4. Use Excel® and SPSS® to analyze the data with and without the extreme case.
5. Create a summary of findings.

REAL-WORLD LAB VIII: SOLUTIONS

Table 12.13 shows the Excel® database that includes the extreme case (case #1 in Group 1 that is shaded). I showed key data analyses that are appropriate for ANOVA.

1. Discuss Whether the Data Meet the Assumptions for ANOVA

Assumption 1: Are Populations Normally Distributed? Tables 12.14 and 12.15 show the respective Excel® and SPSS® descriptive summaries for the FR groups

TABLE 12.13 The Data for Real-World Lab VIII

	Low FR (X_1)	X_1^2	Medium FR (X_2)	X_2^2	High FR (X_3)	X_3^2
	26	665.64	56	3180.96	22	479.61
	62	3782.25	60	3564.09	41	1697.44
	69	4774.81	63	4019.56	46	2070.25
	75	5670.09	65	4225	47	2218.41
	75	5685.16	67	4448.89	49	2430.49
	76	5745.64	70	4900	54	2883.69
	78	6068.41	70	4942.09	61	3672.36
	82	6707.61	70	4956.16	62	3794.56
	83	6822.76	70	4956.16	66	4303.36
	83	6938.89	73	5299.84	71	5069.44
	84	6972.25	77	5882.89	71	5097.96
	84	7022.44	79	6162.25	75	5625
	85	7157.16	79	6193.69	76	5715.36
	85	7225	79	6304.36	80	6400
	85	7259.04	80	6320.25		
	87	7482.25	83	6855.84		
	87	7516.89	92	8500.84		
	87	7534.24	95	8968.09		
	88	7691.29				
	88	7761.61				
	90	8010.25				
	90	8136.04				
	91	8226.49				
	94	8817.21				
	96	9120.25				
	96	9120.25				
	96	9235.21				
ΣX	2,217.90		1,327.60		819.70	
ΣX^2		187,149.13		99,680.96		51,457.93
$(\Sigma X)^2$	4,919,080.41		1,762,521.76		671,908.09	
n	27		18		14	
Mean	82.14		73.76		58.55	

TABLE 12.14 The Excel® Descriptive Analyses

	Low FR (X_1)	Medium FR (X_2)	High FR (X_3)
Mean	82.14	73.76	58.55
Standard error	2.66	2.40	4.36
Median	85.00	71.60	61.10
Mode	95.50	70.40	#N/A
Standard deviation	13.81	10.18	16.32
Sample variance	190.81	103.71	266.50
Kurtosis	10.51	0.04	0.25
Skewness	−2.83	0.41	−0.73
Range	70.30	38.30	58.10
Minimum	25.80	56.40	21.90
Maximum	96.10	94.70	80.00
Sum	2217.90	1327.60	819.70
Count	27	18	14

(including the extreme case). As you can see from both tables, the normal distribution assumption is not met in the FR Low group. Both skewness and kurtosis are well out of bounds (i.e., far beyond our 2–3 guideline). The results are most clear in Table 12.15 because SPSS® reports the standard error for skewness and kurtosis. The graphs in Figure 12.24 provide a clear visual summary of these group one violations.

Assumption #2: Are Population Variances Equal? We will check this assumption by examining the Levene's Test results in SPSS®.

Assumptions #3 (Independent Selection) and #4 (Interval Data) Are Met.

2. (Temporarily) Eliminate the School in the Low FR Category with an Extreme Result and Compare the Results with Your Findings for #1 Above

Both the Excel® and SPSS® results (Tables 12.16 and 12.17, respectively) indicate no skewness or kurtosis violations when the extreme case is removed. The graphs in Figure 12.25 confirm these findings with visual evidence.

TABLE 12.15 The SPSS® Descriptive Analyses: ReadingPercentMetStandard

FR	Mean	N	Standard Deviation	Kurtosis	Standard Error of Kurtosis	Skewness	Standard Error of Skewness
1.00	82.14	27	13.813	10.506	0.872	−2.828	0.448
2.00	73.76	18	10.184	0.037	1.038	0.408	0.536
3.00	58.55	14	16.325	0.249	1.154	−0.728	0.597
Total	73.99	59	16.254	1.616	0.613	−1.186	0.311

REAL-WORLD LAB VIII: SOLUTIONS

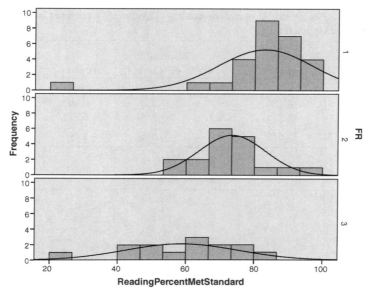

FIGURE 12.24 The SPSS® graphs of FR groups with reading results to assess assumption #1.

TABLE 12.16 The Excel® Descriptive Analyses (Excluding Extreme Case)

	Low FR (X_1)	Medium FR (X_2)	High FR (X_3)
Mean	84.31	73.76	58.55
Standard error	1.60	2.40	4.36
Median	85.10	71.60	61.10
Mode	95.50	70.40	#N/A
Standard deviation	8.16	10.18	16.32
Sample variance	66.57	103.71	266.50
Kurtosis	1.28	0.04	0.25
Skewness	−0.93	0.41	−0.73
Range	34.60	38.30	58.10
Minimum	61.50	56.40	21.90
Maximum	96.10	94.70	80.00
Sum	2192.10	1327.60	819.70
Count	26	18	14

TABLE 12.17 The SPSS® Descriptive Analyses (Excluding Extreme Case): ReadingPercentMetStandard

FR	Mean	N	Standard Deviation	Kurtosis	Standard Error of Kurtosis	Skewness	Standard Error of Skewness
1.00	84.31	26	8.159	1.280	0.887	−0.931	0.456
2.00	73.76	18	10.184	0.037	1.038	0.408	0.536
3.00	58.55	14	16.325	0.249	1.154	−0.728	0.597
Total	74.82	58	15.079	1.615	0.618	−1.090	0.314

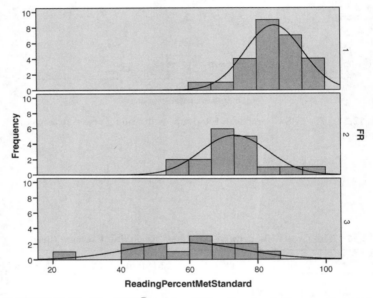

FIGURE 12.25 The SPSS® graphs of FR groups (excluding extreme case).

3. Calculate the One-Way ANOVA Test by Hand Showing Results (*Including* the Extreme Case)

Calculating SS_T ($SS_T = 15,322$)

$$SS = \sum X^2 - \frac{(\sum X)^2}{N},$$

$$SS_T = (187,149 + 99,681 + 51,458) - \frac{(2217.9 + 1327.6 + 819.7)^2}{59},$$

$$SS_T = 338,288 - 322,966$$

REAL-WORLD LAB VIII: SOLUTIONS

Calculating SS_B ($SS_B = 5133$)

$$SS_B = \frac{(\Sigma X_1)^2}{n_1} + \frac{(\Sigma X_2)^2}{n_2} + \frac{(\Sigma X_3)^2}{n_3} (-) \frac{(\Sigma X)^2}{N},$$

$$SS_B = \frac{4,919,080}{27} + \frac{1,762,522}{18} + \frac{671,908}{14} (-) \frac{(2217.9 + 1327.6 + 819.7)^2}{59},$$

$$SS_B = 182,188 + 97,918 + 47993 (-) 322,966$$

Calculating SS_W ($SS_W = 10,189$)

$$SS_W = SS_T - SS_B,$$
$$SS_W = 15,322 - 5133$$

Table 12.18 shows the ANOVA summary table with the calculated results. The calculated F (14.10) is significant ($p < 0.05$) as determined by the critical F value in an F distribution table ($F_{(0.05, 2, 56)} = 3.17$). The effect size can also be determined from Table 12.18 as:

$$\eta^2 = \frac{SS_{Between}}{SS_{Total}}$$

$$\eta^2 = \frac{5133}{15,322}$$

$$\eta^2 = 0.34$$

Tukey's Post Hoc Analysis. The Tukey Range table of values points to $q_{(Range)} = 3.40$ as with our previous (math) example. Using the Tukey formula, we obtain

$$HSD = q_{(Range)} \sqrt{\frac{MS_w}{n}},$$

$$HSD = 3.40 \sqrt{\frac{182}{20}},$$

$$HSD = 10.26$$

TABLE 12.18 The Completed ANOVA Summary Table for Lab VIII

Source of Variance	SS	df	MS	F Ratio
Between	5133	2	2567	14.10
Within	10,189	56	182	
Total	15,322	58	—	

TABLE 12.19 The Group Mean Difference Matrix for Reading

	M_1 (82.14)	M_2 (73.76)	M_3 (58.55)
M_1 (82.14)	—	8.39	23.59
M_2 (73.76)		—	15.21
M_3 (58.55)			—

Recall that this formula requires the sample sizes to be equal. Without introducing a formula that accommodates unequal sample size, we will take the average of the sample sizes $((27 + 18 + 14)/3 = 20)$ as we did with the math example. Table 12.19 shows the results with the reading outcome variable. The two shaded cells are significant because these mean differences exceed the HSD value of 10.26. Therefore, Groups 1 (Low FR) and 2 (Medium FR) are not different from one another, but Group 3 (High FR) is different from both the other groups.

4. Use Excel® and SPSS® to Analyze the Data with and without the Extreme Case

Before we analyze the ANOVA results, note that the Levene's Test results are different, depending on whether the extreme case is included. Table 12.20 provides a comparison from SPSS® printouts. The top panel shows a *significant Levene's statistic when the extreme case is included, which indicates that the sample group variances are not equal*. In this case, we would need to decide how to proceed if we wished to continue with this analysis. The bottom panel indicates that the Levene's statistic is not significant, which indicates that the sample group variances are equal and the equal variance assumption for ANOVA is met.

TABLE 12.20 SPSS® Levene's Test Comparison with and without the Extreme Case

Test of Homogeneity of Variances (with Extreme Case)			
ReadingPercentMetStandard			
Levene Statistic	df1	df2	Significance
5.660	2	55	0.006

Test of Homogeneity of Variances (without Extreme Case)			
ReadingPercentMetStandard			
Levene Statistic	df1	df2	Significance
1.576	2	56	0.216

REAL-WORLD LAB VIII: SOLUTIONS

TABLE 12.21 Excel® Comparison Tables for Reading—FR Lab VIII

ANOVA (With Extreme Case)						
Source of Variation	SS	df	MS	F	P Value	F crit
Between Groups	5133.863	2	2566.932	14.1088	1.09E-05	3.161861
Within Groups	10,188.55	56	181.9383			
Total	15,322.41	58				

ANOVA (Without Extreme Case)						
Source of Variation	SS	df	MS	F	P Value	F crit
Between Groups	6068.697	2	3034.348	24.2158	2.86E-08	3.164993
Within Groups	6891.746	55	125.3045			
Total	12,960.44	57				

Table 12.21 compares the Excel® ANOVA results with (top panel) and without (bottom panel) the extreme case. As you can see, the calculated F ratios (shaded) are quite different. Both are significant. Table 12.22 shows the SPSS® results comparison which are identical to the Excel® results.

Note the different way in which the significance is reported by both programs. SPSS® does not include the tabled F comparison value (Excel® reports it in the final column, "F crit"). SPSS® reports significance as "0.000," whereas Excel® reports the actual probability value in scientific notation.

TABLE 12.22 SPSS® Comparison Tables for Reading—FR Lab VIII

ANOVA (Including Extreme Case)					
ReadingPercentMetStandard					
	Sum of Squares	df	Mean Square	F	Significance
Between Groups	5133.863	2	2566.932	14.109	0.000
Within Groups	10,188.546	56	181.938		
Total	15,322.409	58			

ANOVA (Excluding Extreme Case)					
ReadingPercentMetStandard					
	Sum of Squares	df	Mean Square	F	Significance
Between Groups	6068.697	2	3034.348	24.216	0.000
Within Groups	6891.746	55	125.304		
Total	12,960.443	57			

You can also compare differences in effect sizes, including the extreme case results in $\eta^2 = .34$ versus $\eta^2 = 0.47$ when the case is excluded.

$$\eta^2 = \frac{5133}{15322}, \qquad \eta^2 = .34$$

$$\eta^2 = \frac{6069}{12960}, \qquad \eta^2 = .47$$

Excel® does not provide *post hoc* analyses, so we will use SPSS® to show the Tukey results. Table 12.23 shows the Tukey results both with (top panel) and without (bottom panel) the extreme case. As you can see from the shaded cells, the results are different, depending on whether the extreme case is included.

TABLE 12.23 SPSS® *Post Hoc* Results with and without the Extreme Case

Multiple Comparisons

Tukey HSD

(I) FR	(J) FR	Mean Difference (I−J)	Standard Error	Significance	95% Confidence Interval Lower Bound	95% Confidence Interval Upper Bound
1.00	2.00	8.389	4.104	0.111	−1.49	18.27
	3.00	23.594[a]	4.442	0.000	12.90	34.29
2.00	1.00	−8.389	4.104	0.111	−18.27	1.49
	3.00	15.206[a]	4.807	0.007	3.63	26.78
3.00	1.00	−23.594[a]	4.442	0.000	−34.29	−12.90
	2.00	−15.206[a]	4.807	0.007	−26.78	−3.63

Tukey HSD

(I) FR	(J) FR	Mean Difference (I−J)	Standard Error	Significance	95% Confidence Interval Lower Bound	95% Confidence Interval Upper Bound
1.00	2.00	10.556[a]	3.432	0.009	2.29	18.82
	3.00	25.762[a]	3.711	0.000	16.82	34.70
2.00	1.00	−10.556[a]	3.432	0.009	−18.82	−2.29
	3.00	15.206[a]	3.989	0.001	5.60	24.81
3.00	1.00	−25.762[a]	3.711	0.000	−34.70	−16.82
	2.00	−15.206[a]	3.989	0.001	−24.81	−5.60

[a] The mean difference is significant at the 0.05 level.

When the extreme case is included, groups 1 and 2 are not different from each other, but both are different from group 3. However, when the extreme case is excluded, all three group differences are significant.

5. Create a Summary of Findings

Taken together, the data analyses indicate that groups of FR have an impact on schools' reading achievement results. In particular, as the percentages of students qualified for FR increase in the schools, the percentage of students meeting the reading achievement assessment declines. The results are sharpened by excluding the results of one alternative school that is not structurally comparable to the other sample schools. When the case is excluded, $F = 24.22$ ($p < 0.05$) and the effect size is substantial ($\eta^2 = 0.47$), indicating that 47% of the schools' reading achievement scores are affected by grouping on the FR variable (Low, Medium, and High). A Tukey HSD analysis indicated that each of the three FR groups' reading achievement percentages were significantly different from the other groups.

13

FACTORIAL ANOVA

In Chapter 12, we discussed research studies with one independent variable and one dependent variable. The designation "one-way" ANOVA makes reference to the independent variable. We can extend these procedures by adding a second (or more) independent variables to the design. This is known as *Factorial* ANOVA because the focus is multiple factors (i.e., independent variables). When a second independent variable is added, the procedure is known as a "two-way ANOVA." A shorthand identification of factorial ANOVA is "2×ANOVA" indicating that there are two independent variables. (A 3×ANOVA has three independent variables, etc.)

Before we discuss factorial ANOVA, I want to mention briefly some extensions of the ANOVA procedure. We will not explore these in this book, but you will see how "adaptable" ANOVA is to different research situations.

EXTENSIONS OF ANOVA

There are several ANOVA procedures for special research designs and situations. The basic ANOVA design is very flexible in that it can admit several different sources of variance to compare at the same time.

Within-Subjects ANOVA

An ANOVA procedure designed for use with *repeated measures* is known as Within-Subjects ANOVA. These are designs that use two or more measures from

---

Understanding Educational Statistics Using Microsoft Excel® and SPSS®. By Martin Lee Abbott.
© 2011 John Wiley & Sons, Inc. Published 2011 by John Wiley & Sons, Inc.

the same subjects or matched group subjects. (We will briefly examine one of these, within subjects ANOVA, in a later chapter.) An example might be an ANOVA design with only one group of subjects that is measured more than once, perhaps as a "before–after" design or "pretest–post-test" design.

In these designs, the two measures of the dependent variable are highly related to one another because we are measuring the same person(s) twice. Thus, we need a statistical procedure that will take this relationship into account as the overall ANOVA is calculated.

Two-Way Within-Subjects ANOVA

If the ANOVA design has more than one independent variable on which the same group or subjects are measured twice, the researcher can use this Two-Way Within-Subjects ANOVA. These research situations call for procedures that help ferret out the differences among repeated measures. For example, suppose we used pre–post comparisons on our noise–learning subjects (within subjects), but we also tested them under "non-white noise" as well as "white noise" conditions. In this case, we have more than one independent variable on whose levels the same subjects are measured twice.

ANCOVA

A very useful ANOVA design that researchers use to control extraneous influences is *Analysis of Covariance*. If the researcher cannot truly randomize in the study, ANCOVA might be helpful as a way of limiting the influence of a variable or variables "outside" the design (known as "covariates"), but that might affect the results.

As an example, if we had not been able to randomly select and assign subjects to the noise–learning experiment, we might be concerned that variables other than noise might affect the outcome measure (learning). We might use some existing measure of "learning ability" (GPA, achievement test score) as a "control" on learning so that the post-test measure would measure only the effects of the noise, not the individual subjects' *ability to learn*. A pretest score is a popular covariate.

ANCOVA procedures are controversial when the researcher relies on them rather than randomization for producing valid and meaningful results. Randomization provides the greatest confidence that all the extraneous variables are controlled in a research situation. However, as we have seen, the researcher may not have the luxury of full randomization. In these cases, particularly in quasi-experimental designs, ANCOVA is a helpful tool.

ANCOVA procedures are also (perhaps especially) useful in *post facto* designs when there is no manipulation of the independent variable. The researcher attempts to control as many additional influences as possible on the outcome measure and ANCOVA can be very helpful. ANCOVA is also used in procedures that are not ANOVA-based, like multiple regression procedures.

MULTIVARIATE ANOVA PROCEDURES

The procedures we cover in this book are *univariate* procedures. This means that we have only one dependent or outcome variable in the research design. As you move further into advanced statistical measures, you will see *multivariate* procedures that have more than one outcome variable. MANOVA and MANCOVA are two examples of multivariate procedures.

MANOVA

When a researcher adds one or more *dependent* variables to an ANOVA design, the procedure is called MANOVA or Multivariate Analysis of Variance. To extend my noise experiment example from Chapter 12, I could have added another dependent variable such as performance on a visual recognition task. Thus, I would have had one independent variable (noise) and two dependent variables (learning performance and visual recognition). I would use MANOVA with this design because all the sources of error are contained when you examine all the statistical tests at the same time. (We discussed this as limiting familywise error in Chapter 12.)

MANCOVA

Of course, the statistical procedures can become increasingly sophisticated along with the sophistication and complexity of the design. MANCOVA (Multivariate Analysis of Covariance) is the extension of ANCOVA to designs with *multiple dependent variables*.

FACTORIAL ANOVA

The factorial ANOVA performs the single ANOVA procedure twice (or as many times as you have independent variables in the design) within the same analysis. In this way, the familywise error is contained. In the output, we will learn to recognize the separate effects of these two independent variables (called *main effects*).

Interaction Effects

The factorial ANOVA also performs an additional analysis of the *interaction effects*. As I have stated elsewhere, *an interaction is present when the relationship between one predictor and the outcome variable changes at different levels of another predictor variable* (Abbott, 2010). The key feature of an interaction effect is that both independent variables have effects on each other as well as the dependent variable; different levels of one independent variable affects the levels of the other independent variable.

An Example of 2×ANOVA

To take an example, suppose that in our noise–learning example, we were interested in the difference between men and women as well as the difference between noise conditions in producing learning errors. This 2×ANOVA would include the following primary analyses:

- **Main Effect 1:** Do noise conditions differ in # of learning errors?
- **Main Effect 2:** Do men and women differ in # of learning errors?
- **Interaction Effect:** Do men and women produce different learning errors under different noise conditions?

Figure 13.1 shows the design of an experiment in which we added a second independent variable (men versus women) to our earlier noise–learning design. (In the Figure 13.1 example, I created three groups of noise rather than four groups for a clearer example.) As you can see, there are two main effects indicated by the analyses between rows (noise) and between columns (men versus women). The dependent variable learning errors are represented by the shaded areas where they would be entered according to the independent variable categories.

The other analysis in 2×ANOVA is the interaction effect, represented by the different shades in Figure 13.1. The darker the shade, the higher the learning errors. In Figure 13.1, men and women differ in their patterns of errors under different noise conditions. This is an interaction effect. Figure 13.2 shows how the interaction effect appears with a line chart.

As you can see, Figure 13.2 shows that men produce more errors as the noise levels increase. Women show a different pattern. Their highest errors are produced under no-noise conditions with lower numbers of errors in medium- and high-noise conditions. This is a hypothetical example, so we cannot draw conclusions about the nature of noise and sex on learning errors. However, the example provides insight into the fact that 2×ANOVA conducts several analyses simultaneously. The interaction effect is shown separately in the 2×ANOVA output.

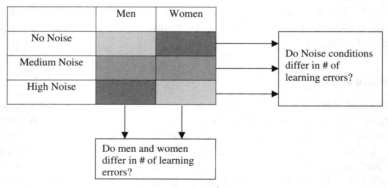

FIGURE 13.1 Main effects analyses in 2×ANOVA.

FACTORIAL ANOVA 311

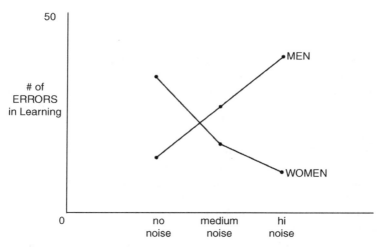

FIGURE 13.2 The interaction effect of sex and noise conditions.

Charting Interactions

When the interaction graph shows that the lines do not cross or intersect, we do not expect a significant interaction. In these cases, the groups may be different on the dependent variable, but the values of one independent variable are consistently parallel across values of the other independent variable.

There are different kinds of interactions that you will recognize by graphing them. *Disordinal interactions* are those in which the lines cross in the plotted graph. Figure 13.2 is an example of the disordinal interaction.

Ordinal interactions are those in which the lines may not cross within the plotted graph, but they are not parallel. In these cases, the lines could intersect or cross if they were carried out beyond the plotted area of the graph. (In these cases, the researcher would need to decide if the extension of the analyses would not be of interest to the research question.) Figure 13.3 shows examples of an ordinal interaction and a study with no interaction.

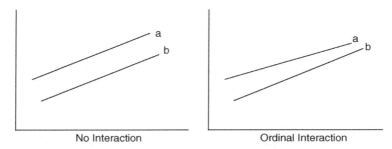

FIGURE 13.3 Ordinal interaction patterns compared to no interaction.

Simple Effects

Generally, a significant interaction effect in the 2×ANOVA takes precedence over the main effects findings. If there is a significant interaction, the main effects dynamics "hide" the unequal cell findings across the factors. Looking at Figure 13.1, for example, you could not detect the different ways that men and women are affected by noise conditions if you were simply comparing the total of men versus women. The different shades would be mixed together and indistinguishable. Likewise, if you were just considering noise differences, the men's and women's shades would be mixed together.

The interaction effects allow you to "disaggregate" the findings in such a way that you can see the differences in one variable at each level of the other. *Simple effects* are the ways you can "see" all these nuances. Simple effects show the differences in categories of one independent variable within single categories of the other independent variable.

In Figure 13.2, both the men's and women's lines represent separate simple effects. The figure shows that if you focus only on men, you can see that the errors increase across noise conditions. For women, the effects of noise are inconsistent on their error production.

There are separate calculations for simple effects that researchers learn to calculate in advanced statistics books or through SPSS® and other statistical programs. I will demonstrate these procedures briefly.

THE EXAMPLE DATASET

The example data I will use for this chapter is taken from the STAR Classroom Observation Protocol™ data provided by The BERC Group, Inc. I included a general description of this database in Chapter 3 as a process for measuring the extent to which Powerful Teaching and Learning™ is present during a classroom observation. The BERC Group, Inc. has performed thousands of classroom observations of all grade levels and subject areas in Washington State. I chose a 5% sample of the 1189 observations in year four of one study in Washington schools and randomly chose five observations per cell to yield the "balanced" data in Table 13.1.[1]

CALCULATING FACTORIAL ANOVA

The factorial ANOVA has several comparisons that recognize the complexity of the data. There are three primary comparisons:

---

[1] Although randomly drawn, I am using this small set of data only to illustrate 2×ANOVA, not to make study conclusions.

CALCULATING FACTORIAL ANOVA

- Main effects (two comparisons)
  Row effects (Level of school in Table 13.1)
  Column effects (Subject in Table 13.1)
- Interaction effects (one comparison)

In order to calculate these comparisons, we need to start by calculating the total, between, and within sums of squares as we did with the one-way ANOVA. These are the "building blocks" of our 2×ANOVA calculations. With the one-way ANOVA, recall that the total SS is the measure of variation of all the individual scores around the grand mean. The between SS is the measure of variation between the group means and the grand mean. The within SS (the error measure) is the measure of variation of the group scores around their own group means.

The "new" measures represent the following sources of variation:

- Row SS measure the variation of the row means (school level means in the example) in reference to the grand mean.
- Column SS measure the variation of the column means (subject means in the example) in reference to the grand mean.
- Interaction SS measure the variations of row means at levels of the column means in reference to the grand mean.

Table 13.2 shows the data in Table 13.1 with the key calculations needed for the 2×ANOVA. As you can see, there are totals for the individual cells as well as separate summaries for columns and rows.

TABLE 13.1 Data for 2×ANOVA Example

	English	Math
Elementary	1	2
	3	2
	3	3
	3	4
	3	3
Middle	4	3
	2	2
	3	4
	4	1
	2	3
High School	4	1
	4	2
	4	1
	3	1
	3	1

TABLE 13.2 Data Summaries for 2×ANOVA Calculations

		ENGLISH		MATH			
		X	X^2	X	X^2		
Elementary		1	1	2	4		
		3	9	2	4		
		3	9	3	9	$n_{(row)}$	10
		3	9	4	16	$\Sigma X_{(row)}$	27
		3	9	3	9	$\Sigma X^2_{(row)}$	79
	n	5		5			
	ΣX	13		14			
	ΣX^2		37		42		
Middle		4	16	3	9		
		2	4	2	4		
		3	9	4	16	$n_{(row)}$	10
		4	16	1	1	$\Sigma X_{(row)}$	28
		2	4	3	9	$\Sigma X^2_{(row)}$	88
	n	5		5			
	ΣX	15		13			
	ΣX^2		49		39		
High School		4	16	1	1		
		4	16	2	4		
		4	16	1	1	$n_{(row)}$	10
		3	9	1	1	$\Sigma X_{(row)}$	24
		3	9	1	1	$\Sigma X^2_{(row)}$	74
	n	5		5			
	ΣX	18		6			
	ΣX^2		66		8		
	$n_{(col)}$	15		15			
	$\Sigma X_{(col)}$	46		33			
	$\Sigma X^2_{(col)}$		152		89		

Calculating SS_T (all six "cells" are included in the equation):

$$SS = \sum X^2 - \frac{(\sum X)^2}{N}$$

$$SS_T = (37 + 42 + 49 + 39 + 66 + 8) - \frac{(13 + 14 + 15 + 13 + 18 + 6)^2}{30}$$

$$SS_T = 241 - 208.03$$

$$\mathbf{SS_T = 32.97}$$

Calculating SS_B (including the six cell means):

$$SS_B = \frac{(\sum X_1)^2}{n_1} + \frac{(\sum X_2)^2}{n_2} + \frac{(\sum X_3)^2}{n_3} + \frac{(\sum X_4)^2}{n_4} + \frac{(\sum X_5)^2}{n_5} + \frac{(\sum X_6)^2}{n_6} (-) \frac{(\sum X)^2}{N}$$

CALCULATING FACTORIAL ANOVA

$$SS_B = \frac{(\sum 13)^2}{5} + \frac{(\sum 14)^2}{5} + \frac{(\sum 15)^2}{5} + \frac{(\sum 13)^2}{5} + \frac{(\sum 18)^2}{5} + \frac{(\sum 6)^2}{5}$$
$$(-)\frac{(13+14+15+13+18+6)^2}{30}$$

$SS_B = 223.80(-)208.03$

$SS_B = \mathbf{15.77}$

Calculating SS_W (including the six cell values):

$$SS_W = SS_T - SS_B$$
$$SS_W = 32.97 - 15.77$$
$$SS_W = \mathbf{17.2}$$

Calculating the $SS_{(Rows)}$ (the main effect of school level):

$$SS_{(Rows)} = \frac{(\sum row1)^2}{n_1} + \frac{(\sum row2)^2}{n_2} + \frac{(\sum row3)^2}{n_3}(-)\frac{(\sum X)^2}{N}$$

$$SS_{(Rows)} = \frac{(27)^2}{10} + \frac{(28)^2}{10} + \frac{(24)^2}{10}(-)208.03$$

$SS_{(Rows)} = 208.9(-)208.03$

$SS_{(Rows)} = \mathbf{0.87}$

Calculating the $SS_{(columns)}$ (the main effect of subjects):

$$SS_{(columns)} = \frac{(\sum col1)^2}{n_1} + \frac{(\sum col1)^2}{n_2}(-)\frac{(\sum X)^2}{N}$$

$$SS_{(columns)} = \frac{(46)^2}{15} + \frac{(33)^2}{15}(-)208.03$$

$SS_{(columns)} = 213.7(-)208.03$

$SS_{(columns)} = \mathbf{5.63}$

Calculating the Interaction

Like the SS_W we can calculate the interaction SS (symbolized as SS_{rXc}) by subtracting the SS for rows and columns from the overall SS between as follows:

$$SS_{rXc} = SS_B - (SS_{(Rows)} + SS_{(columns)})$$
$$SS_{rXc} = 15.77 - (0.87 + 5.63)$$
$$SS_{rXc} = \mathbf{9.27}$$

The 2×ANOVA Summary Table

Just as we did with the one way ANOVA, we can now create the ANOVA summary table for the 2×ANOVA. It has three rows to show the main effects and interaction

TABLE 13.3 The 2×ANOVA Summary Table

Source of Variance	SS	df	MS	F
Between	15.77			
Row main effect (level of school)	0.87			
Column main effect (subject)	5.63			
Interaction effect (level X subject)	9.27			
Within (error)	17.2			
Total	32.97			

effect, but it is otherwise the same as the one way table. Table 13.3 shows the summary table with the calculated SS values included.

Recall that the ANOVA is an inferential process, so we need to estimate the population variances (MS values) from the sample variances (SS values). We need the degrees of freedom to create the estimates appropriately. These are a bit different from the one-way ANOVAs because we are estimating different values, but the basics are the same. Here are the ways to calculate the various degrees of freedom:

$\text{df}_{\text{Total}} = N - 1$ (the overall number of observations minus 1)

$\text{df}_{\text{Row}} = r - 1$ (the number of rows minus 1)

$\text{df}_{\text{Col}} = c - 1$ (the number of columns minus 1)

$\text{df}_{\text{rXc}} = (r - 1)(c - 1)$ (# of rows − 1 × # of columns minus 1; or $\text{df}_{\text{Row}} * \text{df}_{\text{Col}}$)

$\text{dfW} = N - (\text{df}_{\text{Row}} * \text{df}_{\text{Col}})$ (or N − the number of cells)

Creating the MS Values

In the one-way ANOVA, we used the following formula to calculate the F ratio:

$$F = \frac{\text{MS}_B}{\text{MS}_W}$$

This formula simply compared the between to the within sums of squares in order to see if the group means were far apart relative to the (error) variance within each of the sample groups. With 2×ANOVA, we do the same thing, except that we calculate the F values for each of the three effects (main effects and interaction effect) by dividing each by the overall MS_W.

$$F_{\text{row}} = \frac{\text{MS}_{\text{row}}}{\text{MS}_W}, \quad F_{\text{col}} = \frac{\text{MS}_{\text{col}}}{\text{MS}_W}, \quad F_{\text{rXc}} = \frac{\text{MS}_{\text{rXc}}}{\text{MS}_W}$$

Table 13.4 shows the completed 2×ANOVA summary table with the values we calculated above. Now that we have the F ratios for the various components of the study, we can proceed with hypotheses tests to determine whether the F ratios are in the exclusion area of the comparison distribution and therefore not likely to occur by chance.

CALCULATING FACTORIAL ANOVA

TABLE 13.4 The Completed 2×ANOVA Summary Table

Source of Variance	SS	df	MS	F	F-Table Value[1]
Between	15.77	5	3.15	4.38	2.62*
Row main effect (level of school)	0.87	2	0.435	0.61	3.40
Column main effect (subject)	5.63	1	5.63	7.82	4.26*
Interaction effect (level X subject)	9.27	2	4.64	6.44	3.40*
Within (error)	17.2	24	0.72		
Total	32.97	29			

[1] F ratios exceeding F-table values are significant at or beyond the 0.05 level.

The Hypotheses Tests

Table 13.4 shows the F-table values in the last column. These values indicate the 0.05 exclusion region. Thus, any F ratio larger than the tabled values would be considered significant at the $p < 0.05$ level. According to the table, the omnibus test, the main effect for subject and the interaction effect are significant. Thus, their F-ratio values are too high to be expected to occur by chance (at the 0.05 level).

The Omnibus F Ratio

When we tested the one way ANOVA, we created the omnibus F ratio, a transformed ratio of our sample values applied to the comparison distribution in order to assess whether our sample values were likely a chance finding. In 2×ANOVA, we focus on the F ratios of the specific row, column, and interaction effects to determine the same thing. That is why they can be computed by adding the SS values for the main effects and the interaction effects. This will provide an overall $SS_{Between}$ value that can yield the omnibus F ratio. Thus,

$$SS_{Between} = SS_{row} + SS_{col} + SS_{rXc},$$

$$SS_{Between} = 0.87 + 5.63 + 9.27,$$

$$SS_{Between} = 15.77$$

The $MS_{Between}$ can then be calculated by dividing the $SS_{Between}$ by its df measure (in this example it is 5 because it combines the df measures for the three effects).

$$MS_{Between} = \frac{SS_{Between}}{df_{Between}},$$

$$MS_{Between} = \frac{15.77}{5},$$

$$MS_{Between} = 3.15$$

The omnibus F ratio can be calculated by using the $MS_{Between}$ value divided by the MS_W value as we did for the one way ANOVA:

$$F = \frac{MS_B}{MS_W},$$

$$F = \frac{3.15}{0.72},$$

$$F = 4.38$$

Effect Size for 2×ANOVA: Partial η^2

Just as we did with one-way ANOVA, we need to understand the effect size of our study. When we had one independent variable, it was a simple matter to calculate η^2, which expresses the proportion of the total variance explained by the independent variable groups:

$$\eta^2 = \frac{SS_{Between}}{SS_{Total}}$$

We can calculate the "omnibus" effect size as follows:

$$\eta^2 = \frac{SS_{Between}}{SS_{Total}},$$

$$\eta^2 = \frac{15.77}{32.97},$$

$$\eta^2 = 0.48$$

This value expresses the combined impact of the groups of school levels and subjects on the STAR™ overall measure. Thus the combined groupings explain almost half (0.48) of the STAR™ outcome measure of teaching observations.

Because we added another independent variable, however, we can "break the overall explained variance down" into its specific proportions. We have created three proportions of the total variance that can be so measured. The effect size for 2×ANOVA is therefore known as "Partial Eta Squared" or "Partial η^2." It is partial because it is not an overall measure. Rather, it "partials out" the overall variance so that *only the contribution of the specific main effect (or interaction effect) is measured on the total variance explained.*

Because the SS_W is the error measure in ANOVA, we have to take it into account in the overall assessment of calculating the F ratios and the effect sizes as well. The following is a "conceptual equation" explaining how effect size works for partial effects:

$$\text{Partial } \eta^2 = \frac{SS_{Effect}}{SS_{Effect} + SS_W}$$

As you can see, the SS measure for the specific effect (main effect and/or interaction effect) is divided by the same SS measure plus the SS_W. The reasoning is that the bottom part of the equation constitutes the *total SS for the particular effect*

CALCULATING FACTORIAL ANOVA

under consideration. Thus, the formula yields a *percentage of variance explained in the dependent variable that is provided by the proportion of the specific effect relative to its error measure.*

Here is how Partial η^2 is calculated for each main effect:
Effect Size for Row Main Effect (School Level):

$$\text{Partial } \eta^2_{\text{row}} = \frac{SS_{\text{row}}}{SS_{\text{row}} + SS_W},$$

$$\text{Partial } \eta^2_{\text{row}} = \frac{0.87}{0.87 + 17.2},$$

$$\text{Partial } \eta^2_{\text{row}} = 0.048$$

Effect Size for Column Main Effect (Subject):

$$\text{Partial } \eta^2_{\text{col}} = \frac{SS_{\text{col}}}{SS_{\text{col}} + SS_W},$$

$$\text{Partial } \eta^2_{\text{col}} = \frac{5.63}{5.63 + 17.2},$$

$$\text{Partial } \eta^2_{\text{col}} = 0.247$$

Effect Size for Interaction Effect (School Level × Subject):

$$\text{Partial } \eta^2_{\text{rXc}} = \frac{SS_{\text{rXc}}}{SS_{\text{rXc}} + SS_W},$$

$$\text{Partial } \eta^2_{\text{rXc}} = \frac{9.27}{9.27 + 17.2},$$

$$\text{Partial } \eta^2_{\text{rXc}} = 0.35$$

The question of the effect size "meaningfulness" is a matter for the researcher to decide, as we discussed in Chapter 12. If you recall, I suggested the following guidelines for assessing η^2: 0.01 (small), 0.06 (medium), and 0.15 (large). Because these are measures for the "omnibus" effect size, they may be too large for the partial measures. However, they do provide some benchmark for judging the magnitude of partial η^2 results.

Discussing the Results

As we saw in Table 13.4, the omnibus test, the main effect for subject and the interaction effect are significant. Thus, school level and subject (english or math) taken together significantly affect the STAR™ overall observation value (because the omnibus test was significant). The results are inconsistent among the various groups of the factors, however. The researchers task is now to "dig deeper" into the findings to understand the specific contributions of each of the levels of the factors.

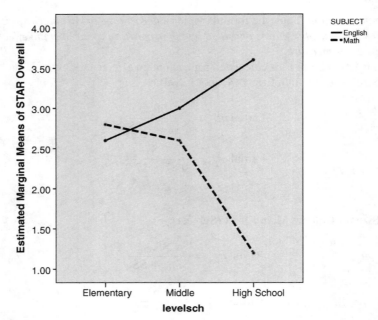

FIGURE 13.4 The interaction graph for the 2×ANOVA example (levelsch by subject).

I mentioned above that *when there is a significant interaction effect in the 2×ANOVA, this finding takes precedence over the main effects findings.* Because there are inconsistent values of one variable on values of the other variable, we need to examine the interaction findings before other values. The two ways to do this are by graphing the interaction and examining simple effects. I will show one of the plots from the SPSS® analysis that shows the interaction, and then I will discuss the simple effects findings in the SPSS® output.

Figure 13.4 shows the two lines for subject (english and math) on the different school levels. As you can see, there is a disordinal interaction between these two factors on the outcome measure of STAR™ (estimated) mean values. The three schools compare differently to one another, depending on the subject. Elementary school classroom observations are similar for both subjects, whereas the observations are quite different at the high school level (with English outperforming math).

The simple effects can be viewed by considering each line separately. To take the example of math in Figure 13.4 (dashed line), you can see that the value of the "overall" scores at the high school level is much lower than at the other two school levels. The simple slope of English shows a different pattern. The values of "overall" change from elementary to high school also, but the difference in "overall" scores at the different school levels is not as extensive as in the case of math. I will discuss the SPSS® simple slope findings below. The simple slope questions for English and math "overall" scores among the different school levels are as follows:

- Are the differences between elementary, middle-school, and high-school math classrooms' overall scores large enough to be considered statistically significant?
- Are the differences between elementary, middle-school, and high-school English classrooms' overall scores large enough to be considered statistically significant?

The trends in both math and English lines appear to show that the values are quite different, depending on the level of the school. But are they "different enough" to be considered significant? We could calculate the simple effects by hand, but most researchers rely on statistical programs to help assess all the findings for 2×ANOVA.

Excel® does not have a straightforward method for assessing 2×ANOVA. One procedure exists for "ANOVA: Two-Factor without Replication," but there is no provision for examining interaction effects.

SPSS® has a very easy way to create the complete 2×ANOVA procedure including interaction effects. Examining simple effects is not immediately apparent through the main menus, however. The researcher must rely on "syntax" statements that create the appropriate output for simple effects. Syntax is akin to programming within SPSS®. It is not as intimidating as it sounds because the program includes "macro-like" statements that the researcher can use to create custom output. I will mention these statements for simple effects in the following section.

USING SPSS® TO ANALYZE 2×ANOVA

Figure 13.5 shows the data file structure for the SPSS® 2×ANOVA analysis. As you can see, the dependent variable (overall) is listed with its raw score values followed by the two factors (i.e., independent variables). The values in the latter variables identify the factor categories. Thus, levelsch categories are: 1, elementary; 2, middle school; and 3, high school. Subject categories are: 1, English; 2, math.

Once the file is created, you can access the 2×ANOVA through the Analyze menu. Figure 13.6 shows the choices leading to the appropriate ANOVA. As you can see, the "General Linear Model" menu is chosen first. This general category includes procedures that allow the researcher to examine the linear effects of one or more independent variables on dependent variable(s). By choosing "Univariate" we are specifying the procedures that include only one dependent variable.

These menus allow you to specify the 2×ANOVA analyses through a series of sub-menus. Figure 13.7 shows the overall menu for specifying the factors and dependent variable measure of the analysis. As you can see, I specified overall as the "Dependent Variable:" and both levelsch and subject as the "Fixed Factor(s):" in this analysis.

Overall	Levelsch	Subject	Overall	Levelsch	Subject
1	1	1	3	2	2
3	1	1	2	2	2
3	1	1	4	2	2
3	1	1	1	2	2
3	1	1	3	2	2
2	1	2	4	3	1
2	1	2	4	3	1
3	1	2	4	3	1
4	1	2	3	3	1
3	1	2	3	3	1
4	2	1	1	3	2
2	2	1	2	3	2
3	2	1	1	3	2
4	2	1	1	3	2
2	2	1	1	3	2

FIGURE 13.5 The SPSS® data file for the 2×ANOVA example.

The Univariate menu allows the researcher several choices for specifying the analysis. The buttons on the upper right side of the callout window can be used to customize the output. In what follows, I will show a series of callouts for a "basic" 2×ANOVA procedure for our example. More complex analyses can be created by researchers who gain experience with 2×ANOVA.

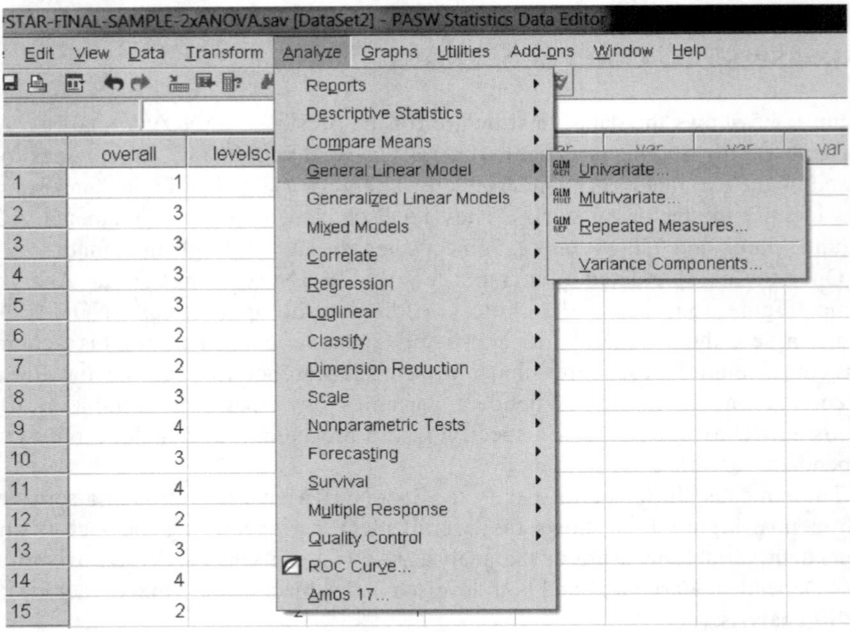

FIGURE 13.6 The SPSS® "General Linear Model" and "Univariate" menus.

USING SPSS® TO ANALYZE 2×ANOVA

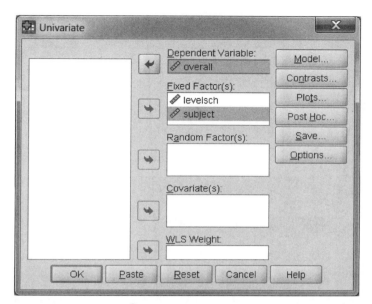

FIGURE 13.7 The SPSS® menus for specifying the 2×ANOVA procedure.

The "Plots" Specification

This choice allows you to create plots of the factors and is a very helpful procedure. Figure 13.8 shows how to use the menus to specify the plot that I produced in Figure 13.4. As you can see, I created the "levelsch*subject" plot by specifying levelsch in the horizontal axis and subject in the "Separate Lines:" window. You must push the "Add" button (located just above the Plots window) once you specify the plot so that it shows up in the Plots window.

You can choose the next button "Post Hoc" from the Univariate window if you wish to examine the group differences. However, since our interaction term was significant, we will need to look at the simple effects.

The "Univariate: Options" button is important for specifying further the kinds of output you wish to see. Figure 13.9 shows some of the choices you can make. For this basic example, I specified descriptive statistics and effect size estimates. The "Estimated Marginal Means" shown in the upper part of the panel are *estimated population means for the cells*. This menu would help with the simple effects analyses. However, SPSS® does not allow simple effects analyses to be specified from this menu. Researchers must use SPSS® "syntax" (a series of pre-programmed commands) in conjunction with the 2×ANOVA procedure. The following syntax specification must be added to the procedure syntax in order to produce simple effects output (readers unfamiliar with SPSS® syntax may want to note these for further study):

/EMMEANS = TABLES (levelsch*subject) comp (subject)
/EMMEANS = TABLES (levelsch*subject) comp (levelsch)

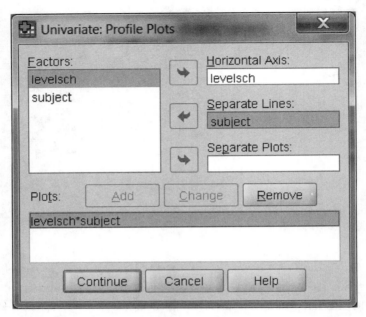

FIGURE 13.8 The Plots window that specifies the results graph.

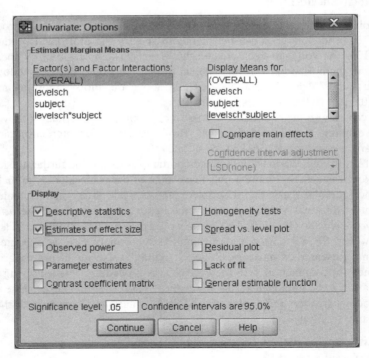

FIGURE 13.9 The choices in the "Univariate: Options" window.

Tests of Between-Subjects Effects

Dependent Variable:overall

Source	Type III Sum of Squares	df	Mean Square	F	Significance	Partial Eta Squared
Corrected model	15.767[a]	5	3.153	4.400	0.006	0.478
Intercept	208.033	1	208.033	290.279	0.000	0.924
Levelsch	0.867	2	0.433	0.605	0.554	0.048
Subject	5.633	1	5.633	7.860	0.010	0.247
Levelsch*subject	9.267	2	4.633	6.465	0.006	0.350
Error	17.200	24	0.717			
Total	241.000	30				
Corrected total	32.967	29				

[a] R squared = 0.478 (Adjusted R squared = 0.370).

FIGURE 13.10 The SPSS® 2×ANOVA summary table.

After the foregoing choices are made, the researcher can create the 2×ANOVA by choosing "OK" on the main Univariate menu. In what follows, I will present some of the main parts of the SPSS® output relevant to our example. Figure 13.10 shows the main summary table. You can compare the values in this table to the values we calculated by hand above.

Omnibus Results

As you can see in Figure 13.10, two columns are added to the summary table that we did not include in our hand calculations. The "Significance" column specifies whether, and how far, the transformed F-ratio value falls into the comparison distribution. Thus, significance values beyond 0.05 indicate that the F results fall further into the exclusion region and are considered significant. All the values except the F ratio for levelsch are significant.

The "Partial Eta Squared" values match those we calculated by hand. One note here is that the Partial Eta Squared for the Corrected Model (0.478) shown in Figure 13.10 is actually the omnibus η^2. The "Corrected Model" results represent the omnibus Between variance measures. The "Corrected Total" variance measure represents our Total variance measure.

Figure 13.10 results indicate that the significant interaction will take precedence in our discussion and findings of these data. Had the interaction not been significant, we could have proceeded to examine the results for each main effect separately. In what follows, I will discuss the simple effects findings because the interaction was significant.

Simple Effects Analyses

When you include the appropriate syntax instructions for simple effects, SPSS® produces simple effects tables showing separate F tests for each level of one

Univariate Tests

Dependent Variable: overall

levelsch		Sum of Squares	df	Mean Square	F	Significance	Partial Eta Squared
Elementary	Contrast	0.100	1	0.100	0.140[a]	0.712	0.006
	Error	17.200	24	0.717			
Middle	Contrast	0.400	1	0.400	0.558	0.462	0.023
	Error	17.200	24	0.717			
High School	Contrast	14.400	1	14.400	20.093	0.000	0.456
	Error	17.200	24	0.717			

[a] Each F tests the simple effects of subject within each level combination of the other effects shown. These tests are based on the linearly independent pairwise comparisons among the estimated marginal means.

FIGURE 13.11 The SPSS® simple effects table for levels of schools on subject areas.

variable across the levels of the second factor. Figure 13.11 shows the simple effects table for each of the levels of the schools across levels of subject. Each main row is a separate F test of that particular level.

As you can see from Figure 13.11, the third category of findings (High School) is a test of the overall scores among high school classrooms to see whether the subject levels (English and math) differ significantly from one another. The F ratio (20.093) is significant ($p = 0.000$), which indicates that English and math are significantly different among high school classes.[2] English and math are not different from each other at the other levels of schools. Elementary schools and middle schools show nonsignificant differences between English and math overall scores ($F = 0.14$ and $F = 0.558$, respectively).

The plot in Figure 13.12 shows these simple effects results. Like the plot in Figure 13.4, this figure shows how the dependent variable scores on each level of one factor change across levels of the other factor. In Figure 13.12, the plot provides a visual description of the simple effects results we saw in Figure 13.11. The English and math overall scores differed significantly at the high school levels as shown in the plot. The other two lines (elementary and middle schools) show very little difference in overall scores between English and math subjects.

Figure 13.13 shows the other simple effects analysis. In this table, only the second subject effect (math) is significant ($F = 5.302$, $p = 0.012$). This analysis indicates that the different school levels show significantly different overall scores among math classrooms. Figure 13.4 is a plot of this simple effect analysis. As you can see, the math subject line is considerably different across levels of school, with the high school level much lower than the elementary school and middle school levels. The English subject line indicates differences among the school levels (with the high school level showing highest overall scores), but not significantly so.

---

[2] Remember that a significance level of 0.000 as reported by SPSS® is shorthand for a very small "actual" probability. You may want to simply report $p < 0.001$ to indicate a very small chance probability.

TERMS AND CONCEPTS

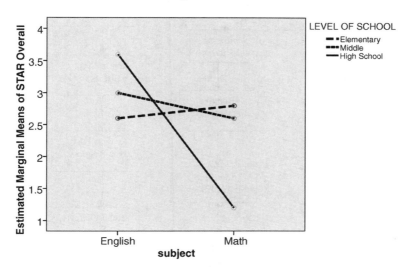

FIGURE 13.12 The interaction graph for the 2×ANOVA example (subject by levelsch).

SUMMARY CHART FOR 2×ANOVA PROCEDURES

The procedures for 2×ANOVA may seem complex compared to those for one-way ANOVA, but it is straightforward. SPSS® provides a thorough series of output to help with the simple effects analyses, which seem to be the most time consuming aspects. Figure 13.14 is a procedural flow chart for 2×ANOVA that you might consider as you prepare to use the procedure with your data.

TERMS AND CONCEPTS

ANCOVA (Analysis of Covariance) An ANOVA type design that limits the influence of a variable or variables outside the design (known as "covariates") that might affect the results.

Univariate Tests

Dependent Variable:overall

Subject		Sum of Squares	df	Mean Square	F	Significance	Partial Eta Squared
1.00	Contrast	2.533	2	1.267	1.767[a]	0.192	0.128
	Error	17.200	24	0.717			
2.00	Contrast	7.600	2	3.800	5.302	0.012	0.306
	Error	17.200	24	0.717			

[a]Each F tests the simple effects of levelsch within each level combination of the other effects shown. These tests are based on the linearly independent pairwise comparisons among the estimated marginal means.

FIGURE 13.13 The SPSS® simple effects table for subject areas on levels of schools.

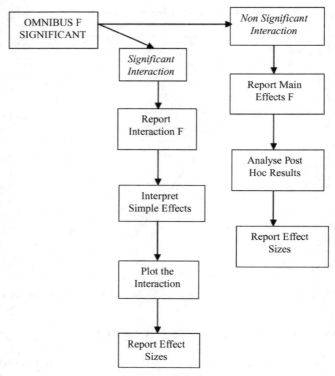

FIGURE 13.14 The 2×ANOVA procedure chart.

Disordinal Interactions Interactions indicated by graph lines that cross in the plot area.

Factorial ANOVA A designation indicating an ANOVA with more than one independent variable (i.e., factor).

Interaction Effects When the relationship between one predictor and the outcome variable changes at different levels of another predictor variable [See Abbott (2010)].

MANCOVA (Multivariate ANCOVA) The statistical procedure that uses ANCOVA with multiple dependent variables.

MANOVA (Multivariate ANOVA) The statistical procedure that uses ANOVA with multiple dependent variables.

Multivariate The designation in statistical analyses indicating more than one dependent variable.

Ordinal Interactions Interactions that may not cross in the plot area, but which may cross if the lines were continued outside the plot area.

Partial Eta Square The effect size measure expressing the proportion of the total variance explained by an independent variable when other influences are controlled.

Simple Effects   Differences (outcome) in the categories of one variable within single categories of another variable.

SPSS® Syntax   Preprogrammed commands used in SPSS® procedures.

Univariate   The designation in statistical analyses indicating one dependent variable.

Within-Subjects ANOVA   Indicating an ANOVA used for repeated measures; either the same group of subjects or subjects in matched groups.

REAL-WORLD LAB IX: 2×ANOVA

We will return to our school database for this lab. The chapter example focused on teaching and learning in math and English classes using the STAR™ data. We will continue the focus on math, but the school data uses aggregated variables to predict/explain school-level math achievement.

In previous labs and exercises, we have explored the importance of FR in education studies. FR, as an indicator of family income, has been shown in the literature to be a very powerful influence on achievement at the school level. We will use FR in this study along with an indicator of ethnicity, "percent white."

Both family income and ethnicity are featured in most all studies of school and student achievement. More infrequently, both variables are included in studies to determine which is the more powerful explanation of achievement. The studies that do include both most often conclude that income is the stronger predictor of achievement [e.g., Abbott (2010)]. This lab provides an opportunity to examine the dual influence of income and ethnicity on school-level achievement using 2×ANOVA.

I have already discussed FR as a predictor of achievement in Chapter 12. Let me add a note about the ethnicity variable, "percent white." The school database provides the percent of students in various ethnic categories at schools in the state. Studies commonly use the percent of Caucasian students at a school as an indicator of "white/non-white" proportion. That is, if you report the percent of white students, this is also a way of reporting the percentage of students who are non-white. Studies often specify ethnic categories for further analysis, but a general "non-white" category is a helpful indicator in some procedures.

Here are the variables we will use in this lab:

- Math: The percentage of students at the school who passed the state math assessment.
- FR: The percentage of students in schools qualified for free and/or reduced lunches. Three equal groups were created using percentiles.
- Ethnicity: The percentage of students in schools categorized as "white." Two groups were created by a median split.

FR	Ethnicity (Low)	Ethnicity (High)
1 (Low)	65 70	56 85
2 (Medium)	24 46	69 51
3 (High)	13 28	47 46

FIGURE 13.15 The balanced data for the 2×ANOVA Real-World Lab.

The study schools were those in the database we used for the Chapter 12 example of one-way ANOVA. In this lab, we will use a balanced 2×ANOVA design comprising 6 cells (similar to the chapter example of STAR™ data). We randomly chose the schools for each cell from the 58 (randomly chosen) schools we used in Chapter 12 (we used the database in which we dropped the extreme score) to provide an example of the 2×ANOVA procedure.

Figure 13.15 shows the data table for this lab. As you can see, in order to balance the design, we are only using 12 schools. This sample is randomly chosen; but because it is so small, we should use caution in generalizing the findings. I will comment further at the end of the lab procedure.

Real-World Lab IX Questions.
1. Calculate the 2×ANOVA by hand.
2. Calculate effect sizes.
3. Interpret the results from your calculated summary table.
4. Perform the procedure using SPSS® (if available) and interpret results.

REAL-WORLD LAB IX: 2×ANOVA SOLUTIONS

The results for this lab are included under each question.

1. Calculate the 2×ANOVA by Hand

Calculating SS_T (all six "cells" are included in the equation):

$$SS = \sum X^2 - \frac{(\sum X)^2}{N},$$

$$SS_T = (9125 + 10,361 + 2692 + 7362 + 953 + 4325)$$
$$- \frac{(135 + 141 + 70 + 120 + 41 + 93)^2}{12},$$

$SS_T = 34,818 - 30,000,$

$SS_T = \mathbf{4818}$

REAL-WORLD LAB IX: 2×ANOVA SOLUTIONS

Calculating SS_B (including the six cell means):

$$SS_B = \frac{(\sum X_1)^2}{n_1} + \frac{(\sum X_2)^2}{n_2} + \frac{(\sum X_3)^2}{n_3} + \frac{(\sum X_4)^2}{n_4} + \frac{(\sum X_5)^2}{n_5} + \frac{(\sum X_6)^2}{n_6} (-) \frac{(\sum X)^2}{N},$$

$$SS_B = \frac{(\sum 135)^2}{2} + \frac{(\sum 141)^2}{2} + \frac{(\sum 70)^2}{2} + \frac{(\sum 120)^2}{2} + \frac{(\sum 41)^2}{2} + \frac{(\sum 93)^2}{2}$$
$$(-)\frac{(135 + 141 + 70 + 120 + 41 + 93)^2}{12},$$

$$SS_B = 33,868(-)30,000,$$

$$\mathbf{SS_B = 3868}$$

Calculating SS_W (including the six cell values):

$$SS_W = SS_T - SS_B,$$
$$SS_W = 4818 - 3868,$$
$$\mathbf{SS_W = 950}$$

Calculating the $SS_{(Rows)}$ (the main effect of school level):

$$SS_{(Rows)} = \frac{(\sum row1)^2}{n_1} + \frac{(\sum row2)^2}{n_2} + \frac{(\sum row3)^2}{n_3} (-) \frac{(\sum X)^2}{N},$$

$$SS_{(Rows)} = \frac{(276)^2}{4} + \frac{(190)^2}{4} + \frac{(134)^2}{4} (-)30,000,$$

$$SS_{(Rows)} = 32,558(-)30,000,$$

$$\mathbf{SS_{(Rows)} = 2558}$$

Calculating the $SS_{(columns)}$ (the main effect of subjects):

$$SS_{(columns)} = \frac{(\sum col1)^2}{n_1} + \frac{(\sum col2)^2}{n_2} (-) \frac{(\sum X)^2}{N},$$

$$SS_{(columns)} = \frac{(246)^2}{6} + \frac{(354)^2}{6} (-)30,000,$$

$$SS_{(columns)} = 30,972(-)30,000,$$

$$\mathbf{SS_{(columns)} = 972}$$

Calculating the Interaction. Like the SS_W, we can calculate the interaction SS (symbolized as SS_{rXc}) by subtracting the SS for rows and columns from the overall SS between as follows:

$$SS_{rXc} = SS_B - (SS_{(Rows)} + SS_{(columns)})$$
$$SS_{rXc} = 3868 - (2558 + 972)$$
$$\mathbf{SS_{rXc} = 338}$$

Source of Variance	SS	df	MS	F^a
Between	3868	5	773.6	4.8859*
Row main effect (FR)	2558	2	1279	8.078*
Column main effect (Ethnicity)	972	1	972	6.139*
Interaction effect (FR × Ethnicity)	338	2	169	1.067
Within (error)	950	6	158.33	
Total	4818	11		

[a]The F ratios indicated by an asterisk are significant at or beyond $p<0.05$ as determined by the F distribution table of values.

FIGURE 13.16 The 2×ANOVA summary Table for Lab IX.

The 2×ANOVA Summary Table. Figure 13.16 shows the 2×ANOVA summary table. Note that you may show slightly different computed values due to rounding. I used an Excel® spreadsheet that retains the actual value of the entries but used the rounded values in the SS calculations above. Any differences due to these rounding procedures should not affect the overall findings much.

2. Calculate Effect Sizes

Omnibus Eta Square:

$$\eta^2 = \frac{SS_{Between}}{SS_{Total}},$$

$$\eta^2 = \frac{3868}{4818},$$

$$\eta^2 = 0.80$$

Effect Size for Row Main Effect (FR):

$$\text{Partial } \eta_{row}^2 = \frac{SS_{row}}{SS_{row} + SS_W},$$

$$\text{Partial } \eta_{row}^2 = \frac{2558}{2558 + 950},$$

$$\text{Partial } \eta_{row}^2 = 0.729$$

Effect Size for Column Main Effect (Ethnicity):

$$\text{Partial } \eta_{col}^2 = \frac{SS_{col}}{SS_{col} + SS_W},$$

$$\text{Partial } \eta_{col}^2 = \frac{972}{972 + 950},$$

$$\text{Partial } \eta_{col}^2 = 0.5057$$

Effect Size for Interaction Effect (FR × Ethnicity):

$$\text{Partial } \eta^2_{rXc} = \frac{SS_{rXc}}{SS_{rXc} + SS_W},$$

$$\text{Partial } \eta^2_{rXc} = \frac{338}{338 + 950},$$

$$\text{Partial } \eta^2_{rXc} = 0.2624$$

3. Interpret the Results from Your Calculated Summary Table

The 2×ANOVA summary table indicates that the omnibus F ratio (4.8859) is significant ($p < 0.05$), indicating that the data show significant differences among the various (cell) mean groups. School-level math achievement is affected by FR and/or ethnicity; both main effects ($F = 8.078$ and $F = 6.139$, respectively) are significant ($p < 0.05$). The interaction effect is not significant.

Eta Squared for the omnibus F ratio is 0.80, indicating that over 80% of the variance in the school math achievement results are attributable to the FR and ethnicity groupings. Partial eta square values for FR and ethnicity are both substantial (0.729 and 0.5057, respectively), with FR indicating the stronger effect.

Note: Because there are so few cases, we must use caution in interpreting the results. However, I performed the 2×ANOVA on the entire dataset (which you can also do with the data files provided in the Wiley website) and confirmed the same trends. Overall ($N = 1039$), FR, ethnicity, and the interaction are all significant (which, with the N size, is not surprising). The effect sizes indicate the various strenths of the findings, however. The partial eta squares for FR (0.261) are much stronger than those for ethnicity (0.012) and the interaction (0.017), indicating that FR is the stronger predictor of math achievement. The findings in Lab IX with $n = 12$ show the same trends, albeit with a stronger ethnicity effect size.

4. Perform the Procedure Using SPSS® (If Available) and Interpret Results

As you can see, the SPSS® results in Figure 13.17 match the hand calculations in Figure 13.16. The actual significance levels in the "Significance" column provide a more specific picture of how far into the exclusion region the transformed F ratios fell. From this point of view, ethnicity was narrowly significant.

Figure 13.18 shows the (Tukey) *post hoc* analyses of FR. Because the interaction was not significant, we can proceed to analyze the main effects and show the *post hoc* analyses. (I did not calculate these by hand because they are readily available from SPSS®; I am presenting the results here for the example Lab data.) Note that the ethnicity *post hoc* analyses are not available because there are only two groups. You can examine the mean values directly to see which levels of ethnicity are higher.

You can see from Figure 13.18 that the differences in the row variable are due to the differences between group 1 (Low) and group 3 (High) FR values. Thus, schools with the highest FR values (i.e., the schools with the greatest percentages of students eligible for FR funding) show a lower math achievement average.

Tests of Between-Subjects Effects

Dependent Variable:Math

Source	Type III Sum of Squares	df	Mean Square	F	Significance	Partial Eta Squared
Corrected model	3868.000[a]	5	773.600	4.886	0.040	0.803
Intercept	30,000.000	1	30,000.000	189.474	0.000	0.969
FR	2558.000	2	1279.000	8.078	0.020	0.729
Ethnicity	972.000	1	972.000	6.139	0.048	0.506
FR*ethnicity	338.000	2	169.000	1.067	0.401	0.262
Error	950.000	6	158.333			
Total	34,818.000	12				
Corrected total	4818.000	11				

[a]R squared = 0.803 (Adjusted R Squared = 0.639).

FIGURE 13.17 The SPSS® summary table for lab IX.

You can use your calculated values to compare the ethnicity group means. The data indicate that the low ethnicity group of schools show a lower math achievement passing proportion (41%) than the high ethnicity group of schools (59%).

You can see the results of the overall analysis by looking at the simple effects plots *despite the fact that the interaction was not significant*. Figure 13.19 shows one such plot. From our discussion of interaction plots above, it appears there is an ordinal interaction in Figure 13.19. Recall that in the ordinal interaction, the lines do not cross in the plotted area, but may cross if they are extended. In this case, however, the extension of the lines would make no sense because there are no other

Multiple Comparisons

Math

Tukey HSD

(I) FR	(J) FR	Mean Difference (I−J)	Standard Error	Significance	95% Confidence Interval Lower Bound	95% Confidence Interval Upper Bound
1.00	2.00	21.5000	8.89757	0.114	−5.8002	48.8002
	3.00	35.5000[b]	8.89757	0.017	8.1998	62.8002
2.00	1.00	−21.5000	8.89757	0.114	−48.8002	5.8002
	3.00	14.0000	8.89757	0.326	−13.3002	41.3002
3.00	1.00	−35.5000[b]	8.89757	0.017	−62.8002	−8.1998
	2.00	−14.0000	8.89757	0.326	−41.3002	13.3002

[a]Based on observed means. The error term is mean square (error) = 158.333.
[b]The mean difference is significant at the 0.05 level.

FIGURE 13.18 The *post hoc* analyses from the 2×ANOVA[a].

REAL-WORLD LAB IX: 2×ANOVA SOLUTIONS

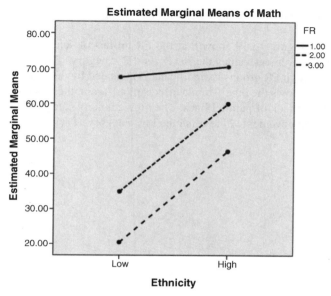

FIGURE 13.19 The FR simple effects plot for the 2×ANOVA lab.

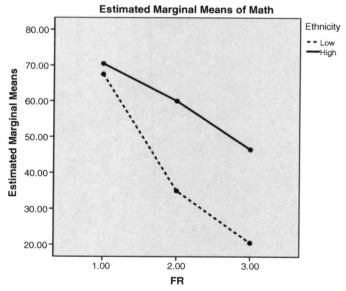

FIGURE 13.20 The ethnicity simple effects plot for the 2×ANOVA lab.

categories of ethnicity. These are categories that do not extend as would continuous data values.

The lines in Figure 13.19 show that the FR groups of schools generally show higher math achievement values the lower the FR category. The distance between the low FR and high FR groups is significant as indicated by the Tukey results.

Figure 13.20 shows the other simple effects plot. Again, there is no ordinal interaction despite the lack of parallel lines. The high ethnicity category is significantly higher in math achievement values than the low category of schools.

14

CORRELATION

Probably everyone has an understanding of the basic principles of correlation; it has an intuitive appeal. Most everyone understands that correlation analyzes whether changes in one thing are linked to changes in something else. Thus, for example, we may observe that students who get the highest reading achievement test scores are also the ones who read the most. Or, stated another way, as the achievement scores change in the upward direction, so does the amount of time spent reading, and vice versa.

In Chapter 5, I introduced correlation as a way of understanding the association between two variables. What does association mean? It refers to "relatedness" or the extent to which two events "vary with one another." To use the example I alluded to in Chapter 5, if the percentage of Caucasian students increases in schools, achievement increases. We can say that the values of the one variable (achievement) change positively as the values of the other variable (ethnicity) increase. Thus, the values of both variables increase together, or "co-vary."

Of course, as I also discussed in Chapter 5, not every correlation is what it seems! There may be additional variables not taken into account in the analysis that give the original two variables the "appearance of co-varying." This is known as spuriousness. The Real-World Lab IX in Chapter 13 hints at the complexity of the ethnicity–achievement correlation by considering the additional impact of family income in the relationship.

Understanding Educational Statistics Using Microsoft Excel® and SPSS®. By Martin Lee Abbott.
© 2011 John Wiley & Sons, Inc. Published 2011 by John Wiley & Sons, Inc.

THE NATURE OF CORRELATION

We will examine correlation in some depth in this chapter. It is a useful procedure for many reasons, and several methods of calculating correlations exist that are "fitted" to the nature of the research question.

Explore and Predict

Evaluators use correlation to *explore* the relationships among a series of variables they suspect may be important to a research question. Other evaluators may use correlation to help *predict* an outcome knowing that the predictor and the outcome variables are related. Explanation and prediction are two important uses of correlation.

Different Measurement Values

Correlation is somewhat unusual in that the researcher can measure the relationship between two variables that are operationally measured differently. For example, in the opening paragraph, I mentioned the relationship between reading achievement and amount of reading. If I operationally define *reading achievement* as the score on a state achievement test and *amount of reading* as the number of books read per month, I will end up with two different sorts of scores. The state reading test may represent a standardized score that ranges from 100 to 600, while the amount of reading may simply be a number that ranges from 0 to 20 (although not many K–12 students read 20 books per month!).

Different Data Levels

In the foregoing example, the researcher can measure the association between two differently measured variables because we are looking at how one score changes as the other score changes. Thus, both variables do not have to be measured with the same *values*. Correlation measures are so powerful that a researcher can also calculate the correlation of two variables measured with different *levels of data* (i.e., interval, ordinal, or nominal). A researcher might correlate reading achievement scale scores (interval level) with students' subjective appraisals of the extent of their reading (ordinal level) as their rating of "A Lot, Some, A Little, or None."

Correlation Measures

In this chapter, I will discuss correlation measures that can be used with two interval or ratio level variables. Table 14.1 shows some *additional* correlation measures that can be calculated with variables of different levels of data.

As you can see, there are several measures of correlation that can be "customized" for the specific nature of the researcher's study question. The first three columns of Table 14.1 show measures of correlation that use different levels of

THE CORRELATION DESIGN

TABLE 14.1 Measures of Correlation

Nominal Level Data	Ordinal Level Data	Nominal by Ordinal	Proportional Reduction in Error (PRE)
Contingency Coefficient	Spearman's Rho	*Biserial*	η^2 Partial η^2 r^2
Phi Coefficient	Kendall's Tau	Point Biserial	Gamma
Cramer's V			(Cohen's) Kappa
Tetrachoric			

data. The correlation measures that are italicized are those that use "artificial" measures. For example, the tetrachoric correlation measures two continuous (interval level) variables that have been *dichotomized*. In some of our examples in this book, we created categories out of continuous scores, like creating FR categories (low, medium, and high) out of a continuous score (percentage of students at a school eligible for FR) to illustrate a particular statistical procedure. Tetrachoric correlation might measure the relationship between FR and another dichotomized variable.

Table 14.1 also shows a column (the fourth column) of measures known as proportional reduction in error (PRE) measures. We have already worked with some of those measures at the interval level (eta squared and partial eta squared), but there are several others at other levels of data I indicated with different shades (i.e., gamma for ordinal and Cohen's Kappa for nominal). These are measures in which correlation is used to "reduce error." We saw in Chapter 12 (one-way ANOVA) that η^2 is a measure of how much variance in the outcome variable is explained by the categories of the predictor variable. That is, how much variation (error) is reduced by knowing the relationship between the two variables? Eta squared and other such measures are expressed in the amount of *variance explained*.

We cannot hope to talk about all of these in this book. I wanted you to see the tremendous variety of correlation measures; and there are yet more measures I didn't include in Table 14.1! In this chapter I will discuss Pearson's correlation coefficient, which is the most common correlation procedure with interval data.

THE CORRELATION DESIGN

The correlation study is a *post facto* design because the researcher is *relating* two sets of scores that have been gathered on an individual case. We do not use correlation in an experimental design because in the latter we are attempting to detect *differences* in group scores after an intervention. Correlation measures "sameness," and an experiment measures "difference." Multiple regression procedures, which are based on correlation, can be used in experimental designs under certain circumstances (see Abbott, 2010).

PEARSON'S CORRELATION COEFFICIENT

Named after Karl Pearson, the Pearson's Correlation Coefficient, symbolized by r, is used to measure the relationship between two interval level variables. I used the above example of reading achievement and number of books read to show how this method can be quite versatile and helpful.

Interpreting the Pearson's Correlation

When we calculate Pearson's r, we need to know what it means! Pearson's r is a number that varies from -1.0 to $+1.0$. The closer the r value is to 0, the less the two variables are related to one another. Here are the two primary "dimensions" of Pearson's r that are helpful for interpreting the relationship:

- *Strength: The closer the r value gets to either -1 or $+1$, the stronger the correlation between two variables.* An r value of 1.00 would indicate that every time one variable increased by one unit, the second variable increases by one unit. It is also the case that a value of 1.00 would indicate that each time a variable decreases by one unit, the second variable also decreases by one unit.
- *Direction: When the variables change their values in the same direction, the r is a positive correlation. Whenever the variables change in opposite directions, the r value is negative.* "Positive" and "negative" do not mean "good" and "bad," they simply indicate the direction of change in both variables. Negative correlations are also called *inverse correlations* because one variable is going up as the other is going down in value.

Figure 14.1 shows some examples of the different possibilities of weak and strong, positive and negative r values. Keep in mind these r values are not real. I introduced some of them as examples in earlier chapters and I made the others up! I simply wanted to show that an r value can take different values and have different "signs." We will discuss later how to calculate the Pearson's r and how to determine whether the r value is strong or weak.

I used shading in Figure 14.1 to indicate strength of relationship. The "Weak" column thus shows lighter shading and correlations (both positive and negative) that are close to 0. The "Strong" column has a darker shade indicating stronger correlations (both positive and negative) closer to 1.00 and -1.00. The arrows in the cell indicate direction, whether positive (both arrows pointing in the same direction) or negative (arrows pointing in opposite directions). You can see from these features that a negative correlation closer to -1.00 is considered strong even though

PEARSON'S CORRELATION COEFFICIENT

		STRENGTH	
		WEAK	STRONG
DIRECTION	POSITIVE	$r = 0.02$	$r = .85$
		↑ Storks–Babies ↑	↑ Family Income–Student Achievement ↑
		There is very little (non-spurious) relationship between the number of storks and the number of babies being born	As we have seen, schools with wealthier families have higher aggregate achievement scores
	NEGATIVE	$r = -.10$	$r = -.73$
		↑ Noise–Human Learning ↓	↑ Lattes Consumed–Success at Threading a Sewing ↓ Needle
		In my undergraduate experiment, the data only weakly reported an association of increasing noise and decreasing learning	Generally, as people ingest more caffeine, they are less dexterous!

FIGURE 14.1 Examples of Pearson's r values.

it is an inverse relationship. Remember that negative does not mean bad, just opposite direction.

The Fictitious Data

I will use a fictitious example to demonstrate how to calculate Pearson's r and then I will introduce a real data example. Table 14.2 shows the fictitious data we will use to show how to calculate Pearson's correlation. The Reading Achievement (RA) variable consists of test scores from eight students on a reading test with 150 points possible. The Books Read (BR) variable is the number of books the students reported reading over the summer break. The research question is whether there is a correlation between the two variables.

TABLE 14.2 Fictitious Data for Correlation Example

Reading Achievement	Books Read
120	10
20	3
50	7
90	5
110	9
100	8
40	6
10	3

Assumptions for Correlation

As with the other statistical procedures, there are assumptions that must be met before we can use Pearson's r correlation. The primary assumptions are as follows:

- Randomly chosen sample.
- Variables are interval level (for Pearson's r).
- Variables are independent of one another. This assumption is somewhat difficult to understand, but it deals with "autocorrelation," which is the tendency for one set of scores to be linked to a second set in a series, like time-related measures. If we measure daily crime rates, for example, there will be a "built-in" correlation because each day is most often related to the next. With correlation, we need to make sure that we have no such "linkages" like a time series in our data.
- Variables are normally distributed.
- Variances are equal. Pearson's r is robust for these violations unless one or both variables are significantly skewed.
- Linear relationship. The two variables must display a "straight line" when plotting their values. Thus, for example, if we were correlating the age of a car with the value of a car, the correlation would probably be a straight line (in a downward direction, indicating an inverse relationship). Violations of this assumption might include "curvilinear" relationships in which plotted data appear to be in the form of a "U." For example, with the age and value of a car, the linear relationship might change over time because really old cars *increase* in value. We can "see" this assumption in a *scattergram*, which I will discuss below. Formally, we can detect these "curvilinear" relationships through SPSS®.

It should be noted that correlation is a "robust" test, which means that it can provide meaningful results even if there are some slight violations of these assumptions. However, some are more important than others in this regard, as we will see.

PLOTTING THE CORRELATION: THE SCATTERGRAM

I mentioned the scattergram in discussing the assumption of linearity above. Variously known as scatter diagram, scatterplot, scatter graph, or simply scattergram, we can create a visual graph that shows the relationship between two variables.

Figure 14.2 shows the scattergram between our two example variables, reading achievement and books read. As you can see, the dots are displayed from the lower left side of the plot to the upper right side. This pattern indicates a positive correlation because as one variable increases in value, the other value also increases.

Reading the plot is straightforward. The values in the table of data are presented in 'pairs' with each pair representing a single student's scores. Thus, the top pair of

PLOTTING THE CORRELATION: THE SCATTERGRAM

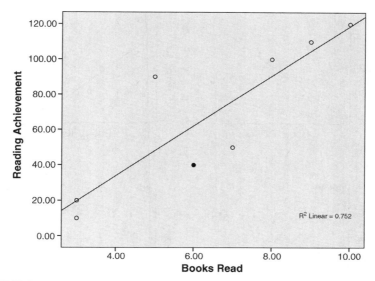

FIGURE 14.2 The scattergram between reading achievement and books read.

values in Table 14.2 indicate that this student scored 120 on the reading achievement test and read 10 books during the summer. I filled in the dot in Figure 14.2 to show the pair of scores for the next-to-last student in Table 14.2 (the student scoring 40 on the reading test and reading 6 books).

As you can see, the pairs of scores are entered into the plot simultaneously so that each dot represents the pair of scores of an individual student. Typically, the outcome variable is placed on the Y axis and the predictor variable is placed on the X axis.

In correlation designs it is not always apparent which variable is the outcome and which is the predictor, you simply have to understand which is which from the research question. The "inherent" research question in our study is whether the amount students read (number of books) will influence their reading achievement. Therefore, the books read is the predictor variable (X) in this study. It could just as easily be the other way around! A researcher could posit that higher reading achievement gives way to increased reading.

Patterns of Correlations

Correlations can be positive like the one in Figure 14.2, or they can be negative and even nonlinear. Figure 14.3 shows some of the possibilities for correlation patterns in scattergrams.

Panel a in Figure 14.3 shows the positive correlation we observed with the actual data in Figure 14.2. However, the results could have been different. Panel b shows a negative (or inverse) correlation in which the more books a student read, the lower the reading achievement score! Panel c shows no correlation at all; the scores do not

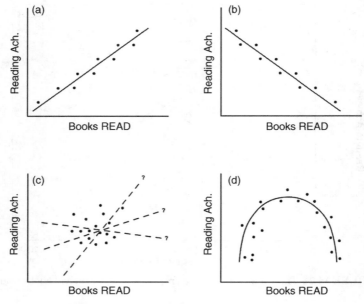

FIGURE 14.3 Correlation patterns in scattergrams.

fall into a recognizable pattern. Panel d shows a curvilinear relationship in which students who read a medium amount of books have high achievement scores, but students who read either a lot or only a few books have poor scores! These panels are only to show the correlation possibilities that might be produced in an actual study.

Strength of Correlations in Scattergrams

The correlation panels in Figure 14.3 showed correlations with different directions or patterns. The dots could extend upward to the right (positive correlation), downward to the right (negative correlation), or in other patterns. Figure 14.4 show scattergrams that indicate the *strength* of the correlation.

The top two panels (a and b) represent positive correlations. The dots are arrayed from bottom left to upper right. But note that the *extent of the scatter around the line* is different. We use a line to represent the scatter and pattern of the dots (in a later chapter, we will learn to calculate the equation for this "line of best fit"). When the dots have a wide scatter, like the scattergram in panel a, the correlation is weaker. This would indicate that as values of one variable (books read) increase, the values of the other also increase, but not "one for one." Panel b shows a positive correlation with a tighter pattern of dots that indicate more of a "one for one" increase in the values of both variables.

Panels c and d of Figure 14.4 show negative correlations. In panel c, the correlation is inverse in that as one variable increases in value, the other decreases, but the

CREATING THE SCATTERGRAM

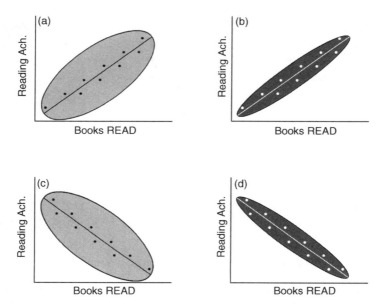

FIGURE 14.4 Strength of correlations in the scattergram.

values do not change in a "one-for-one" relationship. The negative correlation in panel d shows a much higher correlation because the dots are very close to the line, indicating a more "one-to-one" change in values.

The panels underscore the fact that "negative" correlations are not necessarily bad since negative only refers to direction. As you can see in panel d, this negative correlation is very strong. I indicated strength of correlation by the depth of the shading in the panels with stronger correlations showing darker shades.

CREATING THE SCATTERGRAM

You can easily freehand draw a scattergram with pairs of data. However, both Excel® and SPSS® provide simple procedures to create the graphs.

Using Excel® to Create Scattergrams

With the table of data entered in an Excel® spreadsheet, simply highlight the entire table (including labels) and choose the "Insert" menu from the main menu list. Figure 14.5 shows this selection with our table of fictitious data. As you can see, I highlighted the table of values and chose Insert. I can then choose Scatter from the list of possible graphs. I am presented with several choices for the scattergram in a dropdown menu. For the simple, linear correlation the upper left choice is appropriate. As the data increase in complexity, you can experiment with the other scattergram choices.

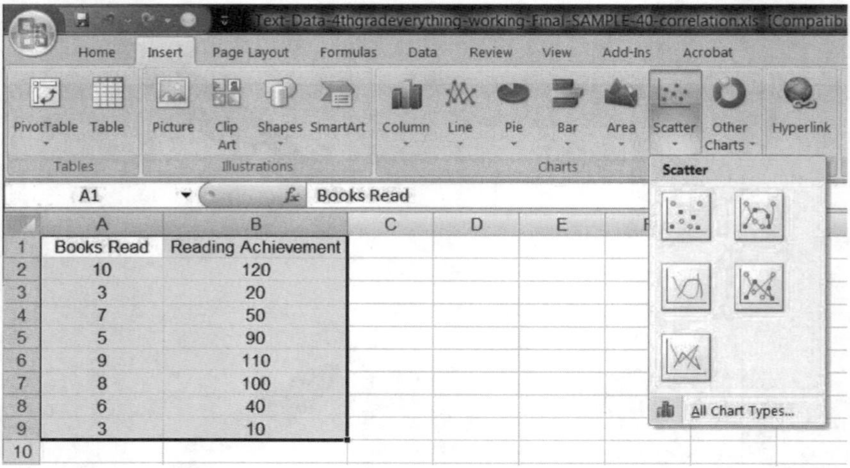

FIGURE 14.5 The Excel® scattergram specification.

Figure 14.6 shows the scattergram created by our menu choices above. As you can see, the dots are displayed according to the pairs in the data table. One note here is that the scattergram will be created according to how you arrange the data. As you can see, the number of books read is shown on the *X* axis because that variable was listed first in the data table. If you reverse the order of the variables, the scattergram will look different because the scales of the variables will be on different axes.

The scattergram in Figure 14.6 is the "plain" graph that Excel® produced by making our menu choices. *You can edit the graph by double clicking on parts of the graph and then right clicking for edit choices.* A simple way to make edits is to use the main menu bar when you have the graph selected. The "Chart Layout" menu provides several different ways that you can make the graph appear. For example, one choice includes axis titles; another includes the raw data table, and so on. I can include the line through the data by using these methods.

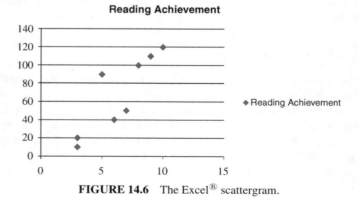

FIGURE 14.6 The Excel® scattergram.

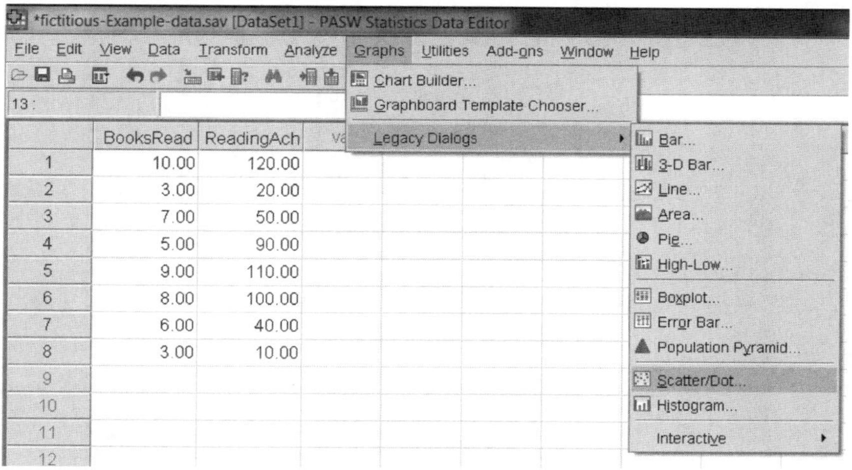

CREATING THE SCATTERGRAM 347

FIGURE 14.7 The SPSS® Graph menu for creating scattergrams.

Using SPSS® to Create Scattergrams

SPSS® has a straightforward way to create scattergrams using the main "Graphs" menu. Figure 14.7 shows how to create the scattergram through the "Graph–Legacy Dialogs–Bar" path of menu choices. Near the bottom of the list is "Scatter/Dot," which will produce a specification box that allows you to design your scattergram. In this system, you do not need to move the data columns so that one of the variables is listed first, as in Excel®.

When you choose "Scatter/Dot" the menu box in Figure 14.8 appears. This box is similar to the Excel® specification in which you can choose the scattergram that matches the complexity of your research data. For our fictitious example, the "Simple Scatter" choice is appropriate.

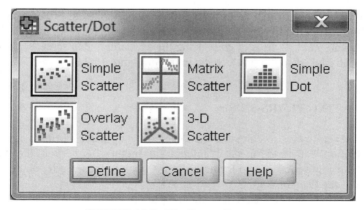

FIGURE 14.8 The Scatter/Dot menus.

FIGURE 14.9 The scattergram specification window in SPSS®.

Choosing the "Simple Scatter" design from the menu produces a specification window like the one shown in Figure 14.9. As you can see, I specified that the Books Read variable should be placed on the X axis. I can select the "Titles" and "Options" buttons in the upper right corner of the window to further specify the graph, but I prefer to edit the scattergram once it is produced.

When you choose "OK" from the window in Figure 14.9, SPSS® produces a graph in the output file like the one shown in Figure 14.2. If you double click on the graph, you can make a series of edits using the available menu screens. I added the line to the "basic" graph to produce the graph in Figure 14.2.

CALCULATING PEARSON'S r

There are several ways to calculate the Pearson's r by hand. Formulas exist using the "deviation" method or the "calculation method" or even the "Z-score method." I want to present two formulas in this book because both point out different facets of the Pearson's r.

The Z-score formula appears to be the most simple until you realize that, in order to use it, you have to transform every score to a Z score. The calculation method is

THE Z-SCORE METHOD

my preferred method because it uses symbols and formulas we have used in past chapters and includes a simple data table of values.

THE Z-SCORE METHOD

The first question most students ask is, Why are there so many different formulas? The simple answer to this is that each formula uses "parts" that express different components of the correlation relationship.

The Z-score formula follows:

$$r_{XY} = \frac{\sum Z_X Z_Y}{N}$$

In this formula, the correlation between two variables X and Y (symbolized by r_{XY}) is calculated by *summing the products* of the X Z scores and the Y Z scores and then dividing by the number of pairs (*in correlation, the N indicates pairs, not the total set of scores*). In order to carry out this formula, you would first need to create Z scores for each of the X and Y values using the formula we discussed in Chapter 7:

$$Z = \frac{X - M}{SD}$$

Creating Z scores by hand is tedious, so if there are many pairs of scores, this is a labor-intensive effort! On the other hand, creating the Z scores in an Excel® spreadsheet is quite simple, as we discussed in Chapter 7. Table 14.3 shows the values as I created them in an Excel® spreadsheet using the embedded Z formula. I had to first calculate the mean and SD values for both variables as shown.

TABLE 14.3 The Data Table Showing Z Scores

Books Read	Z (Books)	Reading Achievement	Z (Achievement)	ZX*ZY
10	1.48	120	1.32	1.95
3	−1.38	20	−1.19	1.64
7	0.26	50	−0.44	−0.11
5	−0.56	90	0.56	−0.32
9	1.07	110	1.06	1.14
8	0.66	100	0.81	0.54
6	−0.15	40	−0.69	0.11
3	−1.38	10	−1.44	1.98
	$\Sigma = 0$		$\Sigma = 0$	
$M = 6.375$		$M = 67.5$	$\Sigma Z_X Z_Y$	6.93
$SD = 2.45$		$SD = 39.92$	r	0.87

Once these calculations are made, you can use the Z-score formula to calculate Pearson's r as follows (I included the relevant calculations in the final column of Table 14.3):

$$r_{XY} = \frac{\sum Z_X Z_Y}{N},$$

$$r_{XY} = \frac{6.93}{8},$$

$$r_{XY} = .866$$

Note: Be careful when you calculate the numerator of the formula. It is not the sum of the Z scores for column Z (Books) times the Z-score sum of column Z (Achievement). That would look like the following:

$$r_{XY} = \frac{(\sum Z_X)(\sum X_Y)}{N}$$

The reason is that the sums of these particular columns would equal 0! Like the deviation method for calculating SD that we discussed in Chapter 6, there will be an equal number of negative and positive Z scores in a normally distributed set of scores. Therefore, you must *multiply each pair of Z scores* to arrive at the values in the last column of Table 14.3. Otherwise, this is the result:

$$r_{XY} = \frac{(0)(0)}{8},$$

$$r_{XY} = 0$$

The reason to use the Z-score formula is to use the transformed X and Y values as Z scores that present the X and Y values in the same scales. The reason to do this will become more apparent when we discuss regression procedures; but for now, look at Figure 14.10 that presents the scattergram with Z-score values for both the

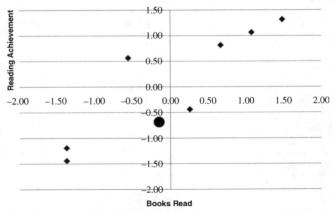

FIGURE 14.10 The Z-score scattergram in Excel®.

THE COMPUTATION METHOD

X and Y values. The X axis is the Z-score values of X (books read) and the Y axis represents the Z values of the Y variable (achievement).

As you can see, all the values of the two scores are now expressed *in the same scale:* as Z-score values, the majority of which (over 99%) range between -3.00 and $+3.00$. Whereas before our highlighted score of the next-to-last student (Books read $=6$, and Achievement score $=40$) was expressed in *raw score units* (that is, the units of the original scores), the transformed scores are expressed in *Z-score units* (Books Read $=-0.15$, and Achievement $=-0.69$). I highlighted the next-to-last student score in both Table 14.3 (shaded) and in Figure 14.10 (as the large round dot) so you can see how the data are arrayed in the scatterdiagram.

Using Z scores thus creates the same scale for raw score variable units of any magnitude. This does not change the correlation or the relationship among the dots, as you can see from the array of dots in Figures 14.6 and 14.10.

THE COMPUTATION METHOD

There are several computation formulas for Pearson's r. I prefer the following because it uses the SS (sum of squares) measures that we have used throughout former chapters. The overall formula looks straightforward. Conceptually, the formula expresses correlation (r) as the proportion of the cross product SS (or the SS of the $X*Y$ pairs) to the square root of the overall product of the X and Y sums of squares.[1]

$$r_{XY} = \frac{SS_{XY}}{\sqrt{(SS_X)(SS_Y)}}$$

Recall that the SS is a general (unstandardized) variance measure of the population of scores. The computation formula thus includes variance (SS) measures of the XY products as well as the individual X and Y SS values. The formula for SS we discussed in Chapter 6 is

$$\sum X^2 - \frac{(\sum X)^2}{N}$$

Each of the SS values (for X and Y) in the Pearson's r formula can be computed with these formulas. The product SS (the numerator in the Pearson's r formula) can

---

[1] This makes conceptual sense if you recognize the denominator as a combined variance measure of X and Y together, because each SS is multiplied (creating a kind of variance measure) and transformed to a combined variance measure. Thus, the formula expresses the "variance" of the XY product measures as a proportion of the overall variance of both sets of scores. This is conceptually what ANOVA does—compares variance components.

be calculated with a variation of the same SS formula by simply using the product column (XY) of values:

$$SS_{XY} = \sum XY - \frac{(\sum X)(\sum Y)}{N}$$

Here is how the example data would be computed using the computation formula:

The SS_{XY}

$$SS_{XY} = \sum XY - \frac{(\sum X)(\sum Y)}{N}$$

$$SS_{XY} = 4120 - \frac{(540)(51)}{8}$$

$$SS_{\text{Books} \times \text{Reading}} = \mathbf{677.5}$$

The SS_Y

$$SS_Y = \sum Y^2 - \frac{(\sum Y)^2}{N}$$

$$SS_{\text{Reading}} = 49,200 - \frac{291,600}{8}$$

$$SS_{\text{Reading}} = \mathbf{12,750}$$

The SS_X

$$SS_X = \sum X^2 - \frac{(\sum X)^2}{N}$$

$$SS_{\text{Books}} = 373 - \frac{2601}{8}$$

$$SS_{\text{Books}} = \mathbf{47.875}$$

Pearson's r

$$r_{XY} = \frac{677.5}{\sqrt{(47.875)(12750)}}$$

$$r_{XY} = \frac{677.5}{781.28}$$

$$r_{XY} = \mathbf{0.867}$$

As you can see, the resulting Pearson's r of 0.867 is equivalent to the r value computed with the Z-score formula. Both formulas yield the same value.

EVALUATING PEARSON'S r

Now that we calculated Pearson's r, what does the value mean? Recall that the (fictional) study asked if reading achievement and the number of books read was correlated. We saw that the scattergrams presented strong visual evidence that there was a relationship, but how can we judge the strength of the relationship simply by looking at the resulting r of 0.867?

The Hypothesis Test for Pearson's r

The r of 0.867 is certainly closer to 1.00 than it is to 0, so we can use this as one gauge of its meaningfulness. However, like the other statistical tests we have discussed in this book, we assess the statistical significance of a test ratio by comparing it to a sampling distribution.

Therefore, the first way to determine whether the r value is significant is to perform a hypothesis test as we have with other statistical procedures. If the calculated value exceeds the tabled value (that is, if the calculated value is further out in the tail of the exclusion area), we concluded that the test value is statistically significant.

We use the same steps as we did with other statistical tests:

1. The Null Hypothesis (H_0): $\rho = 0$. The null hypothesis states that the correlation between the variables in the population from which the sample values came (symbolized by the parameter value of rho or ρ) is 0.

2. The Alternative Hypothesis (H_A): $\rho \neq 0$. The alternative hypothesis states that the population correlation is not 0.

3. The Critical Value $r_{df(0.05)}$. The comparison value for the value of r that defines the exclusion area at a given probability level (I used 0.05 in the above value) is determined by a table of values as are other statistical procedures. The tables for correlation are extensive (as you might imagine since Fisher designed them) and detailed. The exclusion values for Pearson's r can be found in the Appendix.

Like other statistical procedures, testing the null hypothesis for correlation uses degrees of freedom because we are estimating a population parameter from a sample statistic. The degrees of freedom for correlation is $N - 2$, where N is the number of pairs. We "lose" two degrees of freedom because we are estimating a population value with two sample sets of values.

The table for correlation specifies that the two-tailed 0.05 critical (exclusion) value for r at 6 degrees of freedom (8 pairs minus $2 = 6$) is $r_{df(0.05)} = 0.707$. Values that are greater than 0.707 would therefore be considered statistically significant.

4. The Calculated Value (0.867)

5. Statistical Decision. We reject the null hypothesis because the calculated value (0.867) exceeded the critical value of exclusion (0.707). *A significant*

correlation indicates that a calculated value as high as our value is unlikely to occur if there is no correlation in the population.

6. Interpretation. We can consider our calculated r of 0.867 to be statistically significant. This indicates that it is likely not a chance finding. In the words of the (fictitious) research question, the students' reading achievement test scores are positively correlated to the number of books they read during the summer.

The Comparison Table of Values

Because the correlation tables are so extensive and because we will be using Excel® and SPSS®, I include only a brief correlation table in this book. However, there is an alternate way to identify the critical values by using the T table of values that is included. Cohen (1988) provides the following formula that we can use in the T tables as a comparison value. In effect, we are "transforming" the r value into a t ratio that we can then compare against the critical values of T in the T table (using $N-2$ degrees of freedom in the table):

$$t = \frac{r\sqrt{N-2}}{\sqrt{1-r^2}}$$

Therefore, in the present example:

$$t = \frac{0.867\sqrt{8-2}}{\sqrt{1-0.75}}$$
$$t = 4.2474$$

Our r value of 0.867 is "transformed" to a t ratio of approximately 4.2474. When we look in the T table of values using $N-2$ degrees of freedom, we note that the critical value of $t_{0.05,6} = 2.447$. Therefore, our calculated value of r (0.867) is significant because it resulted in a t score that exceeded the critical T value (2.447).

Effect Size: The Coefficient of Determination

A second way to judge the meaningfulness and strength of a correlation is the effect size. With every statistical procedure, we have seen that there is a way to measure the impact of a relationship. With correlation, we can judge the strength of the relationship by the *coefficient of determination* or r^2. *This value is simply the square of r and refers to the amount of variance in one variable explained by the other.* What this means is this: We can consider the fact that a distribution of scores (e.g., reading achievement test scores) vary a certain amount, or are spread out around a mean score.

The question is, Why do reading achievement scores vary? In a world where every child was the same, there would be no variance—everyone would get the

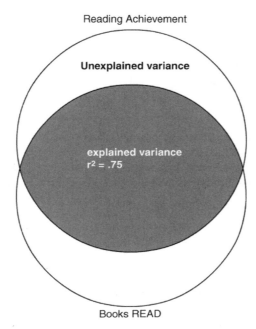

FIGURE 14.11 The effect size of correlation—explaining variance.

same score. But we don't live in that kind of world, so something, or things, are responsible for kids getting different reading achievement scores. Figure 14.11 shows how the variables relate to one another in terms of "explaining variance."

If we establish a correlation is significant (i.e., between achievement test scores and number of books read), the variables' measures overlap (like the Venn diagram in Figure 14.11), and we can understand this overlap to be the amount of variance in the original spread of the scores (reading achievement) that is explained by this new variable (number of books read). Thus, knowing how many books a student reads during the summer is a partial explanation of their reading achievement scores. The number of books read doesn't explain all the variation in reading achievement, because there is still a lot of "unexplained variance" (the amount of the reading achievement circle not overlapping with the books read circle), but it "chips away" at the overall spread of the scores. The r^2 value is this explained variance.

Figure 14.12 shows another way to visualize effect size in correlation. The overall distribution represents the entire variability of reading achievement scores. With a correlation of 0.867, the r^2 of 0.75 reduces the amount of "unexplained variance" in reading achievement by 75%. Thus, there are other variables not in our analysis that might help to further "explain away" the remaining 25% of unexplained variation.

I will discuss further insights into r^2 in Chapter 15. *For now we can understand it as the percentage of the variance in one variable (outcome) contributed by another (predictor).* Knowing how much variance is explained is helpful to the researcher,

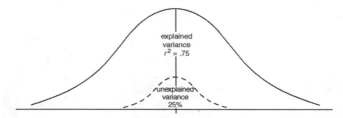

FIGURE 14.12 The effect size components produced by correlation.

but what guidelines do we have that will help us judge the extent of the explained variance? Cohen (1988) provides the following conventions for the r^2:

Small Effect Size: $r^2 = 0.01$ ($r = 0.10$)

Medium Effect Size: $r^2 = 0.09$ ($r = 0.30$)

Large Effect Size: $r^2 = 0.25$ ($r = 0.50$)

These r^2 size conventions may appear to be small, and for some studies they may not be appropriate. The researcher is ultimately in charge of establishing the meaningfulness of the correlation effect size since it is tied to the nature of the research question. Thus, I might not be excited about an r^2 of 0.09 in a small exploratory summer school study of reading achievement and books read, but I might be very excited about the same effect size in a study of the relationship between a collaborative teaching approach developed in several school districts and increased student reading scores.

The r^2 is one of the PRE methods I mentioned at the opening of this chapter. If you look at Figure 14.1 r^2 is listed in the last column with the other PRE methods that we discussed in the ANOVA chapters. All these are related measures in that they focus on explaining variance. I will develop this notion further in Chapter 15.

CORRELATION PROBLEMS

There are several factors that can affect the size of a correlation and therefore its power (i.e., the ability to reject the null hypothesis when it should be rejected). I will mention several such factors in this section.

Correlations and Sample Size

As I stressed in previous chapters, hypotheses tests establish whether a finding is likely a chance finding. (We just discussed the effect size, which is distinguished from the statistical rejection of the null hypothesis.) Be aware that *the sample size dramatically affects the size of Pearson's r.*

For example, with 15 cases, the researcher can reject the null hypothesis (two-tailed at 0.05) with $r = 0.514$. With a sample of 30, the researcher can reject the null hypothesis at 0.361. With a sample of 102, the researcher can reject the null

CORRELATION PROBLEMS

hypothesis at 0.195. It only takes an $r = 0.098$ to reject the null hypothesis with $N = 402$! *Therefore, with the same study variables, but simply larger samples, the researcher will reject the null hypothesis more easily.* (This relates to power analysis, as I noted above.)

Correlation is Not Causation

We have noted this injunction previously. The problem with discovering correlations is the temptation to assume that the variables are *causally* related. Thus, we might be tempted to assume that the fictitious study correlation above means that time reading causes achievement test score increases. It does not necessarily mean this. There are many things that are related to reading achievement test scores other than the number of books one reads during the summer—and some of them may be better explanations of increases in scores. What are some? If you were to list other *potential* causes, you could probably produce a long list (e.g., the nature of *how* the student reads, ability, school curriculum, nature of the environment at home, interest level, parents ability, and so on).

Restricted Range

The strength and size of a correlation is dramatically affected by "restricted range," or the selection of scores that do not display full variability. Figure 14.13 shows this problem with a scattergram.

If a study sample is relatively homogenous on the variables (e.g., all the subjects appear to be similar in achievement or time reading), the resulting correlation can appear to be very low. In the top panel of Figure 14.13, students of all

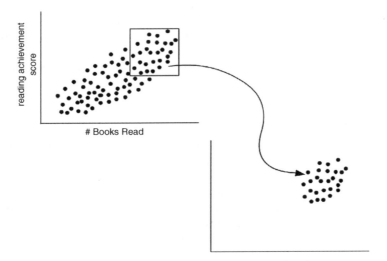

FIGURE 14.13 The correlation problem of restricted range.

different achievement scores and number of books read were correlated, which resulted in a strong correlation. If the study is restricted to students who are very similar to one another, like a group of students from a gifted reading program who were obsessive readers, the resulting correlation would be less pronounced and perhaps not significant. The power of a correlation is related to the variability of the study variables.

Extreme Scores

"Outlier" scores can have a dramatic effect on the calculated r. This is the same problem with any statistical procedure like the one we examined in Chapter 12. As we saw, deleting one extreme score had a surprisingly marked effect on the results. With correlation, extreme scores (especially in studies with small sample sizes) can likewise be problematic.

I offer the same suggestion with correlation studies because with studies using other statistical procedures, the researcher must use their understanding of the data and the research situation in order to "manage" extreme scores. They can transform such scores, delete them, or retain them as legitimate depending on their understanding.

The first step in limiting the effects of extreme scores is to detect them! In other works [e.g., Abbott (2010)] I devote significant attention to diagnostic procedures for correlation and regression studies. The serious researcher should seek these works out before engaging in a correlation study.

Heteroscedasticity

This essentially refers to violating the equal variance assumption among study variables. Homoscedasticity is the condition in which an outcome variable has equal variation across the levels of the predictor variable. If this condition is not present—for example, if the scores of one study variable were markedly skewed—the correlation would show a distortion. Figure 14.14 shows the two conditions of homoscedasticity (a) and heteroscedasticity (b). In some cases, heteroscedasticity may result in a departure from linearity.

Curvilinear Relations

I showed an example of a curvilinear relationship in Figure 14.3 (panel d). In these types of relationships, the pattern of the data may extend in one direction and then break in a different direction creating a nonlinear path. In Abbott (2010), I discussed diagnosis of curvilinear relationships in great detail. These are often not easy to detect, but SPSS® has a straightforward procedure that helps to make the assessments about whether the linearity assumption is violated in a study. I will develop this procedure in Chapter 15 when I discuss bivariate regression.

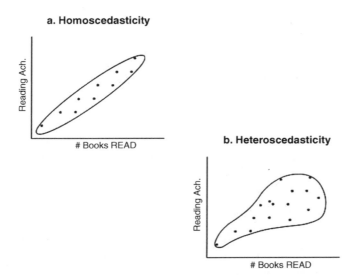

FIGURE 14.14 The correlation conditions of homoscedasticity and heteroscedasticity.

THE EXAMPLE DATABASE

Because I introduced the research question about the relationship among ethnicity, FR, and achievement in Chapter 13, I will use these with a larger dataset to illustrate the correlation procedure. The database is a random sample of 40 schools taken from the Washington State school database that I discussed in Chapter 6. The variables we will consider for the correlation example are the same as those in Real-World Lab IX: 2×ANOVA. Both are continuous variables (interval data):

- Math: the percentage of students at the school who passed the state math assessment.
- FR: the percentage of students in schools qualified for free and/or reduced lunches.

Table 14.4 shows the sample data in two columns (each). I will use this database to provide a real world example of calculating correlation and making appropriate conclusions. For the current example, we will calculate Pearson's r for the relationship between math and FR. We will use the same database for other calculations in the Real-World Lab.

We will use the steps I discussed above as the process for performing the correlation analysis and making conclusions:

1. Check assumptions.
2. Calculate Pearson's r.

3. Evaluate Pearson's r using:
   a. Hypothesis test
   b. Effect size

Assumptions for Correlation

We can use Excel® and SPSS® to assist with the assumptions. Some do not require additional analyses.

- Randomly chosen sample: The sample ($N = 40$) was randomly selected from the total set of schools in the state of Washington with fourth-grade results for the included variables ($N = 1054$).
- Variables are interval level: All data are percentages.
- Variables are independent of one another: The variables are not linked.
- Variables are normally distributed: Figures 14.15 (Excel®) and 14.16 (SPSS®) show the descriptive results for the study variables. The variables appear to be normal from the results. The SPSS® report includes the standard error for skewness and kurtosis, so you can confirm that the values for both variables lay within the boundaries we discussed for both measures.

Figures 14.17 and 14.18 show the histograms for both variables. As you can see, both graphs appear normal, although, due to sample size, the FR graph is a bit horizontal.

TABLE 14.4 The School Database Sample

Cases 1–20		Cases 21–40	
Math	FR	Math	FR
50.0	42.9	39.8	38.9
42.2	13.3	56.1	25.3
37.7	42.4	73.3	62.4
41.3	53.8	40.0	72.1
62.5	24.6	46.0	88.6
69.2	28.9	49.6	35.9
9.7	84.7	53.0	45.1
27.9	91.6	58.9	34.3
24.3	86.1	50.0	70.9
34.2	68.1	72.7	0.0
26.5	73.9	58.2	11.8
36.4	65.4	61.7	19.0
30.2	99.8	65.9	61.6
56.3	76.5	88.5	12.1
50.0	60.9	77.4	1.3
51.1	81.3	50.0	59.0
25.4	87.2	48.5	44.7
36.9	49.4	72.1	15.4
62.9	58.2	73.9	34.7
34.6	32.1	79.3	52.2

THE EXAMPLE DATABASE

	Math	FR
Mean	50.61	50.16
Standard error	2.79	4.25
Median	50.00	50.77
Mode	50.00	#N/A
Standard deviation	17.68	26.88
Sample variance	312.44	722.32
Kurtosis	−0.41	−0.93
Skewness	0.01	−0.07
Range	78.80	99.76
Minimum	9.70	0.00
Maximum	88.50	99.76
Sum	2024.20	2006.47
Count	40	40

FIGURE 14.15 The Excel® descriptive statistics.

Descriptive Statistics

	N	Mean	Stdandard Deviation	Variance	Skewness		Kurtosis	
	Statistic	Statistic	Statistic	Statistic	Statistic	Standard Error	Statistic	Standard Error
Math	40	50.60500	17.676073	312.444	0.008	0.374	−0.412	0.733
FR	40	50.1618	26.87608	722.324	−0.073	0.374	−0.930	0.733
Valid N (listwise)	40							

FIGURE 14.16 The SPSS® descriptive output for the study variables.

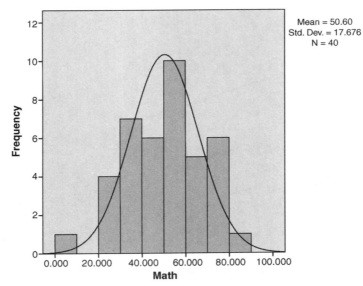

FIGURE 14.17 The SPSS® histogram for Math.

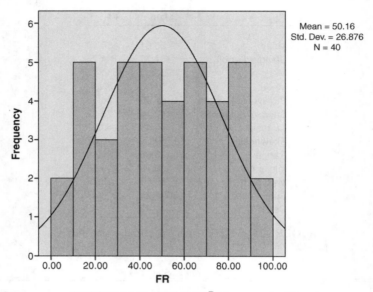

FIGURE 14.18 The SPSS® histogram for FR.

- Variances are equal: We can assume equal variance because neither variable is *markedly* skewed.
- Linear relationship: Figure 14.19 shows the scattergram between Math and FR. As you can see, the pattern of the dots is fairly evenly distributed around the line and the overall shape of the dots is in a straight line.

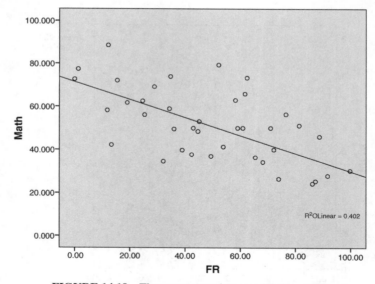

FIGURE 14.19 The scattergram between Math and FR.

THE EXAMPLE DATABASE

Computation of Pearson's *r* for the Example Data

Table 14.5 is the database with the key sums in the columns. After I show the hand calculations, I will use both Excel® and SPSS® to provide the correlation analyses. The general correlation formula using SS values is

$$r_{XY} = \frac{SS_{XY}}{\sqrt{(SS_X)(SS_Y)}}$$

The SS_{XY}

$$SS_{XY} = \sum XY - \frac{(\sum X)(\sum Y)}{N}$$

$$SS_{FR \times Math} = 89,793.8 - \frac{(2024.2)(2006.5)}{40}$$

$$\mathbf{SS_{FR \times Math} = -11,743.7}$$

The SS_X

$$SS_X = \sum X^2 - \frac{(\sum X)^2}{N}$$

$$SS_{FR} = 128,818.7 - 100,651.1$$

$$\mathbf{SS_{FR} = 28,170.62}$$

The SS_Y

$$SS_Y = \sum Y^2 - \frac{(\sum Y)^2}{N}$$

$$SS_{Math} = 114,619.9 - 102,434.6$$

$$\mathbf{SS_{Math} = 12,185.3}$$

Pearson's *r*

$$r_{XY} = \frac{-11,743.7}{\sqrt{(28,170.62)(12,185.3)}}$$

$$r_{XY} = \frac{-11,743.7}{18,527.5}$$

$$\mathbf{r_{XY} = -0.63}$$

TABLE 14.5 The Database with Sums for Calculation

Math	M²	FR	FR²	Math X FR
50.0	2500.0	42.9	1843.8	2147.0
42.2	1780.8	13.3	176.1	560.0
37.7	1421.3	42.4	1795.5	1597.5
41.3	1705.7	53.8	2894.9	2222.1
62.5	3906.3	24.6	606.0	1538.6
69.2	4788.6	28.9	832.9	1997.2
9.7	94.1	84.7	7181.8	822.0
27.9	778.4	91.6	8387.8	2555.2
24.3	590.5	86.1	7411.0	2091.9
34.2	1169.6	68.1	4644.3	2330.7
26.5	702.3	73.9	5461.1	1958.3
36.4	1325.0	65.4	4272.3	2379.2
30.2	912.0	99.8	9952.8	3012.9
56.3	3169.7	76.5	5854.6	4307.8
50.0	2500.0	60.9	3705.1	3043.5
51.1	2611.2	81.3	6607.5	4153.7
25.4	645.2	87.2	7603.8	2214.9
36.9	1361.6	49.4	2438.1	1822.0
62.9	3956.4	58.2	3385.6	3659.9
34.6	1197.2	32.1	1028.6	1109.7
39.8	1584.0	38.9	1514.5	1548.9
56.1	3147.2	25.3	641.0	1420.4
73.3	5372.9	62.4	3893.8	4573.9
40.0	1600.0	72.1	5205.5	2886.0
46.0	2116.0	88.6	7853.4	4076.5
49.6	2460.2	35.9	1291.9	1782.8
53.0	2809.0	45.1	2035.6	2391.2
58.9	3469.2	34.3	1176.3	2020.1
50.0	2500.0	70.9	5033.3	3547.3
72.7	5285.3	0.0	0.0	0.0
58.2	3387.2	11.8	138.2	684.3
61.7	3806.9	19.0	361.1	1172.5
65.9	4342.8	61.6	3800.0	4062.3
88.5	7832.3	12.1	147.3	1074.3
77.4	5990.8	1.3	1.7	100.7
50.0	2500.0	59.0	3483.5	2951.1
48.5	2352.3	44.7	1995.9	2166.8
72.1	5198.4	15.4	237.0	1110.1
73.9	5461.2	34.7	1203.9	2564.1
79.3	6288.5	52.2	2720.8	4136.4
$\Sigma X = 2{,}024.2$		$\Sigma Y = 2006.5$		$\Sigma XY = 89{,}793.8$
	$\Sigma X^2 = 114{,}619.9$		$\Sigma Y^2 = 128{,}818.7$	

THE EXAMPLE DATABASE

Using the table values and the formulas above, I calculated Pearson's r to be -0.63. Because the numbers are so large, I used an Excel® spreadsheet as a calculator by entering the formulas for each SS value in the spreadsheet. Therefore, my totals may be a bit different from yours if you used a calculator because Excel® carries out the values to many decimal points. You should end up with the same, or very slightly different, Pearson's r value.

Evaluating Pearson's r: Hypothesis Test

We will use the same hypothesis test steps as we have with other statistical procedures and as we did above with the fictitious example:

1. The Null Hypothesis (H_0): $\rho = 0$
2. The Alternative Hypothesis (H_A): $\rho \neq 0$
3. The Critical Value $r_{38(0.05)} = 0.312$

Remember that the degrees of freedom for correlation is $N - 2$, where N is the number of pairs. Because $N = 40$, the df $= 38$ ($40 - 2$).

The table for correlation specifies that the two-tailed 0.05 critical (exclusion) value for r at 38 degrees of freedom is 0.312. Values that are greater than 0.312 would therefore be considered statistically significant.

4. The Calculated Value (-0.63)
5. Statistical Decision

We reject the null hypothesis because the calculated value (-0.63) exceeded the critical value of exclusion (0.312). *A calculated value as high as our calculated value (-0.63) is unlikely to occur by chance.*

6. Interpretation

We can consider our calculated r of -0.63 to be statistically significant ($p < 0.05$). This indicates that it is likely not a chance finding. In the words of the research question, the schools' average math achievement test scores are negatively correlated to the FR percentages. Schools with greater numbers of students eligible for free/reduced lunches show lower math achievement scores. Because this is a random sample, we can be reasonably sure that the findings generalize to all schools in Washington State.

Evaluating Pearson's r: Effect Size

The effect size, r^2, is easy to calculate. It is simply the square of the correlation (-0.63) and yields a value of $r^2 = 0.40$. Therefore, FR values account for

40% of the variance in schools' math achievement scores. According to the criteria we discussed earlier, this would represent a large effect size (i.e., it exceeds $r^2 = 0.25$).

CORRELATION USING EXCEL® AND SPSS®

Correlation is an easy calculation for both Excel® and SPSS®. I will show how to create the correlation and discuss the results.

Correlation Using Excel®

Figure 14.20 shows the menu window when you select "Data–Data Analysis" from the main menu in the spreadsheet where the data are located. This should be a familiar menu now that we have used it in past procedures. When you select "Correlation," you are presented with another specification window as shown in Figure 14.21. As you can see, I selected both columns of data at the same time in the "Input Range:" window. Because I included the labels, I made sure to check the "Labels in First Row" box. The default is to create the data in a separate spreadsheet.

When you select "OK" from the button shown in Figure 14.21, you will see the correlation results in a matrix form, even though there is only one correlation pair. Figure 14.22 shows the FR–Math correlation as −0.63 which matches our hand calculated value.

Note that Excel® does not specify the p value of the result. You must consult the Correlation Table of Values to obtain the comparison value for the hypothesis test.

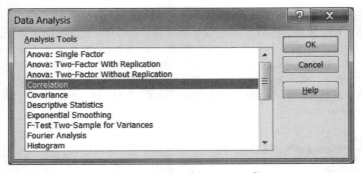

FIGURE 14.20 The "Correlation" window in the Excel® Data–Data Analysis menu.

CORRELATION USING EXCEL® AND SPSS®

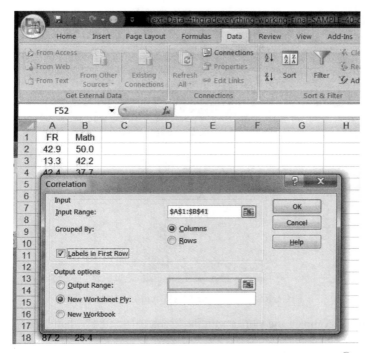

FIGURE 14.21 The "Correlation" specification window in Excel®.

Correlation Using SPSS®

Figure 14.23 shows the SPSS® correlation menu. Note that it is called "Bivariate" because we are specifying a two-variable correlation ("bi" means two). When you make this choice, you are presented with the specification menu shown in Figure 14.24.

As you can see in Figure 14.24, I included both variables in the "Variables:" window. The default values for this specification window are shown. The analysis will produce the Pearson r (this is the only value produced in the default analysis), and SPSS® will create a ("Two-tailed") hypothesis test and "Flag significant correlations" according to the default checks.

If you select the "Options" button in the upper right part of the window, you can request means, standard deviations, and cross-product values. The latter are helpful

	FR	Math
FR	1	
Math	−0.63385	1

FIGURE 14.22 The Excel® correlation matrix.

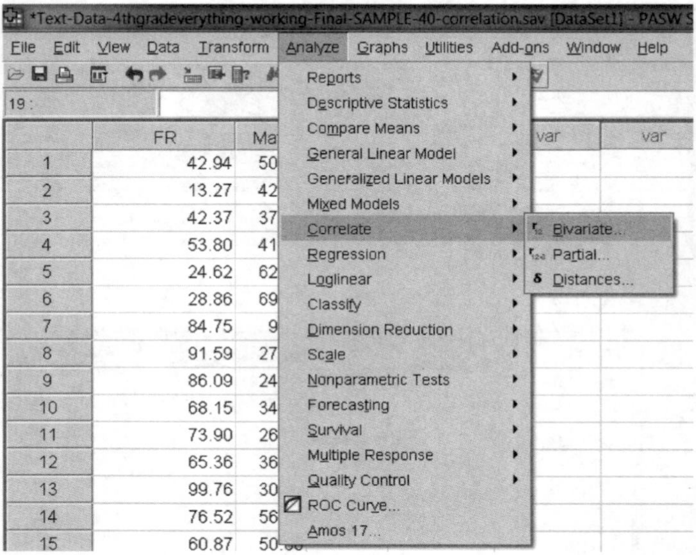

FIGURE 14.23 The SPSS® correlation menu.

if you wish to compare the values with those you calculated by hand. For this example, I did not select the options.

Choosing "OK" will produce the correlation matrix shown in Figure 14.25. As you can see, the same correlation value is produced as in Excel® and in our hand

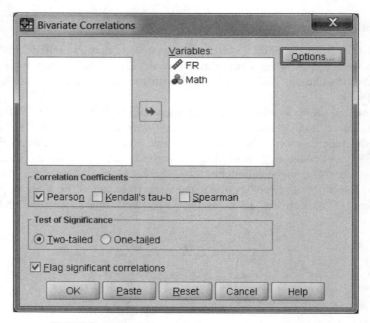

FIGURE 14.24 The correlation specification window.

NONPARAMETRIC STATISTICS: SPEARMAN'S RANK-ORDER CORRELATION (r_S)

Correlations

		FR	Math
FR	Pearson correlation	1	−0.634[a]
	Significant (two-tailed)		0.000
	N	40	40
Math	Pearson correlation	−0.634[a]	1
	Significant (two-tailed)	0.000	
	N	40	40

[a]Correlation is significant at the 0.01 level (two-tailed).

FIGURE 14.25 The SPSS® correlation matrix.

calculations. The additional features of the SPSS® are helpful because they provide the values you need for a hypothesis test. As you can see, the correlation significance level ($p = 0.000$) is listed, which indicates that the correlation is far into the exclusion area and therefore considered a significant correlation; the null hypothesis is rejected.

As with other output we have examined, SPSS® lists the *exact probability* of the finding. When we create hypotheses tests using Correlation Tables for our exclusion values, we report significance as $p < 0.05$ indicating that the calculated value fell into the exclusion area beyond the boundary of the 0.05 level of probability. The exact probability reported in Figure 14.25 (0.000) is actually the following: 1.13E-5, which translates to $p = 0.0000113$! Therefore, our calculated r fell so far into the exclusion region that it has an extremely small probability of representing a chance finding.

NONPARAMETRIC STATISTICS: SPEARMAN'S RANK-ORDER CORRELATION (r_S)

We have not discussed nonparametric correlation procedures thus far, but I listed some examples in Table 14.1. Those listed in the first three columns are nonparametric procedures designed to measure correlations with less than interval data. I will briefly describe one of these in this chapter, Spearman's rho. I will discuss other nonparametric measures in a separate chapter.

Up to now, we have assumed two interval-level variables in calculating Pearson's r. However, researchers often do not have access to interval data. For example, educational researchers might have only data such as "teacher ranking" of students as an operational measure of student achievement. In this case, as we saw in Chapter 5, our rank-ordered data are ordinal data and we can make no assumption that the interval between the numbers is equal.

If we have ordinal data, as in the example of teacher rankings, we can use a variation of correlation called Spearman's r. Also called Spearman's rank-order

TABLE 14.6 Hypothetical Correlation of Rankings by Books Read and Reading Ability

Class Ranking—Books Read	Teacher Ranking—Reading	d	d^2
1	2	−1	1
2	1	1	1
3	5	−2	4
4	3	1	1
5	4	1	1
	Totals	0	8

correlation, or Spearman's rho, this procedure calculates a correlation *using ordinal data* or *interval data that have been ranked*.

Spearman's rho results vary between −1 and +1 as with Pearson's r. You can obtain a negative value, which is interpreted as it is with Pearson's r. However, because this not a parametric statistic, we cannot square the r_s value to arrive at a "variance explained" figure as in Pearson's r. We can perform an hypothesis test to see whether the value is significant.

The hand calculation for Spearman's rho is as follows:

$$r_s = 1 - \frac{6 \sum d^2}{N(N^2 - 1)}$$

The formula appears somewhat strange in comparison to the other procedures we have studied. Like some nonparametric tests, Spearman's rho formula includes a *constant* in the formula (a constant is a number that does not change). The 6 appears mysterious, but the nonparametric formula is created to yield a value that fits the ±1 range, and the value of 6 allows the computed values to fall in this range. There is a longer explanation, but this gives you some insight into nonparametric procedures.

The formula above is used with two sets of ranks like the hypothetical example of data in Table 14.6:

If we use the formula I included above, the following is the calculation for Spearman's rho:

$$r_s = 1 - \frac{6 \sum d^2}{N(N^2 - 1)}$$

$$r_s = 1 - \frac{6(8)}{5(25 - 1)}$$

$$r_s = 1 - \frac{48}{120}$$

$$r_s = 1 - \frac{6 \sum d^2}{N(N^2 - 1)}$$

$$r_s = 0.60$$

Spearman's rho calculations can be part of hypotheses tests using the same process as the hypothesis tests in the parametric procedures. The null and alternative hypotheses state that the correlation will be 0 (or not 0, respectively) in the population the sample data represent. Spearman's rho is calculated with the above formula, and the researcher can compare the calculated value to a critical test value.

Spearman's rho has a special table of comparison values to use for hypothesis tests. In this table, *the researcher does not use degrees of freedom* because no parameters are estimated. However, if a table of values is not available (I do not include such a table in this book), the researcher can use the T Table of values with the same formula I noted above for Pearson's r. *When you use the T Distribution Table for identifying a comparison value, you can use degrees of freedom as we did with Pearson's r* because the table is created for a parametric comparison.

$$t = \frac{r_s\sqrt{N-2}}{\sqrt{1-r_s^2}}$$

$$t = \frac{0.6\sqrt{3}}{\sqrt{0.64}}$$

$$t = 1.30$$

Using the T Table of values, I note that the critical value of rejection at the 0.05 (two tailed) level with $df = 3$ (5 − 2) is 3.182. Therefore, because our calculated t value (from Spearman's rho) is 1.30, we cannot reject the null hypothesis.

The one-tailed 0.05 exclusion value using Spearman's Table of critical values for five data pairs is 0.90. (Spearman's Table does not provide a two-tailed value with $N = 5$.) You would therefore need a 0.90 correlation or higher to reject the null hypothesis. Using either table would therefore indicate the same result: We cannot reject the null hypothesis.

Variations of Spearman's Rho Formula: Tied Ranks

Researchers sometimes use rank-order procedures when they have data that are interval, but the data are not normally distributed, as I mentioned above when I discussed skewed variables. Situations in which this is common involve using financial data, like housing values, teachers' salaries, or school funding. I will provide an extended (actual) example of this in a section below.

More commonly, researchers have studies in which they may have *both* ordinal and interval level variables. In either case, however, the simple Spearman's rho formula above will most likely not be appropriate as I will discuss later. The reason is that larger studies with more cases or studies that involve ranking of interval variables usually result in *tied ranks*, which are problematic to the basic Spearman's rho formula.

TABLE 14.7 Ranking an Interval Variable

Class Ranking-Books Read	Number of Books Read	Books Read Rankings
1	12	2
2	14	1
3	6	3.5
4	6	3.5
5	1	5

If skewed interval-level variables are used along with ranked variables, *the skewed interval data must first be ranked and then correlated with variables already ranked.* In these cases, the Spearman's rho formula above should not be used. Table 14.6 presents a brief example of using interval data that are transformed to rank-order variables.

In our fictitious study, we used Class Ranking—Books Read as one of our ordinal, (ranked) variables. What if we had used the number of books the students read *instead* of the ranking variable? We would then have an interval-level variable. In such a case, we would need to rank order the interval data so that both of our study variables would then be ordinal (remember, the other variable, Teacher Ranking—Reading is already an ordinal variable). Table 14.7 shows the comparison of these variables and how the interval-level variable can be changed to rankings.

I placed the original ranked variable in the first column. The second column is the interval-level variable we are going to use in the study *instead* of the ranked variable. The problem with these data values, however, is that the third and fourth students tied because they read the same number of books (six). If we want to rank this variable, we must *average tied ranks.*

When we rank the students in the order of the greatest number of books read (you can also rank them in the other direction), the top two students are ranked 2 (second) and 1 (first), but the next two students occupy the same rank (third). Therefore, we must *average the rank positions* of the books read (average the rank positions, not the actual values). This means that we average the third and fourth rank positions resulting in two ranks of 3.5 (because $\frac{3+4}{2} = 3.5$). I showed this in Table 14.7 in the third (shaded) column.

Once this variable has been changed to a rank-order variable from an interval-level variable, we can proceed with Spearman's rho calculation with the new ranked variable and the teacher ranking variable in Table 14.6.

Remember, however, the basic Spearman's rho formula I used above for this calculation works well with small samples, but it is affected by variables with *tied ranks.* If we have tied ranks, we must use another formula that takes the tied ranks into account. I will not introduce that formula in this chapter. I just wanted to point out that larger datasets, especially those involving skewed interval data that have been ranked, should be analyzed by SPSS® or similar statistical software. As you can see, ranking skewed data can be quite a process especially if the dataset is large.

NONPARAMETRIC STATISTICS: SPEARMAN'S RANK-ORDER CORRELATION (r_S)

Statistics

Grant$$	
N valid	10,019
N missing	2
Mean	52,463.11
Median	0.00
Skewness	99.984
Standard error of skewness	0.024
Kurtosis	10,004.004
Standard error of kurtosis	0.049

FIGURE 14.26 Descriptive.

A Spearman's Rho Example

I mentioned above that researchers often are faced with using data that are so distorted that they cannot be used with parametric procedures like Pearson's r. In such cases, they might use Spearman's rho because this procedure can be used with skewed data. Creating ranks with these kinds of variables would be very tedious, especially with large datasets. Fortunately, SPSS® includes the Spearman's procedure that can be very helpful.

As an example, one of the databases we have used to demonstrate statistical procedures in this book is the TAGLIT data. One school finance variable in the 2003 TAGLIT data related to instructional technology is how much funding was received from grants by school for benefitting instructional technology. Figure 14.26 shows the SPSS® summary for this variable in one database with over 10,000 cases.

As you can see from Figure 14.26, this is an extremely skewed variable! The median is $0 and the mean is $52,463.11. Therefore, over half of the schools received no grant funding designated for instructional technology in that year. However, there were many schools that exceeded $10,000 and several that received over $100,000. Under these circumstances, the researcher would not want to use these data in a Pearson's correlation despite the fact that they represent interval (even ratio) level data. If the researcher has interval data that depart this far from the assumptions of parametric tests to the extent that they cannot be reasonably used, then Spearman's rho can be used instead of Pearson's r.

I will provide a short example of the data above used in a research study so that you can see how to use Spearman's rho correlation. If you recall, I used a TAGLIT example in Chapter 6 in which I described a measure of Middle/High-School teachers' technology skill. A researcher may wish to correlate such a measure with a

Correlations

			ldr53	mhskills
Spearman's rho	Grant $$	Correlation coefficient	1.000	0.035[a]
		Significance (two-tailed)	—	0.028
		N	10,019	3968
	mhskills	Correlation coefficient	0.035[a]	1.000
		Significance (two-tailed)	0.028	—
		N	3968	3968

[a]Correlation is significant at the 0.05 level (two-tailed).

FIGURE 14.27 The SPSS® Spearman's rho results.

variable like the grants received variable shown in Figure 14.26. The object would be to see whether grant funding was related to teacher technology skill. That is, do MH teacher technology skills increase with increased school grant funding for instructional technology?

Figure 14.27 shows the Spearman's r results for this research question. As you can see, the correlation is 0.035 and is significant at $p = 0.028$. (Figure 14.24 shows how to obtain this SPSS® output. There is simply a box to check in the same specification window in which you call for Pearson's r.)

Spearman's rho is 0.035, which indicates a very low degree of association between the variables. Remember, as the value of Spearman's rho approaches 0, the relationship between the variables is weaker. Note, however, that the significance level is $p = 0.028$! This result underscores our earlier discussion of the importance of sample size. Despite a weak relationship, it is statistically significant.

As a point of interest, look at Figure 14.28 which shows the same variables analyzed using Pearson's r. The results show almost no correlation ($r = -0.003$).

The fact that we were using the extremely skewed variable Grant$$ affected the correlation results using Pearson's r. Even though Spearman's rho was a low correlation, it was nevertheless larger (and significant). These results point to the importance of using the correct correlation procedure with your available data. *They also underscore the point that the power of a study is increased by using the appropriate statistical procedure.*

TERMS AND CONCEPTS

Coefficient of Determination The squared value of r, the Pearson correlation coefficient, which represents the proportion of variance in the outcome variable accounted for by the predictor variable of a study.

Correlations

		ldr53	mhskills
Grant$$	Pearson correlation	1	−0.003
	Significance (two-tailed)		0.855
	N	10,019	3968
mhskills	Pearson correlation	−0.003	1
	Significance (two-tailed)	0.855	
	N	3968	3968

FIGURE 14.28 SPSS® Pearson's r results.

Dichotomized Variables Continuous scores that have been transformed into variables with only two values (e.g., high or low). This is related to, but not the same as, dichotomous variables, which are those variables that naturally have two categories (e.g., male or female).

Heteroscedasticity A violation of the assumption of equal variation. In this condition, the variance of one study variable is not equal at each value of the other study variable.

Homoscedasticity The assumption of equal variance in a study in which the variance of one variable is equal at values of the other variable in a study.

Linear Relationship The relationship among study variables that forms a straight line if plotted on a graph. Violations of linearity might take the form of curvilinear relationships in which the graphed line is not straight but curved.

Outliers Extreme scores in a distribution that may result in a distortion of values of the total set of scores.

Proportional Reduction in Error Measures (PRE) Procedures that measure how much variation is reduced or explained by knowing the relationship among study variables. In correlation, r^2 is such a measure in that it describes the proportion of variance in the outcome variable that is accounted for by the predictor variable.

Restricted Range A problem in correlation studies in which the entire set of scores is not used in an association, but rather a selected group of the scores is used that that do not represent the total variability.

Scattergram A graph that shows the relationship between study variables when they are plotted together. Also called scatter diagram, scatterplot, or scatter graph.

TABLE 14.8 The Real-World X Data

Cases 1–25		Cases 26–50	
Earmark$	mhskills	Earmark$	mhskills
15.00	2.65	0.00	3.33
0.00	3.50	8.00	3.13
15.00	3.19	0.00	3.03
15.00	2.79	0.00	3.11
0.00	3.00	0.00	3.32
2.00	3.08	0.00	3.74
0.00	3.15	5.00	3.07
5.00	3.10	200.00	2.98
27.60	3.28	0.00	3.10
0.00	3.07	7.21	3.12
10.00	2.87	0.00	2.80
0.00	3.47	1.00	3.28
14.00	3.25	0.00	3.40
5.00	2.90	0.00	3.58
35.00	2.96	0.00	3.00
70.00	3.29	30.00	2.87
0.00	2.89	0.00	3.20
3.00	2.97	2.50	2.82
0.00	2.75	5.00	3.09
35.00	2.89	10.00	3.00
10.00	3.16	0.00	3.47
14.00	3.28	2.00	3.40
0.50	2.80	0.00	3.50
0.00	3.06	6.50	3.10
0.00	3.32	3.00	3.00

REAL-WORLD LAB X: CORRELATION

In this lab, we will use TAGLIT data with a research question similar to the example in the chapter. The TAGLIT data in Table 14.8 are a random sample ($N = 50$) of schools with MH Teacher Skills data ($N = 3968$) selected from the national database in 2003. The two variables are:

- Earmark$ is the funding a school earmarks for technology (in thousands of dollars).
- MHSkills is the index of technology skills among middle- and high-school teachers. The higher the index value, the greater the perceived technology skill of the teacher.

Lab X Questions

1. Do the variables meet the assumptions for Pearson's r?

2. Conduct the appropriate correlation procedure with Excel® and SPSS® and explain your results.
3. Calculate the Pearson's *r* by hand and perform the hypothesis test.
4. What is the effect size?
5. Discuss the results.

REAL-WORLD LAB X: SOLUTIONS

1. Do the Variables Meet the Assumptions for Pearson's *r*?

- Randomly chosen sample: The sample ($N=50$) was randomly selected from the total set of school data in the 2003 TAGLIT national ($N=3968$).
- Variables are interval level: All data are either dollars or index scores.
- Variables are independent of one another: The variables are not linked.
- Variables are not normally distributed: Figures 14.29 (Excel®) and 14.30 (SPSS®) show the descriptive results for the study variables. The Earmark$ variable appears to be severely positively skewed from both sets of results. The SPSS® report includes the standard error for skewness and kurtosis, which confirms the extreme skew. The MHSkills variable is normally distributed.

Figures 14.31 and 14.32 show the histograms for both variables. As you can see, the histogram for Earmark$ (Figure 14.31) indicates a severe positive skew in agreement with the numerical data. Figure 14.32 shows a normal distribution for MHSkills.

- Variances are equal: We cannot assume equal variance because one variable is *markedly* skewed.

	Earmark$	MHSkills
Mean	11.13	3.12
Standard error	4.26	0.03
Median	2.00	3.10
Mode	0.00	3.00
Standard deviation	30.13	0.24
Sample variance	907.77	0.06
Kurtosis	32.80	−0.15
Skewness	5.40	0.38
Range	200.00	1.09
Minimum	0.00	2.65
Maximum	200.00	3.74
Sum	556.31	156.07
Count	50	50

FIGURE 14.29 The Excel® descriptive statistics for Lab X.

Descriptive Statistics

	N	Mean	Standard Deviation	Skewness		Kurtosis	
	Statistic	Statistic	Statistic	Statistic	Standard Error	Statistic	Standard Error
Earmark$	50	11.13	30.129	5.397	0.337	32.802	0.662
mhskills	50	3.1214	0.23577	0.383	0.337	−0.151	0.662
Valid N (listwise)	50						

FIGURE 14.30 The SPSS® descriptive output For Lab X.

- Linear relationship: Figure 14.33 shows the scattergram between Earmark$ and MHSkills. As you can see, the pattern of the dots does not indicate a linear relationship due to the extreme skew of Earmark$.

2. Conduct the Appropriate Correlation Procedure with Excel® and SPSS® and Explain Your Results

Given the violation of assumptions, we will conduct the Spearman's rho procedure with SPSS®. Figure 14.34 shows the results.

According to the results in Figure 14.34, Spearman's rho correlation ($r_s = -0.324$) is significant ($p = 0.022$). There is an inverse relationship between Earmark# and MHSkills. The more funding was earmarked for instructional technology, the lower the MHTeacher skills. This is an interesting finding that bears greater study. One possible reason for the inverse relationship is that schools

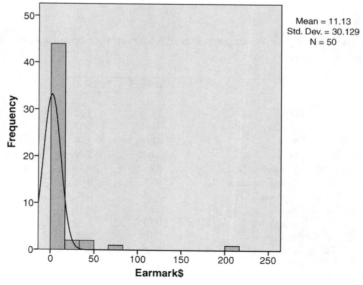

FIGURE 14.31 The SPSS® histogram for Earmark$.

REAL-WORLD LAB X: SOLUTIONS

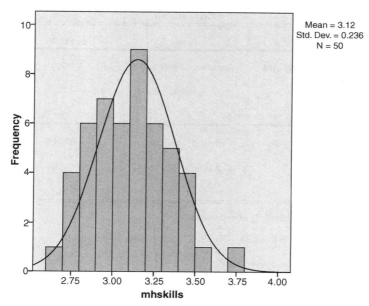

FIGURE 14.32 The SPSS® histogram for MHSkills.

responded to the lack of teacher technology skills by earmarking more funds for instructional technology. This suggestion, if accurate, points to the fact that correlation is not causation; either variable may be considered the predictor and/or the outcome. The other problem is that nearly half the sample schools earmarked no funding, which may have affected the relationship.

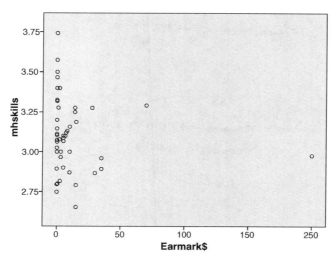

FIGURE 14.33 The SPSS® scatterplot of Earmark$ (predictor) and MHSkills (outcome).

Correlations

			Earmark$	mhskills
Spearman's rho	Earmark$	Correlation coefficient	1.000	−0.324[a]
		Significance (two-tailed)	—	0.022
		N	50	50
	mhskills	Correlation coefficient	−0.324[a]	1.000
		Significance (two-tailed)	0.022	—
		N	50	50

[a]Correlation is significant at the 0.05 level (two-tailed).

FIGURE 14.34 The SPSS® Spearman's rho results.

3. Calculate Pearson's r by Hand and Perform the Hypothesis Test

Table 14.9 shows the data table with sums (not all data are shown). The Pearson's r hand calculations follow the data table. The calculated Pearson's r is −0.135. The correlation does not appear strong, so we will conduct the hypothesis test.

The SS_{XY}

$$SS_{XY} = \sum XY - \frac{(\sum X)(\sum Y)}{N}$$

$$SS_{\text{Earmark} \times \text{MHSkills}} = 1689.47 - \frac{(556.31)(156.07)}{50}$$

$$\mathbf{SS_{\text{Earmark} \times \text{MHSkills}} = -47}$$

The SS_X

$$SS_X = \sum X^2 - \frac{(\sum X)^2}{N}$$

$$SS_{\text{Earmark\$}} = 50,670.52 - 6189.62$$

$$\mathbf{SS_{\text{Earmark\$}} = 44,480.9}$$

The SS_Y

$$SS_Y = \sum Y^2 - \frac{(\sum Y)^2}{N}$$

$$SS_{\text{MHSkills}} = 489.89 - 487.16$$

$$\mathbf{SS_{\text{MHSkills}} = 2.73}$$

REAL-WORLD LAB X: SOLUTIONS

TABLE 14.9 The Data Table with Sums[a]

Earmark$	Earmark$²	mhskills	Skills²	Earmark$ × Skills
15	225	2.65	7.041174	39.80
0	0	3.50	12.25	0.00
15	225	3.19	10.15759	47.81
15	225	2.79	7.79138	41.87
0	0	3.00	9	0.00
2	4	3.08	9.472716	6.16
0	0	3.15	9.895823	0.00
5	25	3.10	9.61	15.50
28	762	3.28	10.73822	90.44
0	0	3.07	9.447535	0.00
10	100	2.87	8.239026	28.70
0	0	3.47	12.01778	0.00
14	196	3.25	10.5625	45.50
5	25	2.90	8.41	14.50
35	1225	2.96	8.7616	103.60
70	4900	3.29	10.83929	230.46
—	—	—	—	—
—	—	—	—	—
—	—	—	—	—
—	—	—	—	—
0	0	3.50	12.25	0.00
7	42	3.10	9.61	20.15
3	9	3.00	9	9.00
$\Sigma X = 556.31$		$\Sigma Y = 156.07$		$\Sigma XY^2 = 1689.47$
	$\Sigma X^2 = 50{,}670.52$		$\Sigma Y^2 = 489.89$	

[a]Not all data are shown due to space considerations.

Pearson's r

$$r_{XY} = \frac{SS_{XY}}{\sqrt{(SS_X)(SS_Y)}}$$

$$r_{XY} = \frac{-47}{\sqrt{(44{,}480.9)(2.73)}}$$

$$r_{XY} = \frac{-47}{348.47}$$

$$r_{XY} = -\mathbf{0.135}$$

The hypothesis results follow:

1. The Null Hypothesis (H_0): $\rho = 0$
2. The Alternative Hypothesis (H_A): $\rho \neq 0$

3. The Critical Value $r_{48(0.05)} = 0.279$
   4. The Calculated Value (-0.135)
   5. Statistical Decision: Do not reject the null hypothesis. The calculated correlation does not fall into the exclusion area and is therefore likely a chance finding.

4. What Is the Effect Size?

The effect size for this Pearson's r is 0.018. The predictor variable explains (accounts for) less than 2% of the variance in the outcome variable. This amount is considered small by the Cohen (1988) criteria.

5. Discuss the Results

Because the Earmark$ variable was extremely skewed, we conducted a Spearman's rho correlation and found the relationship of MHSkills to Earmark$ to have an significant inverse relationship ($r_s = -0.324$, $p = 0.022$). Although the correlation is significant, it is difficult to determine the direction of the relationship. We suggested MHSkills to be the outcome variable. However, this finding may reveal that MHSkills drove the funding for instructional technology and therefore served as the predictor. Additional study with the full range of variables in the database may reveal the nature of the relationship between these study variables.

15

BIVARIATE REGRESSION

This chapter discusses a very important statistical procedure for evaluators. *Regression techniques use correlation to help predict the values of an outcome variable knowing values of the predictor variable.* If you recall, I mentioned at the outset of Chapter 14 that correlation helps researchers to explain and predict. By extension, regression techniques use correlations between variables to *explain variance* in outcome variables and *predict specific outcome values* at different values of predictor variables.

We have already made reference to regression procedures in the scattergrams we created and discussed. For example, Figure 14.2 showed the scattergram of the correlation between the number of books read and student reading achievement. The scattergram used a line to show the pattern of the student scores. We discussed correlation as being stronger when the dots were close to the line and the direction of the line, indicating whether the relationship was positive or inverse.

The line that "captures" the pattern of the correlation relationship is actually the *regression line of best fit*. We will learn in this chapter to calculate the equation for this line so that we can understand the dimensions of the relationship between the two variables. We can use the line to help us predict values of the outcome variable knowing values of the predictor variable, and we can identify the "explained variance" between the two variables knowing the dimensions of the line.

If a teacher created a regression equation for the relationship between the students' number of books read and reading achievement scores, they could use this information to help them understand other students not in the study. For example, if a new student entered the class, the teacher could ask them how many books they read over the summer and then use the regression equation to predict what their

---

Understanding Educational Statistics Using Microsoft Excel® and SPSS®. By Martin Lee Abbott.
© 2011 John Wiley & Sons, Inc. Published 2011 by John Wiley & Sons, Inc.

reading achievement might be. The prediction would likely be a much better estimate of the student's reading achievement than simply guessing!

The teacher could also understand how much of the variance in reading achievement is explained by the number of books read and how much variance is left unexplained. They will have created a model of explanation that will help them in subsequent studies to focus even further on other variables that might help to explain further the unexplained variance.

THE NATURE OF REGRESSION

Technically, regression refers to the *spread of the values of Y for values of X*. This refers to the fact that at fixed values of the predictor variable (X) there are several values of the outcome variable (Y). This spread of Y values will have a mean at each X value. The path of a line that crosses through all the means is what we refer to as the regression line. It is a line of best fit because it passes through the means of the Y points across the values of X. Consider Figure 15.1, a hypothetical example using books read and reading achievement.

In Figure 15.1, you can see that when the number of books read (X) is 4, there are several students represented. This means that several students read four books, but not all of these students had the same reading achievement score. I drew a small distribution around these dots to show that, at any value of X, there will be a spread of values of Y. The mean of all these Y values at each of the X values will be the best score to represent all the others. Thus, the entire regression line will pass through the means of all the Y values that spread out around values of X.

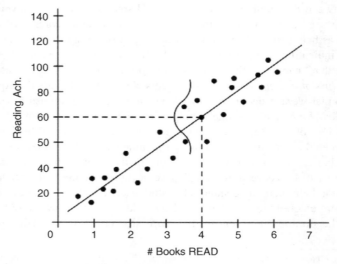

FIGURE 15.1 The line of best fit.

THE REGRESSION LINE

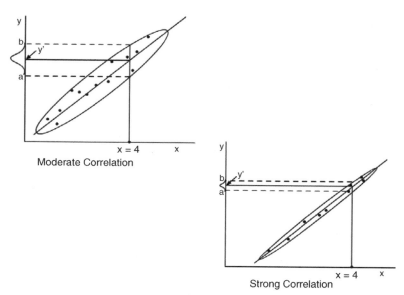

FIGURE 15.2 The effect of correlation on prediction accuracy.

Because the mean of the Y values (at each value of X) is the best representative of the other Y values, we can use this value to predict a value of Y. I showed this in Figure 15.1 by a dashed horizontal line extending from the (mean) Y value (when $X = 4$) to the Y axis. This value (approximately 60) represents the predicted value of Y when $X = 4$. We could extend similar lines to the Y axis at the other values of X.

Figure 15.2 shows another dimension of regression analysis. When we create the regression line, our predicted values of Y represent the mean of the spread of Y values at values of X. However, the *actual* values of Y (which we do not know) might vary from this predicted value. I represented this variability of prediction with a small distribution around the predicted values on the Y axis.

As you can see, when the correlation between the two variables is stronger, the predicted Y values have less variability and can be thought to be more accurate. Because there is less variability of Y scores at each value of X, the "mean of the groups of Y" is a more precise estimate and therefore will produce a more accurate predicted Y value.

Figure 15.3 shows another dimension of regression. When there is no significant correlation, the predictability of Y at values of X is very poor. You can see how this is problematic because with very low correlation, we cannot establish a meaningful regression line that will provide good estimates or predicted Y values.

THE REGRESSION LINE

Just as there were two methods for calculating Pearson's r, there are two methods for calculating the regression line. The first of these methods uses Z scores, just as

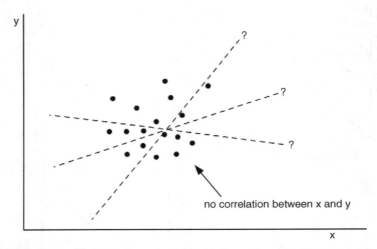

FIGURE 15.3 The lack of meaningful prediction with no significant correlation.

our first correlation formula used Z scores. The logic of this formula is best understood by first looking at the raw score formula, which is the second way of creating the equation for the regression line.

Both the Z-score formula and the raw score formula are variations of the equation for a line that you probably learned in an algebra class:

$$Y = mX + b$$

In this formula, m is the slope of the line (rise over run), and b is the y intercept. Statisticians use the same formula but with different symbols. In the statistical formula, b represents the slope of the line and a is the y intercept. So, the statistical equation for the straight line (i.e., regression line) we use is

$$\hat{Y} = bX + a$$

You will note that $\hat{Y}$ is the symbol for "Y predicted," which is the predicted value of Y at values of X. This is often represented as Y' as you can see in Figure 15.2, or simply as Y_{pred}. I will use the latter (Y_{pred}) in this chapter.

The b and a values in the equation have research meanings beyond slope and intercept. Here is how to use them in research:

- b is the slope of the regression line. When we calculate the equation, the b is the coefficient for the predictor variable that indicates *the unit change in Y with each unit change in X.* That is, for every 1 unit change in the X variable, how many units will the Y variable change?
- a is the Y intercept. This refers to the value of Y when $X = 0$. (You must examine research graphs carefully because some do not show 0 on the X axis. I will discuss this below.)

THE REGRESSION LINE

TABLE 15.1 The Fictitious Data on Books Read and Reading Achievement

Books Read	Reading Achievement
0	45
1	58
3	62
5	76
6	81
8	86
9	90

We can learn best by using a fictitious example and then moving to a real-world example. Table 15.1 is a fictitious dataset that follows the theme we used in Chapter 14 of the relationship between books read and reading achievement. In regression, we are using the correlation to help us better predict the values of one variable from the values of the other variable.

I am using different values than we used in Chapter 14 in order to illustrate the elements of the regression relationship, but the research question is similar. With regression, we are using one variable as the predictor (X = Books Read) and an outcome variable (Y = Reading Achievement). Thus, we are using regression to create a model that will accomplish two things:

1. Explain as much variance as possible in the outcome variable.
2. Predict values of the outcome variable with values of the predictor.

Figure 15.4 shows the scattergram between the two variables (from SPSS®). As you can see, the regression line (line of best fit) has been calculated and drawn through the dots in such a way as to minimize the distance of each dot from the line. As we discussed above, you may also think of the line representing the mean of all the Y values that lie at each value of X. Because we have so few observations, this isn't visually apparent, but you can see it more clearly with more observations as we will in our next example.

Our task is to calculate the regression line so we can understand better the properties of regression. We will calculate both b and a for this equation.

Note that in Figure 15.4 the Y intercept is shown (i.e., the regression line is shown crossing the Y axis), but the intercept value (a) is the value of Y when $X = 0$. *This value lies directly above the 0 on the X axis where the regression line would intersect the Y axis.* I indicated this with the vertical and horizontal lines in the scattergram. Since the $X = 0$ is not directly on the Y axis, the value of a is deceptive. On the graph, a appears to be about 47 (the point of intersection of the regression line and the Y axis). However, as we will see, the calculated value of a is 49.41, which is seen more clearly by the added vertical and horizontal lines. I wanted to show an actual scattergram result so you can understand the meaning of the a value.

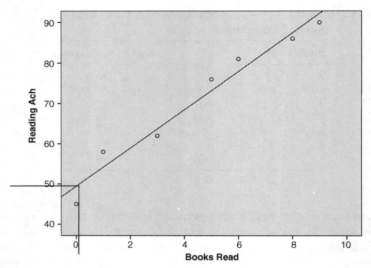

FIGURE 15.4 The scattergram between books read and reading achievement.

CALCULATING REGRESSION

As I noted above, we will begin by calculating the raw score formula for regression. This formula uses the actual values of X and Y as they appear in the raw score data table. Both variables are measured in their own scales. If you look at the data in Table 15.1, you will see that both X and Y are interval-level variables, but the magnitudes of the scale values are different. X values range from 0 to 9, whereas Y values range from 45 to 90. Our raw score formula uses the values as they exist in the table.

$$Y_{\text{pred}} = bX + a$$

If you look carefully at the formula, there are four elements, two of which are dependent upon the calculation of the other two.

- Y_{pred} is the outcome that depends on the values of the other calculations;
- X takes the values from the data the researcher uses as the predictors;

The two values above are "resident" in the equation, they are there but do not show a value in the equation. The other two elements are calculated: b, the slope, is calculated by the following equation:

$$b = \frac{r\sqrt{SS_Y}}{\sqrt{SS_X}}$$

a, the intercept is calculated by the following equation:

$$a = M_Y - b(M_X)$$

The Slope Value b

Note that the slope is simply the r value times the relationship between the variance (SD) of Y to that of X (in the formula, I use the SS value that represents the SD when the square root is taken). Here are some observations about the slope:

- When r *is large*, the slope value will be larger, indicating a larger slant to the regression line and a *greater impact of X on Y*.
- When r *is small*, the slope will be flatter. The r value thus modifies the slope by influencing the size of the rise (Y) over the size of the run (X).
- When $r = 1.0$, a perfect correlation, the slope is the direct relationship between the SD values of Y and X.
- When $r = 0$, or no correlation, the slope will be flat, *indicating no predictive value of X for Y values*.
- Intermediate values of r "weight" the impact of the slope.

The Regression Equation in "Pieces"

The regression equation looks difficult but it is manageable if you think of it in "pieces." Figure 15.5 shows how you might consider the calculation formula.

A Fictitious Example

Table 15.2 shows the values in Table 15.1 needed for calculating the pieces of the regression equation.

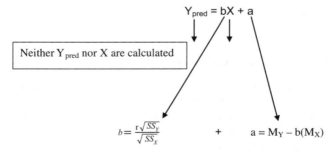

FIGURE 15.5 The formula in pieces.

TABLE 15.2 The Calculated Sums of the Fictitious Data

Books Read (X)	X^2	Reading Achievement (Y)	Y^2	XY
0	0	45	2025	0
1	1	58	3364	58
3	9	62	3844	186
5	25	76	5776	380
6	36	81	6561	486
8	64	86	7396	688
9	81	90	8100	810
$\sum X = 32$	$\sum X^2 = 216$	$\sum Y = 498$	$\sum Y^2 = 37{,}066$	$\sum XY = 2{,}608$
$M = 4.57$		$M = 71.14$		
$SD = 3.156$		$SD = 15.29$		

Here are the pieces of the regression equation. Remember, we must calculate r as we did in Chapter 14. The value of regression depends upon a significant and meaningful r value:

r value:

$$r_{XY} = \frac{SS_{XY}}{\sqrt{(SS_X)(SS_Y)}}, \qquad r_{XY} = \frac{331.43}{\sqrt{(69.71)(1636.86)}} = .98$$

b value:

$$b = \frac{r\sqrt{SS_Y}}{\sqrt{SS_X}}, \qquad b = \frac{.98\sqrt{1636.86}}{\sqrt{69.71}}, \qquad b = 4.75$$

a value:

$$a = M_Y - b(M_X), \qquad a = 71.14 - 4.75(4.57), \qquad a = 49.43$$

Now that we have calculated all the pieces, we can put them in the equation as shown in Figure 15.6. As you can see, the X value simply stays in the equation without a value.

Interpreting and Using the Regression Equation

Once we arrive at the final regression formula, we can interpret the parts and use it for prediction. The b is a coefficient of X and indicates the unit change in Y with every unit change in X. Thus, when X changes by one unit (i.e., when a student reads one additional book), the Y value increases by 4.75. If reading achievement is measured in percentage on the reading test, then each book read results in a gain of almost 5% on the reading test!

EFFECT SIZE OF REGRESSION

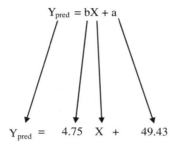

FIGURE 15.6 The completed regression formula.

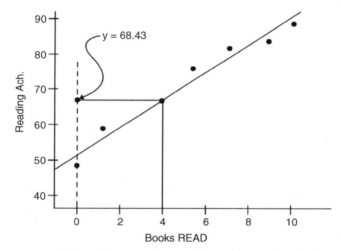

FIGURE 15.7 Using the regression formula to predict a value of Y at $X=4$.

If we wished to predict a value of Y for a certain value of X, we can simply put the X value in the equation and solve for the Y_{pred}. For example, if we want to predict a Reading Achievement score for a student who read four books, the predicted reading achievement score would be 68.43:

$$Y_{pred} = 4.75(4) + 49.43,$$
$$Y_{pred} = 68.43$$

Figure 15.7 shows the predicted reading achievement value (68.43) when the student read four books.

EFFECT SIZE OF REGRESSION

Regression is based on correlation, so the effect size is r^2 as in correlation. In this instance we can conclude, as we did with correlation, that with this sample the number of books read accounts for approximately 96% (0.96) of the variance in reading achievement. (If it were only this simple!). As with correlation, we can use the same criteria from Cohen (1988): r^2 values of 0.01, 0.09, and 0.25 for small, medium, and large effects, respectively.

THE Z-SCORE FORMULA FOR REGRESSION

If you recall, I noted earlier that there is another way to calculate the regression formula using Z scores. Like correlation, this would involve creating Z scores for each of the scores in the data table. The resulting formula is very simple, however, because with bivariate regression the slope is equal to the correlation (slope $= r$).[1] You can see how this works if you consider that the SD of Z scores is 1. Thus, the following would apply:

$$b = \frac{r\sqrt{SS_Y}}{\sqrt{SS_X}}, \qquad b = \frac{r(1)}{(1)}, \qquad b = r$$

If both X and Y were transformed to Z scores, their SDs would be 1. Thus, as you can see from the formula for calculating b, this would mean that we would be dividing the Y SD (which equals 1) by the X SD (which also equals 1) to yield 1. (Remember that the $\sqrt{SS}$ value in the equation is a measure of SD.)

When both X and Y values are Z scores, the scattergram would show the change in scales for both variables. Because both would be in Z-score scales, there would be no Y intercept; the regression line would cross the Y axis at the origin (where $X = 0$ and $Y = 0$). Figure 15.8 shows this. As you can see, the vertical and horizontal lines create the Y and X axes at values of 0 for both. Where they cross is the origin, which represents $X = 0$ and $Y = 0$. (Note that both the Y and X axes have negative values because Z scores are negative as well as positive.) Because the regression line passes through the origin, the Y intercept is always 0, unlike the intercept with the raw score formula.

The Z-score formula therefore changes due to the revised values for b (i.e., $b = r$) and a ($a = 0$ since there is no Y intercept). The following shows the raw-score formula and the resulting Z-score formula that incorporates the changes in values from using Z scores.

Raw-Score Formula : $Y_{pred} = bX + a$
Z-Score Formula : $Z_Y = rZ_X$, which in final form is $Z_Y = \beta Z_X$

Using the Z-Score Formula for Regression

If you use the Z-score formula to predict Z_Y values, you need to remember to use Z_X values. Thus, if you wanted to predict a Z_Y value, you would need to obtain Z values for the X scores.

[1] This is only true for regression studies with one predictor variable (bivariate regression). Multiple regression creates a different value for the slope value with Z-score formulas.

THE Z-SCORE FORMULA FOR REGRESSION

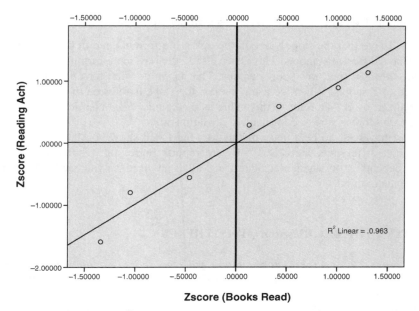

FIGURE 15.8 The SPSS® Z-score scattergram of books read and reading achievement.

For example, if you wanted to use the Z-score formula for the previous problem (i.e., What is Y_{pred} when $X=4$?), you would first need to create the Z score for $X=4$. This would yield the Z_X value of $=-0.181$ $\left(\frac{4-4.57}{3.156}\right)$. Using the Z-score formula, the resulting predicted Z_Y score is -0.177.

$$Z_Y = rZ_X$$
$$Z_Y = 0.98(-0.181)$$
$$Z_Y = -0.177$$

Therefore, when $Z_X = -0.181$, $Z_Y = -0.177$. If you did not want to report the resulting Z_{pred} as a Z score, you could transform it to a raw score using the formula you learned to transform Z scores back to raw scores (Chapter 8). *Just remember to use the M and SD of the Y variable because you are transforming a Z_Y score to a Y raw score*:

$$\text{Raw score } (Y_{pred}) = Z(SD) + M,$$
$$(Y_{pred}) = -0.177(15.29) + 71.14,$$
$$(Y_{pred}) = 68.436$$

This predicted value (68.436) is the raw score (Y) predicted by a Z_X of -0.181. Compare this to the raw score values above calculated by the raw-score formula because these are the Z-score values corresponding to predicting Y from an X score of 4.

Unstandardized and Standardized Regression Coefficients

You will note that the slope value in the raw score formula and in the Z-score formula have different symbols. This is because in the raw-score formula we are using *unstandardized* (i.e., raw score) values. That is, values that have not been transformed to Z scores. When we use Z scores, the slope is referred to as "Beta" and symbolized by β. We refer to this value as the *standardized* coefficient because Z scores are scores expressed in standard deviation units.

This may seem to be a petty distinction, but I will show you later that it is an important difference. Statistical software reports regression results in different ways, depending on whether standardized values are used, so you need to keep this difference in mind.

TESTING THE REGRESSION HYPOTHESES

When you calculate a regression equation, you can use it to explain variance and to predict values. But the same question remains with regression as with the other procedures we have discussed thus far: Are the results statistically significant?

There are two considerations for statistical significance in bivariate regression:

1. The *Omnibus* test is the statistical significance of the *overall regression model*.
2. The *individual predictor(s) test* assesses the statistical significance of each predictor in the model (in bivariate regression there is only one predictor).

We will examine the results for both of these tests by examining the results from Excel® and SPSS®. I will note here that the *omnibus test is an F test* because the F test assesses significance by comparing components of variance. Our raw-score formula uses SS values that will allow you to see the relationships among the variables in the analysis. The *individual predictor test is a T test of whether the slope differs from 0* (in effect, a T test of a single sample). We have discussed both of these tests in past chapters, so you will be prepared to understand the results.

THE STANDARD ERROR OF ESTIMATE

We have examined several inferential statistical tests that use sampling distributions as comparisons for our transformed test values. The regression analysis creates a sampling distribution of sorts, the standard deviation of which is called the standard error of estimate, symbolized by s_{est}.

When we use the regression equation to make predictions, our estimate will infrequently match exactly the actual value of Y. Rather, our Y_{pred} value will be somewhat different than an actual value. This difference is called the error of prediction or simply error. The differences between Y_{pred} values and actual Y values are also

THE STANDARD ERROR OF ESTIMATE

known as *residuals*. When all the prediction errors from the model are placed in a distribution, we can calculate a mean and standard deviation of the entire group. The s_{est} is the standard deviation of all the prediction errors. We can use this value as a comparison of all individual prediction errors, and we can also use it to create a confidence interval of scores within which the actual value of Y will fall a given percentage of time (e.g., 95% of the time).

Calculating s_{est}

Calculating s_{est} is straightforward. The prediction errors are in reference to the variance measure of Y, so we simply adjust this measure by using the correlation measure and the degrees of freedom for regression. This will yield the standard deviation of the distribution of prediction errors.

In the formula, we start with SS_Y, the variance measure of Y. Then we divide it by the degrees of freedom of regression (and correlation). As you recall, the degrees of freedom for correlation is $N - 2$ because we are estimating two parameters (predictor and outcome variables). This is the same in bivariate regression where we have two variables. When you divide the variance measure by the degrees of freedom, we create an estimate of the SD of the sampling distribution (i.e., the standard error). Taking the square root creates an SD measure from the variance measure (SS_Y) we used to create the estimate.

The s_{est} measure is made more precise by taking into account the correlation between the predictor and outcome variables. Thus, in the formula, we modify our estimate by multiplying by $(1 - r^2)$, which is the unexplained variance. As you can see, when there is a large correlation (as there is in our fictitious sample), this unexplained variance will be very small and will therefore result in a s_{est} that is small. However, when the original correlation is weak (i.e., approaching 0), there will be a greater amount of unexplained variance. This will result in a larger standard error of estimate.

$$S_{est} = \sqrt{\frac{SS_Y}{N-2}(1 - r^2)}$$

In our fictitious example, s_{est} is calculated as follows. Your results may be slightly different, depending on whether you round off the elements of the formula before you use it or complete the formula in one step (as with Excel® and SPSS®).

$$S_{est} = \sqrt{\frac{SS_Y}{N-2}(1 - r^2)},$$

$$S_{est} = \sqrt{\frac{1636.86}{7-2}(1 - 0.963)},$$

$$S_{est} = \sqrt{327.372(0.0374)},$$

$$S_{est} = 3.499$$

The result is 3.499. What does this mean? *It is the estimated standard deviation of all the prediction errors of our regression equation.*

Our original correlation ($r = 0.981$) was very large, so the standard error is not large. If the correlation had been small, for example $r = 0.10$, look what would happen to the s_{est} calculation:

$$S_{est} = \sqrt{\frac{1636.86}{7-2}(1-0.01)},$$

$$S_{est} = \sqrt{327.372(0.99)},$$

$$S_{est} = 18.00$$

You can see by comparing these s_{est} results that a stronger correlation reduces the prediction error. The range of s_{est} is therefore:

$s_{est} = 0$, when there is perfect correlation, $r = 1$.
s_{est} = The maximum value is equal to SD_Y when the correlation $= 0$.

CONFIDENCE INTERVAL

Earlier, we predicted a value of Y (68.43) when X was 4. We can have confidence that this value is *close* to the actual value of Y, but it will likely not be exactly the same. Therefore, we can estimate a range of values that will capture the true value of Y (a parameter estimate) by using the s_{est} value. This is the same process we followed with other statistical procedures when we created confidence intervals that identified the boundary values within which the true (population) value fell a given percent of the time (e.g., 95%).

We can use s_{est} to identify the interval within which we can expect the true (parameter) value of Y to fall. Recall that in Chapter 10 we used the following formula to establish the CI for the single sample population estimate:

$$\text{Confidence interval} = \pm t(s_M) + M$$

With regression, we use the same approach, but substituting our different elements. The CI for the (estimated) true value of Y is

$$\text{Confidence interval} = \pm t(s_{est}) + Y_{pred}$$

In this formula, we still use the tabled value of T because our sample size is very small. In larger samples, you can use the Z value (1.96 at the 0.05 level, etc.). Please remember, however, that when you identify the exclusion value with the T table for regression, you need to use the df value of $N - 2$ (since our regression procedure estimates two parameters). The s_{est} replaces the standard error of the mean (s_M) because we are estimating values of Y rather than the population mean. We are establishing boundary values around our predicted value of Y (Y_{pred}), so this value replaces M in the single-sample CI formula.

Using the CI formula, we can estimate the true value of Y when X = 4 as follows:

$$\text{Confidence interval} = \pm t(s_{est}) + Y_{pred}$$
$$CI_{0.95} = \pm 2.571(3.50) + 68.43$$

Lower Boundary = 59.43	Upper Boundary = 77.43
$CI_{0.95} = -2.571(3.50) + 68.43$	$CI_{0.95} = +2.571(3.50) + 68.43$

According to these estimates, we are confident at the 0.95 level that the true value of Y will fall between 59.43 and 77.43. The s_{est} has helped us to identify these values which are much more accurate because we have used the value of the correlation to help us estimate more precisely. Without the knowledge of the correlation, we would estimate the boundaries somewhere between approximately 25.97 and 110.90! This is because we would only have the s_Y (16.52) to use in the estimation. The high correlation between X and Y has helped us to create a more precise estimate.

EXPLAINING VARIANCE THROUGH REGRESSION

I mentioned earlier that the main functions of regression are prediction and explanation. Correlations greater than 0 (i.e., significant) improve the predictions of the outcome variable at various levels of the predictor variable. We saw this in the figures that showed "tighter" patterns of dots around the line having stronger correlations and, therefore, more precise predictions.

We can also use regression relationships to explain the proportion of variance in the outcome variable resulting from its relationship to the predictor variable. The r^2 value is a numerical expression of this explained variance, as we saw in Chapter 14. Now we can delve a bit more deeply into the relationships to show how this works. We can use sum-of-square calculations to measure components of variance that will partition the variance or break the variance down into recognizable parts.

The following formula expresses these parts of the variance in the relationship between the number of books read and the reading achievement of the students in the study. The equation has three parts.

$$\sum (Y - M_Y)^2 = \sum (Y_{pred} - M_Y)^2 + \sum (Y - Y_{pred})^2$$

The part of the equation on the left of the equals sign is the *total variance measure of Y*. As the equation shows, there are two parts on the right side of the equation produced by the regression relationship that explains the total variance in the outcome variable. Here are the three parts and how they are calculated:

$$\sum (Y - M_Y)^2 = \text{the Total sum of squares of } Y, \text{ or simply } SS_Y$$

The next component (immediately to the right of the equals sign) is the portion of variance in the outcome variable that we can identify through the predictive power of the correlation with the predictor variable. It is the known portion of variance in the outcome variable and is called the regression portion, or the sum of squares of regression.

$$\sum (Y_{\text{pred}} - M_Y)^2 = \text{sum of squares of regression, or } SS_{\text{reg}}$$

The last portion represents the part of the variance in the outcome variable that is unknown or that which we cannot identify through the variables in our regression equation. It is called error because its origin cannot be determined. This part is measured by creating residual values that exist between the predicted Y values and the actual Y values. This "residual sum of squares" value is a combination of random error and unexplained variance.

$$\sum (Y - Y_{\text{pred}})^2 = \text{sum of squares of residual, or } SS_{\text{res}}$$

In regression, we are accounting for total variance by partitioning it, or breaking it up, so that we can understand better where it comes from. The following is a simplification of the larger regression equation above:

$$SS_Y = SS_{\text{reg}} + SS_{\text{res}}$$

Stated differently:

Total variation in Y = Known variation (SS_{reg}) + Unknown variation (SS_{res})

These parts of total Y variation are similar to the sources of variance we discussed in the ANOVA chapter (Chapter 12). Recall that we determined the following with respect to the portions of variance in an ANOVA study:

$$SS_T = SS_B + SS_W$$

Like the regression equation, we explained total variance as a combination of between (known) and within (unknown or error) variance. There is a very close relationship between ANOVA and regression that I will discuss further below.

We learned in Chapter 12 that you can calculate eta square (an effect size measure) as follows:

$$\eta^2 = \frac{SS_{\text{Between}}}{SS_{\text{Total}}}$$

Like ANOVA, you can create a similar measure, r^2, from the proportions of variation identified in regression. Consider the following formula and compare it to the eta square formula above:

$$r^2 = \frac{SS_{reg}}{SS_Y}$$

Hopefully, you can see the parallels between these two formulas. Both measure the proportion of known variance (regression) to total variation. Both express the amount of explained variance in an outcome variable accounted for by a predictor variable.

USING SCATTERGRAMS TO UNDERSTAND THE PARTITIONING OF VARIANCE

The parts of the regression equation are easier to understand through a visual examination of the scattergram. Remember that residuals are the prediction errors that result from using our regression equation to predict values of the outcome variable Y. Each prediction yields a value that is a certain distance from the actual value of Y.

Figure 15.9 shows the regression relationship between our two variables in the fictitious example. The regression line is established between the X variable (number of books read) and the Y variable (reading achievement).

I have shown three lines in Figure 15.9 that help to show the parts of variation:

1. The dotted line in the scattergram (the lowest of the three lines) is the *actual value of Y*. (There was no actual value of Y in the study data, so we can assume this to be 65 for illustration purposes. That is, we are assuming that a student not in the study has read four books and received a 65 on the reading achievement assessment.)
2. The solid black line just above the (dotted) mean line is the predicted value of Y at a certain value of X. (The value shown in the figure is $X = 4$, which, as we calculated in an earlier section of this chapter, predicted a value of Y of 68.43. Remember, there was no student who read four books in the study; we simply

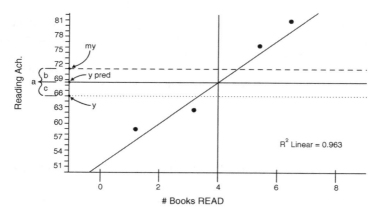

FIGURE 15.9 The parts of the Y variance.

demonstrated earlier that the regression equation could help us predict values of Y at certain values of X.)

3. The dashed line above the predicted value line is the mean of Y (71.14). This is a primary reference point for our analyses because the M_Y is the best guess of the value of Y if we had no other information. That is, without any help from the regression with X, if someone asked you to guess a reading achievement value for a certain student, the best guess would be the mean of the reading achievement scores because most student scores would fall there.

I added three identifiers to Figure 15.9 to point out the parts of the variation we are attempting to understand:

- $a = SS_Y$. This is the SS measure for the differences between the mean of Y and the actual value of Y. Remember that variance is a measure of how far from the mean all the scores in a distribution "scatter out." Thus, in Figure 15.9, this represents the large distance of the actual Y value from the mean of Y.
- $b = SS_{reg}$. This distance represents the "known" variation. Think of this part as representing the distance that a predicted value moves *away from the mean* and *closer to the actual value* as a result of the correlation between X and Y. Thus, the correlation has the effect of identifying the known relationship between X and Y, resulting in the predicted value being moved closer to the actual value. The stronger the correlation, the greater this portion of the variance, because the predicted values will fall further from the mean and closer to the actual values.
- $c = SS_{res}$. This part represents the distance between the predicted value of Y and the actual value. It is the unknown portion of the variation because we have no information about what might cause it to be as large as it is. *The SS_{reg} moved the predicted value closer to the actual value (due to the influence of X), but couldn't get it all the way!* The SS_{res} is what is "left over" or error.

Now, examine the r^2 equation again. As you can see, we are comparing the parts of variation from the regression analysis. The r^2, or effect size, increases as the SS_{reg} increases relative to the overall variation in Y. Another way of saying this is that as the predicted scores move further away from the mean toward the actual values of Y, the more variation we explain in the Y variable.

$$r^2 = \frac{SS_{reg}}{SS_Y}$$

A NUMERICAL EXAMPLE OF PARTITIONING THE VARIATION

Now that we have examined the variation parts in the scattergram, we can use the actual calculations from the fictitious study to measure the proportions of variation. Table 15.3 shows these calculations.

TABLE 15.3 The Calculations for the Components of Variance

X	Y	$(Y-M_Y)$	$(Y-M_Y)^2$	Y_{pred}	$(Y_{pred}-M_Y)$	$(Y_{pred}-M_Y)^2$	$(Y-Y_{pred})$	$(Y-Y_{pred})^2$
							Residuals	
0	45	−26.14	683.46	49.41	−21.73	472.33	−4.41	19.45
1	58	−13.14	172.74	54.16	−16.98	288.29	3.84	14.72
3	62	−9.14	83.59	63.67	−7.47	55.81	−1.67	2.80
5	76	4.86	23.59	73.18	2.04	4.15	2.82	7.95
6	81	9.86	97.16	77.93	6.79	46.12	3.07	9.40
8	86	14.86	220.73	87.44	16.30	265.68	−1.44	2.08
9	90	18.86	355.59	92.20	21.05	443.26	−2.20	4.83
$M_Y=71.14$								
			$SS_Y=1636.86$			$SS_{reg}=1575.64$		$SS_{res}=61.21$

Remember that the equation for these parts is the one we discussed above. I have shown this equation with the actual values Table 15.3:

$$SS_Y = SS_{reg} + SS_{res},$$
$$1636.86 = 1575.64 + 61.21$$

As you can see from the equation, the total variation in Y is a combination of known (SS_{reg}) and unknown (SS_{res}) variation. Comparing these two sources, you can see that the unknown proportion is much smaller than the known proportion because the correlation was so large ($r=0.98$). You can calculate r^2 from the variance sources, as we discussed above:

$$r^2 = \frac{SS_{reg}}{SS_Y},$$
$$r^2 = \frac{1575.64}{1636.86},$$
$$r^2 = 0.96$$

I hope you can see from the previous sections that partitioning the variance is an important avenue for understanding the relationships between an outcome and a predictor variable.

USING EXCEL® AND SPSS® WITH BIVARIATE REGRESSION

We have explored the value of regression for prediction and explanation in the sections above. We now turn to exploring the regression output to illustrate these procedures. I will use the same fictitious data in this section, but then will introduce a "real" data example in a later section.

The Excel® Regression Output

We will use the data in Table 15.1 to demonstrate the Excel® regression procedure. At the Data Analysis window that results from using the main menu options "Data—Data Analysis," you can choose "Regression," which will produce the window in Figure 15.10.

As you can see, I entered the spreadsheet locations of the Y (reading achievement) and X (books read) data. I checked the "Labels" box because I included the variable labels so that the output would identify the variable data. I asked "Confidence Level 95%," but this applies to the regression coefficients (b and a) rather than any specific prediction. I also asked for "Residuals" so you can see how these values appear. Recall that we used them above to partition the variance.

When I run this specification, Excel® returns four panels of results. I show these below in Figures 15.11 to 15.14.

Figure 15.11 confirms the values that we produced in our hand calculations. Note that the Pearson's r is called "Multiple R" when it is used in regression. The "Adjusted R Square" measure is affected by the sample size among other considerations and we can use it with real-world data as I will discuss below. You can see that the "Standard Error" is the same value as our s_{est} of 3.50.

Figure 15.12 shows the data that will help us test the null hypothesis for regression. If you recall, I mentioned that there are two kinds of significance tests with

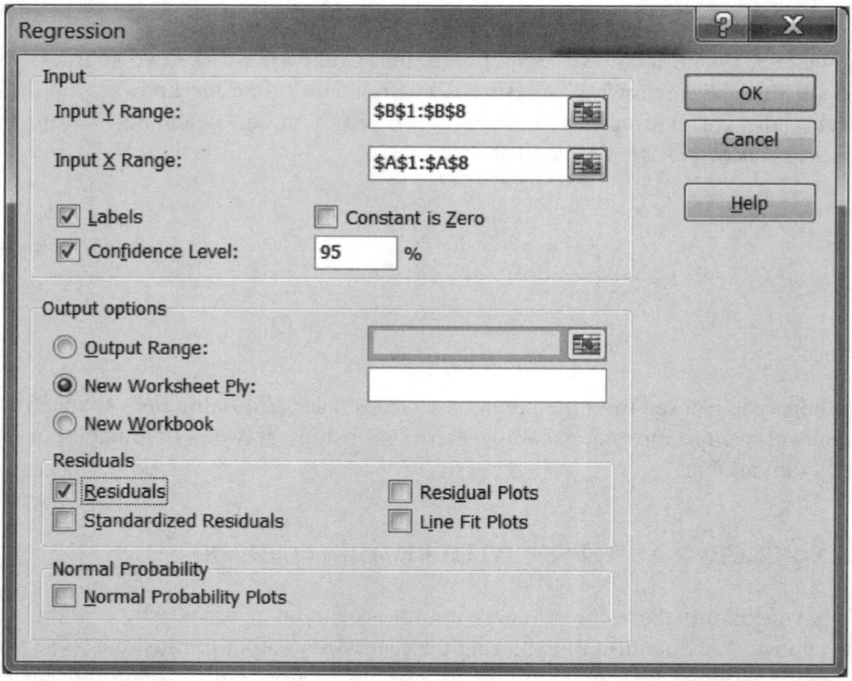

FIGURE 15.10 The Excel® regression specification window.

Regression Statistics	
Multiple R	0.98
R square	0.96
Adjusted R square	0.96
Standard error	3.50
Observations	7

FIGURE 15.11 The Excel® regression statistics output.

regression. One is the *omnibus test* that determines whether the overall results establish a non-chance predictive model between our two study variables. Note that the omnibus test results are provided in an ANOVA table. I discussed above the close relationship between ANOVA and regression because both procedures use sum of squares calculations to understand how the components of variation between the study variables relate to each other. The following are two important elements of the ANOVA results:

1. The omnibus significance test is provided by the F ratio (128.70, significant beyond $p < 0.000$), which indicates that the proportion of total variance explained by the SS of regression is beyond the value expected by chance (according to the F distribution).
2. The total explained variance, or R^2 (0.96) can be obtained by dividing the SS_{reg} (1575.64) by the SS_Y (1636.86).

Figure 15.13 shows the table of values that allows us to perform the second significance test, the test of the individual predictors. Recall that I mentioned that, once the omnibus test was performed, we would need to perform a test on the individual predictor(s) to determine whether the slope was significantly different from 0. This test is primarily important for Multiple Regression, which has more than one independent (predictor) variable. Thus, we would need to perform individual significance on the separate predictors to see which added most to the overall prediction.

As you can see, the t Stat for the predictor variable X (Books Read) is 11.34 and is determined to be significant (in the "P value" column at $p < 0.00$). This test uses the single-sample t ratio to determine whether the derived slope coefficient (4.75) is significantly different than a population value of 0. Therefore, we can conclude that the individual predictor books read is a significant predictor of the reading achievement of this sample of students.

ANOVA					
	df	SS	MS	F	Significance F
Regression	1	1575.64	1575.64	128.70	0.00
Residual	5	61.21	12.24		
Total	6	1636.86			

FIGURE 15.12 The Excel® regression model output: ANOVA.

	Coefficients	Standard Error	t Stat	P-value	Lower 95%	Upper 95%
Intercept	49.41	2.33	21.23	0.00	43.43	55.39
Books Read (X)	4.75	0.42	11.34	0.00	3.68	5.83

FIGURE 15.13 The Excel® regression output showing the regression coefficients.

	RESIDUAL OUTPUT	
Observation	Predicted Reading Achievement (Y)	Residuals
1	49.41	−4.41
2	54.16	3.84
3	63.67	−1.67
4	73.18	2.82
5	77.93	3.07
6	87.44	−1.44
7	92.20	−2.20

FIGURE 15.14 The Excel® predicted values and residuals for the study data.

Figure 15.13 also provides the calculated value of the slope (b) and intercept (a) values we used to create the regression equation. As you can see, the values in the table for a and b (49.41 and 4.75, respectively) correspond to the values we calculated by hand as shown in the following formula from our earlier calculations (the a value is slightly different due to rounding):

$$Y_{pred} = 4.75X + 49.43$$

Figure 15.13 also shows standard error; but this is not s_{est}, which is shown in Figure 15.11. These standard errors are those connected to both the a and b coefficients (that is, the standard deviation of the sampling distributions of estimates for intercept and slope) to create CI values for each. The confidence intervals shown in the last two columns (Lower 95% and Upper 95%) result from estimates using the standard errors.

The last Excel® panel is shown in Figure 15.14 and lists the predicted and residual values for each case in the dataset. You can compare these values to those in Table 15.3. Remember, the s_{est} is the estimated standard error of these residuals. Excel® has an option for "standardized residuals" that I did not request which transforms the residual values to standardized values (i.e., Z values) so you can see at a glance which ones exceed 2 or 3 and might therefore be considered extreme or outlier values. In our fictitious database, we would expect no such outliers, but you can use this feature with actual data in your studies.

The SPSS® Regression Output

SPSS® creates bivariate regression output very similar to the Excel® output in the former section. I will note below some unique features from SPSS® that are helpful

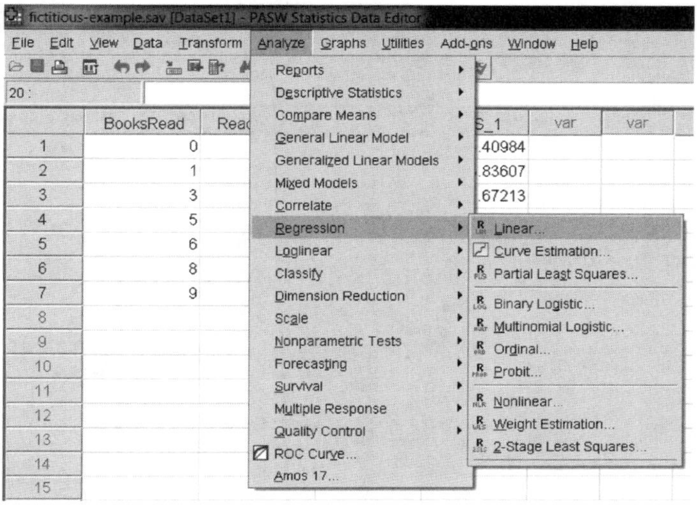

FIGURE 15.15 The regression options in SPSS®.

for bivariate regression and that establish a model for interpreting multiple regression results.

The SPSS® procedure is available through the main "Analyze–Regression" menu. When you select this option, you will be presented with several regression procedures from which to choose. You can see several options for regression (Figure 15.15). For bivariate regression, we can select "Linear" because this procedure applies to regression equations that manage one (bivariate) or many predictors (multiple). Note that this regression procedure assumes a linear relationship that we discussed above. I will review all the regression assumptions in a later section in this chapter.

Figure 15.16 shows several ways to specify the bivariate regression study. The main specification window "Linear Regression" requires that you place the predictor variable in the "Independent(s)" window and the outcome variable in the "Dependent" window. As you can see, I have specified both variables using our fictitious data.

Once the variables are specified, the user can choose from a number of buttons on the right side of the Linear Regression window. All of these are important for different kinds of studies, but the "basic" bivariate regression can be created by including only a few additional specifications from the "Statistics" button. I have shown this window in Figure 15.16.

The "Linear Regression: Statistics" window allows the user to create descriptive information on the study variables ("Descriptives") as well as to produce the ANOVA table ("Model Fit") results and identify changes to the R^2 from predictors added to the analysis (R-squared change). The "Estimates" and "Confidence intervals" choices produce the a and b coefficients with their confidence intervals.

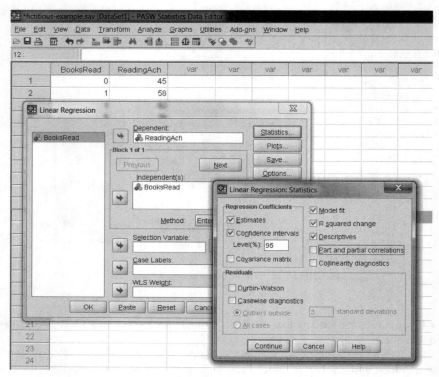

FIGURE 15.16 The SPSS® regression specification windows.

The ANOVA table in SPSS® is identical to that produced in Excel®, so I will not include it here. The same values are represented, which can be used to create the omnibus significance test.

The results panel in Figure 15.17 ('Model Summary') show the outcomes in a bit different format from Excel®. You will find the same values for R, R_2, Adjusted R_2, and s_{est} in this panel as in the Excel® results. There are some new results that are important in this panel, however. I will list these below.

Model Summary

Model	R	R Square	Adjusted R Square	Standard error of the Estimate	Change Statistics				
					R Square Change	F Change	df1	df2	Significance F Change
1	0.981[a]	0.963	0.955	3.499	0.963	128.702	1	5	0.000

[a]Predictors: (Constant), BooksRead

FIGURE 15.17 The SPSS® model summary results panel.

- "R-Square Change" shows the impact of the predictor variable on the overall regression model. There is only one predictor, so this is a bit harder to see; but if there were additional independent variables, there would be separate lines of data and the R-Square-Change value would show how the addition of each predictor would change the overall R-Square value. Because there is only one predictor in the current study, the R-Square-Change value is the same as the R-Square value in the second column. You might think of this value as the predictor variable (books read) increasing the explained variance in Y (reading achievement) from 0% to 96%.
- "F Change" similarly shows the impact of adding the predictor (only one in this study) to the F value.
- "Significance F Change" registers the change in the significance of the F value from the addition of the predictor variable. Both this and the former will change when additional predictors are included in the model.

Figure 15.18 shows the "Coefficients" panel that provides the information necessary to create the regression equation and assess the significance of the individual predictor. As you can see, this panel looks almost identical to the Excel® results (see Figure 15.13). There is one important addition to this output that you should note, however.

Figure 15.18 includes the column "Standardized Coefficients" just before the T-test results are provided. This column of data includes the Beta (β) coefficient, which is the *standardized coefficient* for the regression equation. Recall that at the outset of this chapter I discussed two ways of creating the regression equation. The method we used involved the regression equation that used the actual (raw score) scales of the variables. This method produced the regression equation we noted above that uses the a (49.41) and b (4.754) values located in the "Unstandardized Coefficients" columns in Figure 15.18. Using these coefficients produces the same equation as noted above using Excel®:

$$Y_{pred} = 4.75X + 49.43$$

The other method we discussed was the Z-score method where we transformed the raw score values of both X and Y variables and then created the regression

Coefficients[a]

Model		Unstandardized Coefficients		Standardized Coefficients	t	Significance	95.0% Confidence Interval for B	
		B	Standard error	Beta			Lower Bound	Upper Bound
1	(Constant)	49.410	2.328		21.226	0.000	43.426	55.394
	BooksRead	4.754	0.419	0.981	11.345	0.000	3.677	5.831

[a]Dependent Variable: Reading Ach

FIGURE 15.18 The SPSS® coefficients panel results for the bivariate regression.

equation:

$$Z_Y = \beta Z_X$$

Recall that with bivariate regression, the slope coefficient is equal to the Pearson's r value if we use standardized X and Y values. Thus, Figure 15.18 shows the Beta (β) coefficient, which can be used in the standardized equation:

$$Z_Y = \beta Z_X,$$
$$Z_Y = 0.981 Z_X$$

If you check the values in Figure 15.18, you will see that the standardized coefficient is equal to the Pearson's r because we only have one predictor variable.

Many students are curious as to *why* SPSS® produces both sets of coefficients. There are several answers to this question, but the most common answer is that researchers often wish to work with both of their variables on the same scale of measurement. Thus, in our fictitious study, both books read and reading achievement would be expressed as standardized (Z) scores. SPSS® can save the raw score values as Z scores as we saw in Chapter 8 if you wish to use the standardized values in scattergrams as I did in Figure 15.8.

ASSUMPTIONS OF BIVARIATE LINEAR REGRESSION

As in the case of the other statistical procedures we discussed, there are assumptions for linear regression that, when met, will likely yield the best predictions and explanation of variance. Different authorities look at these assumptions differently, so I will list some common assumptions here.

- Variables are interval level. Regression procedures can use ordinal and even nominal data *as the predictor variable* but not the outcome variable. In advanced courses, you will learn how to create and use categories of data to predict interval level outcomes. I treat this in detail in other publications [see Abbott (2010)]. There are other regression procedures for studies with categorical outcomes (logistic regression, for example), but we will use interval level variables for our bivariate regression examples.
- Variables are normally distributed.
- Variances are equal. As with Pearson's r, regression is robust for these violations unless one or both variables are significantly skewed. You can use scattergrams to detect patterns that may indicate violations of this assumption. If the pattern of dots is not generally evenly distributed around the regression line, the variances may not be equal, for example. Technically, this assumption means that *the variance of one variable should be generally the same at different levels of the other variable*.

- Linear relationship. With linear regression, the two variables must display a "straight line" when plotting their values. Thus, for example, if we were correlating the age of a car with the value of a car, the correlation would probably be a straight line (in a downward direction, indicating an inverse relationship), but then the line would change in an upward direction because really old cars increase in value. Formally, we can detect these curvilinear relationships through SPSS®.
- Cases are independent of one another. As I mentioned with correlation, this assumption is somewhat difficult to understand, but it deals primarily with not using variables in which there are linkages among the participants. The pairs of data from each participant must not be connected.

CURVILINEAR RELATIONSHIPS

Thus far, we have discussed bivariate *linear* relationships. As you check the assumptions for your study, however, you might find that the variables are not related to one another in linear fashion. Scattergrams are always helpful to detect these possible violations of assumptions, but SPSS® has a procedure that provides a numerical analysis.

As a brief example, I will demonstrate these issues with a database we have used previously in this book. This is the school-level sample ($N = 40$) of reading and math achievement scores and FR percentages. The first bivariate study I will use as an example of curve estimation to test the linear assumption is the relationship between reading achievement and FR.

In order to test the assumption of linearity, use the "Analyze–Regression–Curve Estimation" menu to get the screen shown in Figure 15.19.

As you can see, I placed the reading achievement variable in the "Dependent" window and placed FR in the "Independent" window. In the "Models" box, there are several choices for curve "fitting" that might be applicable to our data. I have chosen two of these ("Linear" and "Quadratic") to show how each method fits the data to the regression line. In essence, these methods (and the others listed) are attempts to provide the best fit of the line to the dots. As you can see, there are several different models that might provide a better regression equation.

Figure 15.20 shows the resulting scattergram with two curves placed on the data. The linear (solid) line is shown along with the quadratic (dashed) line. The pattern of the dots *appear* to favor the quadratic line indicating a curvilinear relationship. That is, low and high FR values are related to lower reading assessment scores while middle FR levels are related to higher reading assessment scores. The two lines are close to one another, so how do you decide which is a better fit?

The curve estimation procedure also provides a numerical analysis of the two lines so that you can compare how they fit the data. Figure 15.21 shows the results of this analysis.

You interpret this output as you would a given linear regression output such as the example above with our fictitious data. In the output table, however, there are

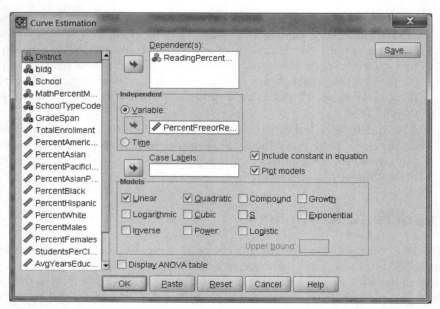

FIGURE 15.19 The Curve Estimation procedure in SPSS®.

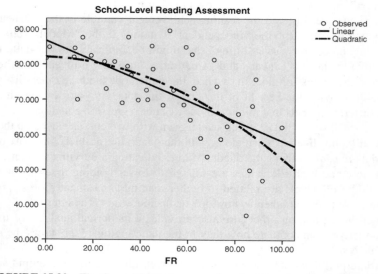

FIGURE 15.20 The Curve Estimation procedure for FR and reading assessment.

Model Summary and Parameter Estimates

Dependent Variable is ReadingPercentMetStandard

Equation	Model Summary					Parameter Estimates		
	R Square	F	df1	df2	Significance	Constant	b1	b2
Linear	0.407	26.098	1	38	0.000	86.779	−0.285	
Quadratic	0.434	14.182	2	37	0.000	82.118	−0.018	−0.003

The independent variable is PercentFreeorReducedPricedMeals.

FIGURE 15.21 The curve estimation model summary.

two lines of data that correspond to two different statistical models. The first line (Linear) shows that the regression equation between FR and Reading would result in an R^2 of 0.407, along with the following linear equation:

$$Y_{pred} = -0.285X + 86.779$$

The second line (Quadratic) shows the regression equation with an R^2 of 0.434, along with an equation of

$$Y_{pred} = -0.018X_1 + -0.003X_1^2 + 82.118$$

This equation looks a bit different because it adds a component ($-0.003X_1^2$) that corresponds to the quadratic pattern. The quadratic model adds the squared values of FR to the regression equation along with the unsquared values in predicting reading assessment. In this way, we can see if the squared values *add information to estimating the curve*. Predicted Y values are created by *combining* two estimated coefficients: (a) the coefficient for the values of the X variable and (b) the coefficient for the squared values of the X variable. *The resulting equation would therefore represent a line that curves or that contains two different trajectories.*

The quadratic model (curvilinear) explains 2.7% more of the variance in reading assessment than the linear equation (0.434 − 0.407 = 0.027). Along with the visual evidence from the scattergram, we might conclude that the quadratic model is a better fit for these variables. However, as a researcher, you might decide that the linear relationship, although explaining a bit less variance, may be the better model for your study.

Compare the reading results above with the curve estimation results for FR with school-level math assessment. Figure 15.22 shows the scattergram and Figure 15.23 shows the numerical results.

As you can see in Figure 15.22, the two lines (linear and quadratic) are almost indistinguishable. The quadratic model explains a fraction more of the variance than does the linear model (0.404 − 0.402 = 0.002, or 0.2 of 1% of explained variance). Both sources of information suggest that the FR–math assessment analysis generally fits the assumption of a linear model.

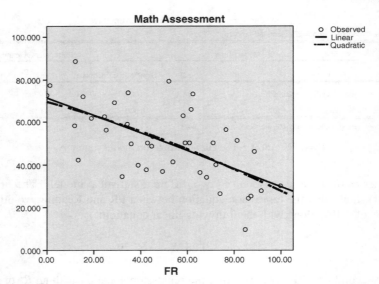

FIGURE 15.22 The curve estimation for FR with math assessment.

DETECTING PROBLEMS IN BIVARIATE LINEAR REGRESSION

I mentioned earlier that I have dealt with detecting problems with statistical assumptions in other publications [see Abbott (2010), in particular about regression procedures]. You can see that performing a regression study can be quite complex, but that *your results can be improved if you meet the assumptions of the procedure.* This is a general rule in statistics: The power of a study is improved by using the proper procedure and meeting the assumptions for each.

SPSS® provides several diagnostic procedures to assist researchers with their attempts to meet the assumptions of their chosen statistical measures. You may want to explore these further as you advance in your use of statistics in problem solving. We have discussed some in this chapter on bivariate regression (i.e., using residual analyses and curve fitting). I encourage you to seek other diagnostic techniques to help with your real-world studies.

Model Summary and Parameter Estimates

Dependent Variable is MathPercentMetStandard

Equation	Model Summary					Parameter Estimates		
	R Square	F	df1	df2	Significance	Constant	b1	b2
Linear	0.402	25.520	1	38	0.000	71.516	−0.417	
Quadratic	0.404	12.524	2	37	0.000	69.679	−0.312	−0.001

The independent variable is PercentFreeorReducedPricedMeals.

FIGURE 15.23 The numerical curve estimation data for FR with school-level math assessment.

A REAL-WORLD EXAMPLE OF BIVARIATE LINEAR REGRESSION

I will use the data from our earlier study to show the various aspects of interpreting Excel® and SPSS® results. This database is the sample of schools I used to show the curve estimation procedures above: the sample ($N = 40$) of schools with FR and math assessment scores. The study question is whether FR can significantly predict math assessment scores and be helpful in explaining the variance in math assessment.

In this example I will not provide the individual data for hand calculations. I will provide the Excel® and SPSS® outcome tables so you can learn to interpret the results. First, we will address the assumptions:

- Variables are interval level. Both are percentages and therefore interval level.
- Linear relationship. We discussed this above in our curve estimation analysis of the study variables and found that the linear model was acceptable.
- Cases are independent of one another. Met by nature of the data.
- Variables are normally distributed. We will examine the descriptive data below and discuss the findings for this assumption.
- Variances are equal. We will discuss this assumption with the normal distribution assumption.

Normal Distribution and Equal Variances Assumptions

Figures 15.24 and 15.25 show the descriptive data from Excel® and SPSS® with respect to these study variables. As you can see from both Figures, the study variables are normally distributed as indicated by skewness and kurtosis values that are within acceptable boundaries. Because neither variable is extremely skewed, the scattergram should show an even distribution of dots around the regression line.

	Math Assessment	FR
Mean	50.61	50.16
Standard error	2.79	4.25
Median	50.00	50.77
Mode	50.00	#N/A
Standard deviation	17.68	26.88
Sample variance	312.44	722.32
Kurtosis	−0.41	−0.93
Skewness	0.01	−0.07
Range	78.80	99.76
Minimum	9.70	0.00
Maximum	88.50	99.76
Sum	2024.20	2006.47
Count	40	40

FIGURE 15.24 The Excel® descriptive summaries for the study variables.

Descriptive Statistics

	N	Mean	Standard deviation	Skewness		Kurtosis	
	Statistic	Statistic	Statistic	Statistic	Standard error	Statistic	Standard error
Math Assessment	40	50.60500	17.676073	0.008	0.374	−0.412	0.733
FR	40	50.1618	26.87608	−0.073	0.374	−0.930	0.733
Valid N (listwise)	40						

FIGURE 15.25 The SPSS® descriptive summaries for the study variables.

Figure 15.22 shows this to be the case, so we can accept that the equal variance assumption is met.

The Omnibus Test Results

Recall that the omnibus test assesses the statistical significance of the overall regression model. We can use both Excel® and SPSS® to examine the omnibus test results through the ANOVA table. I will show the Excel® results in Figure 15.26, although SPSS® provides identical results. As you can see, the F (25.52) is significant ($p = 0.00001$) which indicates that the overall model significantly predicts school-level reading assessment from FR scores.

Effect Size

You can calculate the effect size by using the SS values from the table in Figure 15.26. As you can see, r^2 is 0.402, which is considered a large effect according to the guidelines we discussed earlier.

$$r^2 = \frac{SS_{reg}}{SS_Y},$$

$$r^2 = \frac{4895.65}{12,185.30},$$

$$r^2 = 0.402$$

ANOVA

	df	SS	MS	F	Significance F
Regression	1	4895.646	4895.646	25.52036	0.00001
Residual	38	7289.653	191.833		
Total	39	12185.3			

FIGURE 15.26 The Excel® omnibus test results: ANOVA.

A REAL-WORLD EXAMPLE OF BIVARIATE LINEAR REGRESSION

Model Summary[b]

Model	R	R Square	Adjusted R Square	Stdandard error of the Estimate	Change Statistics				
					R Square Change	F Change	df1	df2	Significance F Change
1	0.634[a]	0.402	0.386	13.850378	0.402	25.520	1	38	0.000

[a]Predictors: (Constant), PercentFreeorReducedPricedMeals
[b]Dependent Variable: MathPercentMetStandard

FIGURE 15.27 The SPSS® model summary results for the reading assessment—FR study.

The Model Summary

Both Excel® and SPSS® provide overall model summary data that provide further information for the omnibus test and effect size. I will show the SPSS® results in Figure 15.27 because of the additional information contained in the output as I described earlier in this chapter. As you can see, the effect size is as we calculated it above, and both the F-Change and R-Square-Change values are significant. The s_{est} is 13.85, which will be helpful if we wish to predict specific values of reading assessment given values of FR.

The Regression Equation and Individual Predictor Test of Significance

I will show the individual coefficient results from both Excel® and SPSS® in order to point out the unique results of the outputs. Both report identical findings for the test of the predictor slope and for the coefficient values.

Both Figures 15.28 (Excel®) and 15.29 (SPSS®) show that the t value (-5.05) for FR is significant ($p < 0.000$). This indicates that the slope for FR is significantly different from 0. That is, the slope is nonzero. The confidence intervals for the coefficients indicate that the estimated population slope falls between -0.58 and -0.25 at the 95% level. If this interval had included 0, we would have to conclude that a 0 population slope was possible and, therefore, that the slope is not significant. However, the values do not include 0, so we can be confident (at the 0.95 level) that the slope is significantly different from 0.

Figures 15.28 and 15.29 show the coefficients we can use to create the regression equation:

$$Y_{pred} = -0.42X + 71.52$$

	Coefficients	Standard Error	t Stat	P-value	Lower 95%	Upper 95%
Intercept	71.52	4.68	15.27	0.00	62.04	81.00
FR	−0.42	0.08	−5.05	0.00	−0.58	−0.25

FIGURE 15.28 The Excel® regression output for the reading assessment—FR study.

Coefficients[a]

Model		Unstandardized Coefficients B	Standard error	Standardized Coefficient Beta[a]	t	Significance	95.0% Confidence Interval for B Lower Bound	Upper Bound
1	(Constant)	71.516	4.683		15.272	0.000	62.036	80.996
	FR	−0.417	0.083	−0.634	−5.052	0.000	−0.584	−0.250

[a]Dependent Variable: MathPercentMetStandard

FIGURE 15.29 The SPSS® Coefficients output for the reading assessment—FR study.

Thus, we can identify that with a change of 1% in FR, the reading assessment value decreases by 0.42%.

Figure 15.29 also includes the standardized regression coefficient (β) of −0.634, which is also r in this analysis. If we wanted to create the standardized regression equation, it would be as follows:

$$Z_Y = -0.634 Z_X$$

Either of these regression formulas can be used for prediction. However, you must remember to use the Z-score formula to transform the outcomes if you wish to create raw-score values using the standardized (Z) formula.

The Scattergram

Figure 15.30 is the (SPSS®) scattergram between math assessment and FR. As you can see, the dots are evenly spaced around the line, indicating that the equal

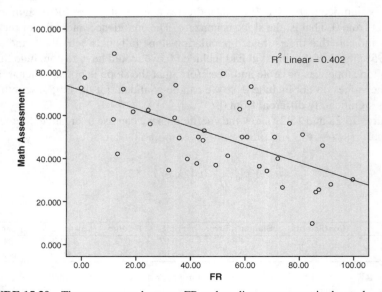

FIGURE 15.30 The scattergram between FR and reading assessment in the study schools.

variance and normal distribution assumptions are met. The intercept is at 71.5, which reflects the a coefficient value in our regression equation. The -0.42 slope is evident in the downward direction of the line. The scattergram also includes the R^2 value in the legend so that you can see the explained variance resulting from the relationship.

ADVANCED REGRESSION PROCEDURES

Correlation and regression are very useful statistical techniques. However, they are somewhat limited in that we are only using one predictor variable to understand the variance in the outcome variable. Real-life research is much more complex. FR does explain a good deal of the variation in school-level math achievement, but there is still a lot of the variance unexplained. In our last example, we saw that the effect size was 0.40. That is a very large effect size, but it means that 60% of the math achievement variance is not explained. Fortunately, there are several statistical procedures that are helpful for these complex problems.

Multiple Correlation

If you recall, we discussed a fictitious study in which we used books read as a way of explaining the variance in reading achievement test scores. However, we know that no one variable by itself will explain all of the variance of another; life is not that simple. Therefore, we might use several predictor variables, analyzed at the same time, to explain more of the variance of a dependent variable. *Multiple correlation is a technique that correlates several predictors to a dependent variable.* Will the combined influence of the set of independent variables explain more of the variance in a dependent variable than will a single predictor?

We can extend our fictitious example to understand multiple correlation. If you look at Figure 14.11, you will see that a Venn diagram illustrates how a correlation results in explained variance (i.e., through r^2). Recognizing the complexity of the research situation, we might add "ability level" to our original correlation to see what *additional* explanation it provides. Figure 15.31 shows how this might appear through Venn diagrams.

As you can see, the addition of the second predictor, reading ability, results in additional overlap with the dependent variable reading achievement. Therefore, the amount of overlap illustrated by the hatching (of both predictors) shows that this multiple correlation explains more of the variance in reading achievement than using a single predictor.

We may never be able to explain all of the variation in reading achievement in this manner, because there are probably an infinite number of potential explanations. But it helps us to get closer to understanding the variance in the distribution of dependent variable scores.

We will not calculate multiple correlation in this book. I wanted to introduce you to the procedure so that you can understand how correlation and regression are used

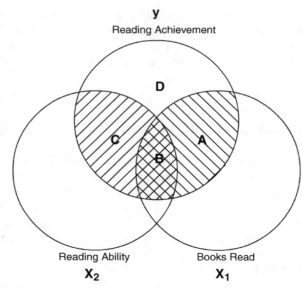

FIGURE 15.31 The multiple correlation relationship in the fictitious study example.

in research. We add (and subtract) variables from the models we build so that we can increase our understanding of a study (i.e., outcome) variable.

With a single predictor, r^2 is the effect size indicator. With multiple correlation, the effect size (the "multiple coefficient of determination") is indicated by the capital r as R^2. You will note that the regression summaries presented in both Excel® and SPSS® show the multiple R^2 because they are designed to describe the effects of multiple predictors.

Partial Correlation

Partial correlation is a process whereby the researcher assesses the correlation between an outcome variable Y and one independent variable (X_1) while holding a second independent variable (X_2) constant. What this means in practice is that we are taking into account all of the various pieces of the overall set of correlations.

Recall the Venn diagram in Figure 15.31. As you can see I labeled the various overlapping sections of the circles with A through D. The partial correlation of books read (X_1) and reading achievement (Y) is like saying we are going to correlate these two variables *after we remove the influence of reading ability (X_2) from both of the other variables*. This is like saying that, among everyone of the same reading ability, the correlation of books read and reading achievement is a certain value.

In this situation, we can understand the partial correlation by looking at the proportion of variance explained in Y after the X_2 influence is removed from both of the other variables. In terms of the Venn diagram in Figure 15.31, the explained variance would be equivalent to the proportion of A to D (i.e., A/D).

Multiple Regression

If multiple correlation can explain additional variance in Y as a result of adding more predictors, multiple regression can use the additional variables in a formula measuring the influence of all the predictors on the outcome variable. The incremental gain of each predictor in explaining more of the variance in Y is identified in a multiple regression formula. The slope values of each independent variable measure the effects of that particular predictor on Y *when the influence of the other predictors is controlled.* These regression coefficients can be tested for significance to provide a view of how important each of the independent variables are in predicting Y.

We will cover multiple regression in greater depth in Chapter 16. I encourage you to pursue an understanding of this procedure, which I believe is one of the most valuable statistical techniques for researchers. I explore the topic in depth in Abbott (2010).

Additional Considerations

In advanced books and courses, you will learn about several statistical techniques that are built on the correlation–regression procedures. Path Analysis, Structural Equation Modeling, Canonical Correlation, Discriminant Analysis and Hierarchical Linear Modeling are just a few of many such powerful techniques.

TERMS AND CONCEPTS

Beta The *standardized* regression coefficient. Sometimes called "beta weight" and "beta coefficient."

Confidence Interval In regression, the confidence interval is the range of y values within which the "true" population value of a predicted y value is likely to fall.

Line of Best Fit This is the term for the regression line calculated from the slope and y intercept values. It is called the line of best fit because a line is drawn through the bivariate scatter of values in such a way that distances from the line are minimized.

Multiple Correlation A statistical procedure that measures the association of several predictors with a single outcome variable.

Multiple Regression A statistical procedure assessing the influence on an outcome variable of more than one predictor variable.

Partial Correlation *Generally*, the process of assessing the correlation between an outcome and predictor variable when holding a second predictor constant.

Slope In a regression equation, the slope is the "angle" of the line which is measured by comparing "rise over run."

Standard Error of Estimate The standard deviation of the distribution of estimates in a regression study (i.e., the distribution of prediction errors).

Y Intercept In a regression equation, this value refers to the y value when $X = 0$. This value will fall at the intersection of the regression line and the y axis.

REAL-WORLD LAB XI: BIVARIATE LINEAR REGRESSION

In this lab, we will use regression with the TAGLIT data I described in Chapter 3.[2] In particular, we will use the Middle/High-School teacher (MHTeacher) technology factors to determine whether the data support a regression model whereby we predict technology impact from technology skills.

You may recall that these factors came from a national study in which I used factor analysis to create indices of various technology uses in the classroom. Four factors emerged from that study:

- Technology Skills—the extent of MH teachers' technology skills.
- Technology Impact—the extent to which MH teachers believed that technology impacted teaching and learning in the classroom.
- Technology Access—The extent to which MH teachers had access to technology in their schools.
- Technology Support—The extent of technology support in MH teachers' schools.

The data for this lab represent a random sample ($N = 60$) from the database ($N = 3444$) of schools from 2003 in which the factors represent aggregated MH teacher factors from schools with at least 10 students and 5 teachers. The response categories from which the factors were created were as follows:

MHImpact. The eight items measured teacher perception of the result of their use of technology in the classroom in various ways with the following response categories:

"1"—No

"2"—Yes, somewhat

"3"—Yes, quite a bit

"4"—Yes, very much

---

[2] The author acknowledges the kind approval of T.E.S.T., Inc., the owner and manager of TAGLIT data, for the use of TAGLIT databases in this book (http://www.testkids.com/taglit/).

MHSkills. The many items forming this factor were based on MHTeacher perceptions of their technology skills in various areas according to the response categories:

"1"—I don't know how to do this
"2"—I can do this but sometimes I need help
"3"—I can do this independently
"4"—I can teach others how to do this

The research question for this lab is whether we can predict Technology Impact (MHImpact) from Technology Skills (MHSkills). That is, do MH teachers rate the impact of technology on the teaching and learning in their classrooms to be higher depending on the level of their technology skills? Table 15.4 shows the data we will use in this lab (in two sets of columns).

TABLE 15.4 Lab XI Data

mhskills	mhimpact	mhskills	mhimpact
2.88	2.64	3.26	2.79
3.00	2.71	2.90	2.83
2.87	2.61	2.90	2.45
3.01	2.75	3.12	2.56
3.24	2.93	3.12	3.08
3.11	2.84	3.08	3.05
2.96	2.65	2.81	2.39
3.14	2.88	3.50	3.23
3.10	2.75	3.48	2.96
3.05	2.71	2.93	2.94
2.92	2.61	3.01	2.43
3.08	2.85	2.86	2.36
3.02	2.66	3.36	3.20
3.00	2.64	2.90	2.35
2.89	2.57	3.08	3.15
2.98	2.61	3.16	2.49
3.17	2.76	3.35	3.27
3.16	2.74	2.71	2.76
2.98	2.84	2.60	2.21
2.96	2.57	3.33	2.65
2.93	2.56	3.14	2.33
3.46	3.14	3.56	3.38
3.04	2.62	2.73	2.87
2.83	2.49	3.37	2.66
2.75	2.63	3.40	3.38
3.20	3.01	2.99	3.42
3.02	2.90	3.47	2.69
2.60	2.50	3.40	3.52
3.15	2.68	2.64	1.99
2.72	2.61	2.63	3.02

Lab-Questions

1. Do the variables meet the assumptions for correlation and regression?
2. What is the correlation between MHSkills and MHImpact?
3. Is the correlation significant?
4. What is the effect size?
5. What is the regression equation?
6. Is there a significant regression between MHSkills and MHImpact?
7. Predict a value of MHImpact when MHSkills is 3.29 and identify the $CI_{0.95}$ values.

REAL-WORLD LAB XI: SOLUTIONS

1. Do the Variables Meet the Assumptions for Correlation and Regression?

- Variables are interval level. The variables represent aggregated factor index scores and are interval. It could be argued that the response categories are not interval data, but I have treated them as interval for creating the index scores. As we have discussed in earlier chapters, some researchers may not make this assumption.
- Variables are normally distributed. Figures 15.32 and 15.33 represent the Excel® and SPSS® descriptive statistics, respectively. As you can see from both figures, the study variables appear to be normally distributed. The histograms for these variables are shown in Figures 15.34 and 15.35.

	mhskills	mhimpact
Mean	3.05	2.76
Standard error	0.03	0.04
Median	3.02	2.71
Mode	2.60	3.38
Standard deviation	0.24	0.31
Sample variance	0.06	0.09
Kurtosis	−0.42	0.28
Skewness	0.19	0.37
Range	0.96	1.54
Minimum	2.60	1.99
Maximum	3.56	3.52
Sum	183.00	165.85
Count	60	60

FIGURE 15.32 The Excel® descriptive statistics for the study variables.

REAL-WORLD LAB XI: SOLUTIONS

Descriptive Statistics

	N	Mean	Standard deviation	Variance	Skewness		Kurtosis	
	Statistic	Statistic	Statistic	Statistic	Statistic	Standard error	Statistic	Standard error
mhskills	60	3.0501	0.23809	0.057	0.185	0.309	−0.418	0.608
mhimpact	60	2.7642	0.30706	0.094	0.365	0.309	0.284	0.608
Valid N (listwise)	60							

FIGURE 15.33 The SPSS® descriptive results for the study variables.

- Variances are equal; linear relationship. We can use a scattergram to show visual evidence for these two assumptions. I will use the SPSS® procedure for curvilinear relationships ("Curve Estimation") to show these results.

As you can see in Figure 15.36, the two best-fit lines are similar, making it difficult to decide about which model to use. Figure 15.37 shows the curve estimation data which can help in our decision. Both the linear and quadratic models are significant. However, the R^2 difference is very slight at 0.008.

Making a decision to assume linearity is often difficult in these circumstances, since there is a slight curvilinear trend in the scattergram. However, I will keep the sample data as an example of a linear relationship according to the following reasoning. An outlier analysis of these data identified one data point, which, if eliminated, would result in (a) an identical R^2 value for both linear and quadratic models

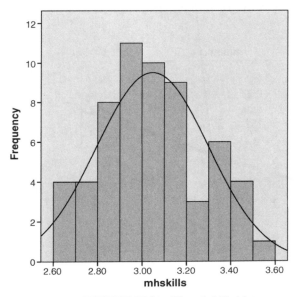

FIGURE 15.34 The mhskills histogram.

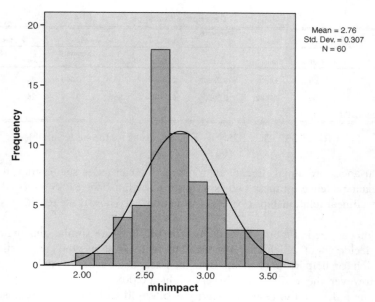

FIGURE 15.35 The mhimpact histogram.

and (b) nearly identical regression lines. I believe it is necessary to leave the outlier score in the analysis because there is no reason to exclude it as being an error. Beyond this, however, a residual analysis, which advanced researchers often conduct, shows no violations of linearity. Therefore, I believe the best course of action is to

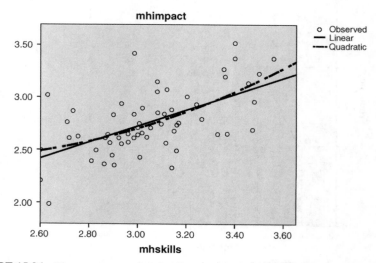

FIGURE 15.36 The scattergram showing linear and quadratic regression lines for the study variables.

REAL-WORLD LAB XI: SOLUTIONS

Model Summary and Parameter Estimates

Dependent Variable is mhimpact

Equation	R Square	Model Summary				Parameter Estimates		
		F	df1	df2	Significance	Constant	b1	b2
Linear	0.353	31.585	1	58	0.000	0.429	0.766	
Quadratic	0.361	16.125	2	57	0.000	4.367	−1.814	0.420

The independent variable is mhskills.

FIGURE 15.37 The Curve Estimation Model Between mhskills and mhimpact.

leave the score in the analysis and treat the data as linear, but note the potential influence of the one outlier score.

- Cases are independent of one another. This assumption is met.

2. What Is the Correlation Between MHSkills and MHImpact?

Table 15.5 shows the sums of square values derived from the data table that you can use to calculate the Pearson's r value. The correlation is positive, indicating that as MHSkills increase, MHImpact scores also increase. Use the formula to calculate Pearson's r by hand, and then confirm your calculation by using Excel® and SPSS® procedures. Figures 15.38 and 15.39 show the correlation tables from SPSS® and EXCEL®, respectively.

$$r_{XY} = \frac{SS_{XY}}{\sqrt{(SS_X)(SS_Y)}},$$

$$0.594 = \frac{2.561}{\sqrt{(3.345)(5.563)}}$$

3. Is the Correlation Significant?

As you can see, both correlation tables (Figures 15.38 and 15.39) report the same Pearson's r value (0.594) and SPSS® notes the significance level ($p = 0.0000006$). Using the Pearson's r table of exclusion values, we compare the calculated $r = 0.594$ to the $r_{0.05,58} = 0.25$. Thus, our calculated r is well into the exclusion range defined by 0.25. Thus, our correlation is significant, $p < 0.05$.

TABLE 15.5 The Data Summaries for Calculations

$\sum X$	$\sum X^2$	$\sum Y$	$\sum Y^2$	$\sum XY$
183.004	561.521	165.85	464.01	508.42
	SS_X	SS_Y	SS_{XY}	
	3.345	5.563	2.561	

Correlations

		mhskills	mhimpact
mhskills	Pearson Correlation	1	0.594[a]
	Significance (two-tailed)		0.000
	N	60	60
mhimpact	Pearson Correlation	0.594[a]	1
	Significance (two-tailed)	0.000	
	N	60	60

[a]Correlation is significant at the 0.01 level (two-tailed).

FIGURE 15.38 The SPSS® correlation matrix for the study variables.

4. What Is the Effect Size?

The coefficient of determination $(r^2) = 0.353$. Therefore, 35% of the variance in MHImpact is explained by MHSkills. This effect size is considered large according to the guidelines we discussed in the chapter above.

5. What Is the Regression Equation?

The hand calculations for the bivariate regression are as follows. We can compare the equation with the Excel® and SPSS® output that follows.

b value:

$$b = \frac{r\sqrt{SS_Y}}{SS_X}, \quad b = \frac{0.594\sqrt{5.563}}{\sqrt{3.345}}, \quad b = 0.766$$

a value:

$$a = M_Y - b(M_X), \quad a = 2.764 - 0.766(3.05), \quad a = 0.428$$

The Regression Equation:

$$\boxed{Y_{\text{pred}} = 0.776X + 0.428}$$

The regression equation indicates that the value of MHImpact increases by 0.776 units when MHSkills increases by 1 unit. The regression line crosses the Y axis at

	mhskills	mhimpact
mhskills	1	
mhimpact	0.593775	1

FIGURE 15.39 The Excel® correlation matrix for the study variables.

	Coefficients	Standard Error	t Stat	P-value	Lower 95%	Upper 95%
Intercept	0.43	0.42	1.03	0.31	−0.41	1.26
mhskills	0.77	0.14	5.62	0.00	0.49	1.04

FIGURE 15.40 The Excel® coefficients table showing the b and a values.

0.428 when $X = 0$. Figures 15.40 and 15.41 show the tables with the a and b coefficients. Figure 15.41, the SPSS® table, shows the standardized β coefficient as well (0.594).

6. Is There a Significant Regression Between MHSkills and MHImpact?

The Omnibus Test. The omnibus test ($F = 31.585$, $p = 0.0000006$) indicates that the regression model (MHSkills predicting MHImpact) is significant. Figure 15.42 shows the Excel® ANOVA table that includes this finding. (The SPSS® ANOVA model is identical.)

The Individual Predictor Test. Both Figures 15.40 and 15.41 show that MHSkills, the predictor, is significant ($t = 5.620$, $p = 0.0000006$). The $CI_{0.95}$ for b (0.493 to 1.039) do not include the value of 0, confirming a significant slope.

Figure 15.43 shows the SPSS® Model Summary table with additional indicators of significance. As you can see, the F Change (31.585) is significant ($p = 0.0000006$).

7. Predict a Value of MHImpact When MHSkills = 3.29 and Identify the $CI_{0.95}$ Values

The following formulas provide the calculated values for s_{est} and the $CI_{0.95}$ values for the predicted value of MHImpact when MHSkills is 3.29. As you can see, the predicted value for MHImpact is 2.981.

$$Y_{pred} = 0.776X + 0.428,$$
$$Y_{pred} = 0.776(3.29) + 0.428,$$
$$Y_{pred} = 2.981$$

Coefficients[a]

Model		Unstandardized Coefficients		Standardized Coefficients[a]	t	Significance	95.0% Confidence Interval for B	
		B	Standard error	Beta[a]			Lower Bound	Upper Bound
1	(Constant)	0.429	0.417		1.028	0.308	−0.406	1.263
	mhskills	0.766	0.136	0.594	5.620	0.000	0.493	1.039

[a]Dependent Variable: mhimpact

FIGURE 15.41 The SPSS® Coefficients table showing the b and a values.

ANOVA					
	df	SS	MS	F	Significance F
Regression	1	1.96	1.96	31.58	0.00
Residual	58	3.60	0.06		
Total	59	5.56			

FIGURE 15.42 The Excel® ANOVA summary table for the MHSkills—MHImpact study.

Model Summary

					Change Statistics			
Model	R	R Square	Adjusted R Square	Standard error of the Estimate	R Square Change	F Change	df1	df2	Significance F Change
1	0.594[a]	0.353	0.341	0.24919	0.353	31.585	1	58	0.000

[a]Predictors: (Constant), mhskills

FIGURE 15.43 The SPSS® Model Summary table showing a significant F-Change value.

The CI.95 values are calculated as follows following the s_{est} calculation. The estimated population value for the Y_{pred} falls between 2.483 and 3.479 at the $CI_{0.95}$ level.

$$S_{est} = \sqrt{\frac{SS_Y}{N-2}(1-r^2)},$$

$$S_{est} = \sqrt{\frac{5.563}{58}(1-0.353)},$$

$$S_{est} = 0.249,$$

$$\text{Confidence Interval} = \pm t(S_{est}) + Y_{pred},$$

$$CI_{0.95} = \pm 2.000(0.249) + 2.981$$

Lower Boundary = 2.483	Upper Boundary = 3.479
$CI_{0.95} = -2.000(0.249) + 2.981$	$CI_{0.95} = +2.000(0.249) + 2.981$

16

INTRODUCTION TO MULTIPLE LINEAR REGRESSION

At the end of Chapter 15, I mentioned that Multiple Linear Regression (MLR) is an extension of bivariate linear regression in which more than one predictor variable is added to an equation that predicts values of an outcome variable. Because MLR is such an important statistical procedure in research, I wanted to discuss the topic in this brief chapter.

MLR is a complex procedure and well beyond the scope of this book. However, I believe it is important, and necessary, for all students of statistics to understand at least the basics of the procedure. I will identify some of the primary elements of MLR and provide a simple example based on real world research. We will not discuss the hand calculations but will instead examine the Excel® and SPSS® output for the research example.

You will need to explore more advanced treatments of MLR in order to gain depth of understanding this versatile process. I encourage you to explore my more comprehensive treatment of MLR in Abbott (2010) as a starting point.

THE ELEMENTS OF MLR

We can start our treatment of MLR by noting the elements common to all regression procedures and especially to bivariate regression that we discussed in Chapter 15. The following are some of the similarities.

Understanding Educational Statistics Using Microsoft Excel® and SPSS®. By Martin Lee Abbott.
© 2011 John Wiley & Sons, Inc. Published 2011 by John Wiley & Sons, Inc.

Same Process as Bivariate Regression

MLR, like bivariate regression, is a process that creates a model for *predicting* values of an outcome variable from predictor variables, along with a way of *explaining* the variance in an outcome variable. Although the hand calculations are more complex, the overall procedure is the same. *We are examining the linear relationship among variables (predictors and outcome) that will improve on our prediction of values of an outcome variable knowing the correlation between the outcome and predictors.*

Similar Assumptions

Generally speaking, MLR shares the assumptions of bivariate regression. Statisticians make a distinction in assumptions when the researcher uses MLR for research in which a predictor can take only certain values ("fixed effects" models) and when MLR is used for research in which the variables can take any number of values ("random effects" models). Both of these approaches can be used for experimental and nonexperimental studies; but you should explore further discussion of these approaches, depending on your particular research question.

Statistical Significance

Like bivariate regression, MLR can be examined for statistical significance by examining the *omnibus* test and the *individual predictor(s)* test(s). Both procedures make use of the ANOVA table because both examine how the various components of variance relate to one another. Bivariate regression has only one individual predictor test while MLR has several, but the statistical significance of each individual predictor can be tested with an individual T test.

Effect Size

Effect size is measured the same way in MLR as in bivariate with the omnibus test of the model. In both cases, R^2 is an indicator of how much variance is explained in the outcome variable as a result of the predictor(s). There are more specific effect size indicators for individual predictors in MLR, but they are based on the same principle.

Coefficients

Like bivariate regression, there is a regression equation with coefficients that help to explain the relationship among the study variables. In MLR, of course, there are additional coefficients because there are additional predictors. The beta coefficients express the same thing: the changes in Y_{pred} when the beta coefficients change one unit. Because MLR has multiple beta coefficients, this interpretation changes to "the changes in Y_{pred} when the beta changes one unit and the effects of the other predictors are controlled." This last restriction allows the researcher to understand how a particular predictor affects the outcome *at similar levels of all the other predictors*. This is the way in which the other predictors are spoken of as being controlled.

Scatterdiagrams

We have seen how useful scattergrams are for expressing visually the relationship between a predictor and an outcome. MLR can use scattergrams for models that include a second predictor by creating a 3-D graph that includes a Z axis as well as axes for X and Y. More complex MLR designs can use graphical methods that capture the additional influences of other variables, but these are beyond the scope of this book.

SOME DIFFERENCES BETWEEN BIVARIATE REGRESSION AND MLR

There will obviously be some differences between the two approaches given the added complexity of additional predictors. These differences are not related to the structure of the process, but are due to the added complexity.

Multiple Coefficients

I mentioned this difference above. Again, added predictors do not change the essential nature of regression, but rather call for additional methods for interpretation of output.

Multicollinearity

Multicollinearity is a facet of MLR that relates to the *relationship among the predictors* as well as the relationship between predictors and outcome. In our fictitious example of books read predicting reading achievement in Chapter 15, we discussed adding a second predictor, reading ability, as a way of understanding multiple correlation Figure 15.31 showed the "pieces" of variance in Y explained by the two predictors. MLR takes into account that predictors probably have some relationship *to one another* as well as to the outcome variable.

The MLR output can be examined to understand the size and nature of the components of variance produced in a model with several predictors. Statisticians have devised guidelines for detecting problems in which "too much" of the predictor-to-predictor variance clouds the understanding of explained variance in the outcome variable and makes it difficult to understand the contribution of individual predictors.

Explanation of R^2

The explanation of components of explained variance in the outcome is related to the complexity of multiple coefficients and multicollinearity. MLR can be used to pinpoint the contribution to the R^2 of individual predictors. This is one of the chief contributions of MLR to research.

Although we cannot pursue this matter very far, I will mention that interpretation of the components of R^2 rests with some procedural elements ("order of entry schemes") as well as with the very important correlation procedures of *partial* correlation and *semi-partial* (or "part") correlation. I introduced these concepts in Chapter 15, and I will demonstrate how to use one of these for interpreting MLR results.

Entry Schemes

MLR results can be affected by the order in which predictors are added to the equation by statistical software. Some software procedures add (and remove) the predictors to the overall model in stepwise fashion according to a set of predetermined statistical guidelines that judge whether the predictors meet certain numerical thresholds. Other schemes allow the researcher to add predictors to a MLR model in the order they determine based on *a priori* or other grounds.

The researcher can use several approaches to identify the influence of predictors on an outcome variable. (I will show a couple of these in the example below.) "Hierarchical" regression is simply an MLR study in which the researcher adds predictors according to an indication of importance to the study and observes the changes to the overall R^2 as a result of adding the predictors.

STUFF NOT COVERED

Because of the complexity of MLR, we cannot hope to discuss all the dimensions in this book. I mentioned some of these areas above (e.g., partial and semi-partial correlation). Here are some other areas of MLR we cannot cover here, but which I encourage you to pursue.

Using MLR with Categorical Data

MLR is quite versatile and allows the researcher to use categorical as well as continuous predictors. It can even be used in experimental studies; there is a statistical connection between MLR and ANOVA, the latter of which is typically used for experimental study.

MLR uses categorical data through procedures in which the researcher transforms each of the categories of a predictor into separate subvariables, which together comprise the predictor. Each of these subvariables can then be understood in their relationship to the outcome variable. (An example might be creating four subvariables from our treatment variable "noise" in predicting human learning in our fictitious study in an earlier chapter.) Transforming predictors to sets of subvariables can be done using dummy, effect, or contrast coding according to the needs of the researcher.

Curvilinear Regression

Like bivariate regression, the relationship of the predictors to the outcome variable in MLR may be nonlinear as well as linear. Statistical software, like SPSS®, have ways of identifying these trends so the researcher can express the most efficient model for predicting the outcome variable.

Multilevel Analysis

Multilevel analyses are procedures that recognize different levels of data in a regression study. For example, we have discussed the relationship between technology skills and technology impact among MH teachers (TAGLIT data). We need to recognize that we could look at this relationship in more than one way.

Typically, researchers simply want to see if one variable (i.e., skills) can predict the other (i.e., impact). This approach understands the data as a teacher-level phenomenon. However, might the school within which the teacher works also influence both of these variables? Some schools may recruit teachers with more advanced technology skills, for example. Or, some schools may place more of an emphasis on using technology in the classroom.

In any case, multilevel analysis is a way of understanding what the relationships are at one level of analysis (teacher level) by recognizing the influence of another level of analysis (school level). These are very powerful tools for a researcher, but they are not easy to learn. You might explore Bickel (2007) for a very straightforward approach to understanding this procedure. Raudenbush and Bryk (2002) discuss this analysis in their treatment of Hierarchical Linear Modeling (HLM).

MLR EXTENDED EXAMPLE

In this section, I will present an MLR study to show some of the ways to interpret the findings from Excel® and SPSS®. Both will perform MLR analyses, but the Excel® capacity is more limited than SPSS®. In addition, because SPSS® is specifically designed for these kinds of complex analytical procedures, there are more custom features provided in the menus.

In Chapter 15 (Real-World Lab XI) I presented a bivariate regression analysis using a sample ($N = 60$) of TAGLIT data in which I predicted the MH teachers' perceived impact of technology on the classroom from the teachers' technology skill values. These were aggregated teacher scores from schools in which at least 10 teachers submitted TAGLIT questionnaires. The current extended example uses the sample database (TAGLIT 2003-Middle-HighSchool-Teachers) available through the Wiley Publications ftp website for my 2010 book ($N = 589$). I will use this database to analyze the same research question, but with an additional predictor.

The following factor scores will be used in this example:

- **mhimpact.** This is the outcome variable that is a combined measure of teachers' perceptions of the impact of technology on their classrooms (teaching and learning).
- **mhskills.** This is the single predictor I used in the Chapter 15 study that measured the teachers' perceptions of their technology skills.
- **mhaccess.** This factor is the teachers' perception of their access to their schools' technology resources.

The study question is whether middle/high-school teachers' perceptions of their technology skills and access to technology significantly predict their perceptions of the impact of technology on teaching and learning. I will treat mhimpact as the outcome variable, mhskills as the first predictor, and mhaccess as the second predictor.

ARE THE ASSUMPTIONS MET?

Before we proceed, we need to ensure that the assumptions are met. I discussed the assumptions for the outcome and first predictor in Chapter 15. I will summarize those findings and add the information for the second predictor.

- Variables are interval level. As in Chapter 15, I am treating all the factor index scores as interval data.
- Variables are normally distributed. Figures 15.32 through 15.35 showed both mhskills and mhimpact were normally distributed. Figure 16.1 shows the descriptive information for mhaccess.

As you can see, the skewness and kurtosis values indicate that there are no violations of kurtosis, but the skewness value appears to be quite high. The standard skewness value (-5.13) is well beyond our suggested guideline of 2 to 3. However, recall that with large sample sizes (beyond 200 or so) we need to rely on visual evidence as well as the numerical evidence. Figure 16.2 shows the histogram for mhaccess. As you can see, the histogram shows a slight negative skew, but only slight. It is reasonable to conclude that if there are normal distribution violations of this predictor, they are slight. We will address other ways below of adding to our decision to use this variable in the MLR analysis.

- Variances are equal; linear relationship. We will use a scattergram to show visual evidence for these two assumptions as we did in Chapter 15 for mhskills and mhimpact. I will use the SPSS® procedure for curvilinear relationships (Curve Estimation) to show these results.

As you can see in Figure 16.3, the two best fit lines (linear and quadratic) are similar, adding to our confidence that this predictor meets the MLR

Statistics

MHTeacher access to technology factor

N valid	589
N missing	0
Mean	2.8411
Median	2.8750
Mode	2.50[a]
Standard deviation	0.38700
Variance	0.150
Skewness	−0.518
Standard error of skewness	0.101
Kurtosis	0.358
Standard error of kurtosis	0.201

[a]Multiple modes exist. The smallest value is shown.

FIGURE 16.1 The descriptive data for mhaccess.

assumptions. Figure 16.4 shows the SPSS® summary table showing the numerical curve fit analysis. Although the quadratic model is significant, it does not increase the R^2 over the linear model. The curve fit analysis between the two predictors is similarly very close.

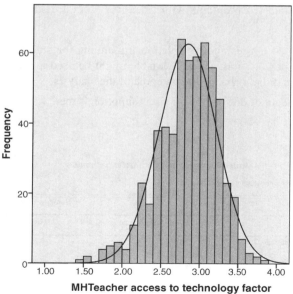

FIGURE 16.2 The mhaccess factor histogram.

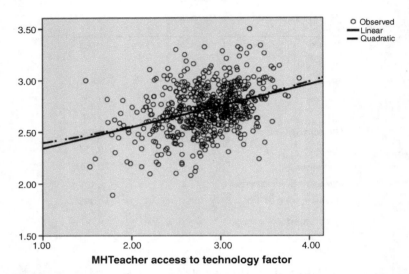

FIGURE 16.3 The SPSS® curve estimation scattergram with mhaccess and mhimpact.

One additional piece of evidence for using mhaccess as it exists is that, even if transformed, there is no impact on the overall R^2 of the model. *Variable transformation* is a process for changing the data in various ways to account for (a) outliers and (b) problems with linearity, normal distribution, and so on.

These transformations do not change the relationship among the study variables, they just help to show how a "normalized" variable might be used in the analysis. Transforming mhaccess did not change the outcome of the analysis.

- Cases are independent of one another. This assumption is met.

Model Summary and Parameter Estimates

Dependent Variable is MHTeacher technology impact factor.

Equation	Model Summary					Parameter Estimates		
	R Square	F	df1	df2	Significance	Constant	b1	b2
Linear	0.128	85.793	1	587	0.000	2.133	0.207	
Quadratic	0.128	42.951	2	586	0.000	2.273	0.104	0.019

The independent variable is MHTeacher access to technology factor.

FIGURE 16.4 The SPSS® curve fit summary for mhaccess and mhimpact.

THE FINDINGS

In what follows, I will present the Excel® and SPSS® findings for the study including comments so that you can understand the output files. I will start with the Excel® findings and then proceed to the SPSS® output to demonstrate the additional features of the latter.

Figure 16.5 presents the general findings including the R^2 and the standard error of estimate. According to these findings, mhskills and mhaccess together explain about 25% of the variance in mhimpact.

Figure 16.6 is an ANOVA summary table that we can use for the omnibus test of the model. As you can see, the F test indicates a significant model ($F = 99.07$, $p < 0.001$) showing that the F ratio is far into the exclusion area. There is only an extremely small probability that this is a chance finding. You can reproduce R^2 (0.25) by calculating the proportion of total variance explained by the regression (7.51/29.72) as we discussed in Chapter 15.

Figure 16.7 shows the individual predictor tests. From this output, you can examine the t tests for the various predictors and identify the (unstandardized) regression equation. The t tests indicate that *both predictors are significant predictors* of mhimpact when the other predictor is held constant. The X_1 coefficient ($t = 9.91$) and X_2 coefficient ($t = 4.55$) are both significant at $p < 0.05$.

The following is the MLR equation as it appears in "parameter language," but each of the elements are the same as in the bivariate equation:

$$Y_{\text{pred}} = \beta_0 + \beta_1 X_1 + \beta_2 X_2$$

β_0 indicates the intercept, or value of the coefficient a. Formally, it is the value of Y when the predictors equal 0.

Regression Statistics	
Multiple R	0.50
R square	0.25
Adjusted R square	0.25
Standard error	0.19
Observations	589

FIGURE 16.5 The overall MLR findings from Excel®.

ANOVA

	df	SS	MS	F	Significance F
Regression	2	7.51	3.75	99.07	0.00
Residual	586	22.21	0.04		
Total	588	29.72			

FIGURE 16.6 The Excel® ANOVA summary table.

	Coefficients	Standard Error	t Stat	P-value	Lower 95%	Upper 95%
Intercept	1.17	0.11	10.20	0.00	0.94	1.39
mhskills	0.41	0.04	9.91	0.00	0.33	0.49
mhaccess	0.11	0.02	4.55	0.00	0.06	0.15

FIGURE 16.7 The Excel® individual predictor coefficients.

$\beta_1 X_1$ indicates the slope (β_1) of the first predictor (X_1).
$\beta_2 X_2$ indicates the slope (β_2) of the second predictor (X_2).

Replacing these values with the actual data from the results in Figure 16.7, we obtain the (unstandardized) regression equation for this study.

$$Y_{\text{pred}} = 1.17 + 0.41 X_{\text{mhskills}} + 0.11 X_{\text{mhaccess}}$$

Researchers can use this equation to predict values of mhimpact when they know the values of mhskills and mhaccess. You would use the same process we discussed in Chapter 15 to make the prediction. You could then construct confidence intervals around the predictions using the (multiple) standard error of estimate shown in Figure 16.5.

Note also in Figure 16.7 that the confidence intervals are shown for the regression coefficients. The confidence limits for neither predictor include the likelihood of a 0 slope, which is another indication of a significant predictor.

THE SPSS® FINDINGS

The SPSS® results include all of the Excel® findings we discussed above. In what follows, I will point out some of the unique features of SPSS® that help to explain some of the nuances of the model.

Figure 16.8 shows the Linear Regression specification window that I presented in Chapter 15 (Figures 15.15 and 15.16). Recall that this series of menus is available through the main Analyze–Regression–Linear set of choices. This selection yields the option window in Figure 16.8. I reproduced this window because I want to point out a couple of very important specifications for MLR analyses.

In Figure 16.8, you see that I placed mhimpact in the "Dependent" window and placed the first predictor, mhskills, in the "Independent(s)" window. What you cannot see in this figure is that I can place the second predictor, mhaccess, in a second "Block" of predictors. Note that mhaccess is highlighted in the list of variables on the left of the screen. When I choose the "Next" button (located just below the "Dependent" window), I can enter this second predictor by itself. In this way, I have instructed SPSS® to create an MLR model with two predictors, but to add the predictors hierarchically.

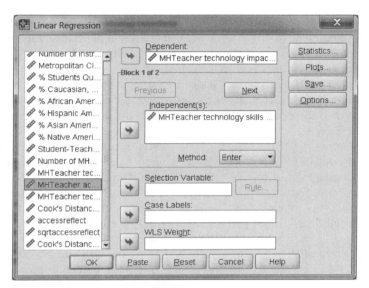

FIGURE 16.8 The SPSS® Linear Regression specification menu.

Thus, our output will show two models of data: one with the first predictor only, and the second with both predictors. In this way, *we can compare how the model summary data change with the introduction of the second predictor.*

Figure 16.8 also shows another important feature that I discussed above as an "order of entry scheme." Note that in the middle of the figure is a dropdown menu entitled "Method," which by default shows "Enter." This is the place where researchers can determine how they want the SPSS® program to manage the predictors in the model. As shown ("Enter"), I have directed the program to treat the introduction of the predictors in my own fashion. Because I used two "Blocks" of predictors, I have instructed SPSS® to enter the first predictor (mhskills) and then enter the second predictor (mhaccess) separately so that I can see the changes when the second predictor variable is added. This is the essence of hierarchical regression.

Other choices for entry include "Stepwise," which allows SPSS® preset entry and removal protocols to enter and retain each variable in a somewhat "mechanical" fashion. For example, if I had a study with five predictors, the stepwise procedure would add and retain (or delete) predictors, depending on the size of their impact on the model. It might be the case that only two predictors would be retained and the others excluded even though the researcher has good reason to include them.

The stepwise procedure can be useful at times, but most researchers avoid the method because it takes the choice for building the model out of the researcher's control. If the researcher is including each predictor on some a priori grounds for inclusion, then it is important to see the impact on the results despite the variable not "making the cut" established by the program.

The menu of entry schemes includes other methods that exceed the boundaries of our treatment of MLR. However, most research studies can profitably use hierarchical regression ("Enter") for their analyses. Depending on how you use the output, it does not matter which scheme you use!

As I showed in Chapter 15, the "Statistics" button in the upper right corner of the window shown in Figure 16.8 allows the researcher to choose several kinds of MLR output. I showed the default choices for bivarite regression in Figure 15.16. These also work fine for MLR, but in Figure 16.9 I show some additional specifications.

First, you can see that in Figure 16.9 I *did not* include confidence intervals because I showed those above in the Excel® output (see Figure 16.7). The two new choices shown in Figure 16.9 are "Part and partial correlations" and "Collinearity diagnostics." I cannot explain these choices exhaustively, but I did want to note them and what they provide so that you can begin to use them in your own research.

I introduced "part" (semi-partial) and "partial" correlations above. These can be used in many ways in the MLR analysis. One very helpful use is to show the "squared part correlation" as the contribution to R^2 for each variable added to the model. I will demonstrate this briefly below. I discussed multicollinearity above as a potential problem resulting from a high "intercorrelation" among the predictors. Choosing collinearity diagnostics helps to ensure that our study avoids serious violations of this assumption.

Figure 16.10 shows the model summary information from SPSS® resulting from our specification. Note that the program actually creates two summaries. The first

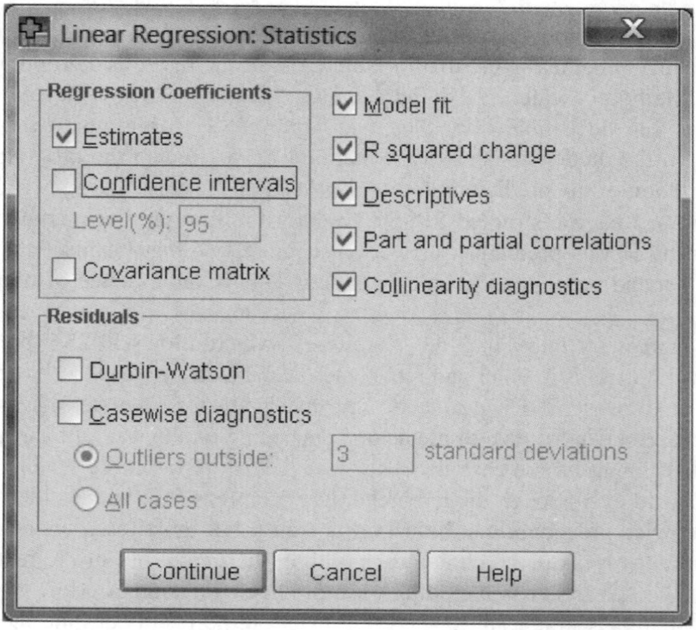

FIGURE 16.9 The SPSS® Linear Regression specification for MLR output.

THE SPSS® FINDINGS

Model Summary[c]

Model	R	R Square	Adjusted R Square	Standard Error of the Estimate	Change Statistics				
					R Square Change	F Change	df1	df2	Significance F Change
1	0.476[a]	0.226	0.225	0.19792	0.226	171.693	1	587	0.000
2	0.503[b]	0.253	0.250	0.19468	0.026	20.690	1	586	0.000

[a]Predictors: (Constant), MHTeacher technology skills factor.
[b]Predictors: (Constant), MHTeacher technology skills factor, MHTeacher access to technology factor.
[c]Dependent variable: MHTeacher technology impact factor.

FIGURE 16.10 The SPSS® MLR model summary.

"Model" is the one that only contains the first predictor and the outcome variable (in this case, only mhskills is included as a predictor of mhimpact). The second model is the one that includes the second predictor (mhaccess) as well. The reason for getting two models is that I specified that each of the two predictors should be entered separately in different blocks so that I could see how the model summary changed when I added the second predictor. This process would be the same no matter the number of predictors; adding each predictor separately would provide a number of hierarchical models.

In Figure 16.10, the last model (Model 2) is the same as the overall model shown in the Excel® output in Figures 16.5 and 16.6 because Excel® combined the two predictors in the same model. Compare these results and you will see that the Model 2 SPSS® results are the same as the Excel® results (R^2, s_{est}, etc.).

I shaded one cell in Figure 16.10 to explain. This "R-Square Change" represents the additional variance of the dependent variable (mhimpact) explained by adding a second predictor. As you can see, the second predictor (mhaccess) explains an additional 2.6% percent of variance (0.026) in mhimpact.

Now, look at the R^2 values in the "R-Square" column for Models 1 and 2 (0.226 and 0.253, respectively). If you subtract these values, you will get the 0.026 shown in the "R-Square Change" column. Either way, you can see how much the second predictor adds to the explanation of the variance in the outcome variables.

This brief example shows the value of hierarchical regression. You can see how the results change by adding subsequent predictors and thereby have a more comprehensive view of the relationship among the study variables. You can see, for example, that this second predictor also resulted in a "Significance F Change" value, which is important to the omnibus test.

The SPSS® ANOVA summary table shown in Figure 16.11 is similar to that from Excel® (see Figure 16.6). As I noted above, the ANOVA results for Model 2 in Figure 16.11 is the same as in Figure 16.6 because Excel® combines the individual predictor data. By examining the results in the two models, you can see the influence of the second predictor.

The next SPSS® output concerns the tests for the individual predictors. Figure 16.12 shows the "Coefficients" output, which specifies the slope and intercept

ANOVA[c]

Model		Sum of Squares	df	Mean Square	F	Significance
1	Regression	6.726	1	6.726	171.693	0.000[a]
	Residual	22.994	587	0.039		
	Total	29.719	588			
2	Regression	7.510	2	3.755	99.071	0.000[b]
	Residual	22.210	586	0.038		
	Total	29.719	588			

[a]Predictors: (Constant), MHTeacher technology skills factor.
[b]Predictors: (Constant), MHTeacher technology skills factor, MHTeacher access to technology factor.
[c]Dependent variable: MHTeacher technology impact factor.

FIGURE 16.11 The SPSS® ANOVA summary table showing two models.

values for each of the models. Once again, the SPSS® results in Model 2 are the same as those from Excel®. (Compare these to those shown in Figure 16.7.)

The Unstandardized Coefficients

The unstandardized coefficients shown in their own columns are the same in Model 2 as those shown in Excel®. Thus, you can create the same unstandardized regression formula as shown above.

The Standardized Coefficients

Figure 16.12 includes a column for "Standardized Coefficients." These values represent what the coefficients would be if we had used Z scores instead of raw scores

Coefficients[a]

Model		Unstandardized Coefficients		Standardized Coefficients	t	Significance	Correlations			Collinearity Statistics	
		B	Standard error	Beta			Zero-order	Partial	Part	Tolerance	VIF
1	(Constant)	1.209	0.116		10.434	0.000					
	MHTeacher technology skills factor	0.497	0.038	0.476	13.103	0.000	0.476	0.476	0.476	1.000	1.000
2	(Constant)	1.166	0.114		10.200	0.000					
	MHTeacher technology skills factor	0.413	0.042	0.395	9.907	0.000	0.476	0.379	0.354	0.802	1.247
	MHTeacher access to technology factor	0.105	0.023	0.181	4.549	0.000	0.357	0.185	0.162	0.802	1.247

[a]Dependent Variable: MHTeacher technology impact factor

FIGURE 16.12 The SPSS® Coefficients output.

THE SPSS® FINDINGS

in the data. This is a very useful set of information because it gives you the standardized (Z) information without the researcher having to transform all the raw scores to Z scores. As you can see in Model 1, the standardized beta coefficient ($\beta = 0.476$) is equal to the R when only the first predictor is in the model. This is the Pearson's r between mhskills and mhimpact because, as I explained in Chapter 15, in bivariate regression (where there is only one predictor) the beta coefficient is equal to the r value. As you can see, this is not true when you have more than one predictor. In the latter case, the standardized coefficients represent how many Z-score units the outcome variable changes with one-unit Z-score changes in the predictors (when the other predictors are held constant).

Collinearity Statistics

The "Collinearity Statistics" columns (last two columns on the right side of the figure) show the results researchers can use to determine if the intercorrelation among the predictors is too high. You can consult advanced statistical treatments of this information as you begin to build your understanding of MLR. I include an explanation in my (Abbott, 2010) book dealing with program evaluation that may be helpful. Generally, the lower the "Tolerance" value, the greater the intercorrelation complications. The "VIF" values are derived from the tolerance values, so higher numbers are problematic. The values in Figure 16.12 do not indicate problems of multicollinearity.

The Squared Part Correlation

I shaded the "part correlation" column in Figure 16.12 to show a very important aspect of the SPSS® output. If you recall Figure 15.34, we can show how multiple correlation helps to explain "pieces" of the variance in the outcome variable as a result of the two predictors in a MLR analysis. I recreated Figure 15.31 as Figure 16.13 in order to show the results for the current study problem.

If you look at Figure 16.13, you will see that I shaded portion A, which represents the "unique" explanation of the variance in the outcome due to the mhskills predictor. Portion B includes the intercorrelation (overlap) between the two predictors (mhskills with mhaccess). This is the proportion of the variance in the outcome that is problematic because we cannot "assign" it to one or the other predictors. Part C represents the unique contribution to the variance in the outcome based on the mhaccess predictor. Part D is the unexplained variance in the outcome.

Part correlation, when it is squared, represents the unique explanation of the variance of an outcome variable based on a single predictor. It is like comparing (dividing) Part A in Figure 16.13 to all the variance in the outcome (Parts A + B + C + D). You can see how this works by looking at the shaded cells of Figure 16.12.

- In Model 1, the part correlation of mhskills is 0.476.

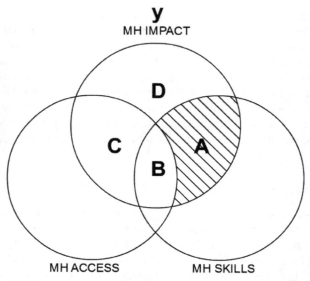

FIGURE 16.13 The part correlation segment of MH Skills.

- The part correlation squared ($0.476^2 = 0.2265$) equals the R-Square-Change value in Figure 16.10.
- In Model 2 the part correlation *of the second predictor added to the model*, mhaccess (0.162), when squared ($0.162^2 = 0.026$), equals the R-Square Change in Model 2 of Figure 16.10.

The general principle of the squared part correlation is that you can use it to understand the explanation of the variance in the outcome variable of a single predictor *no matter how it is entered into the equation*. If you add more than one variable to each "block," this is a bit more complex to see, but it is a powerful feature presented in the SPSS® choice of output.

Conclusion

The example study concludes that the two predictors, mhskills and mhaccess, are significant predictors of mhimpact. Both predictors explain about 25% of the variance (i.e., $R^2 = 0.253$) in mhimpact, a finding considered large according to the guidelines we discussed in Chapter 15. Each predictor is a significant predictor (as determined by the *t* values), but mhskills is the more powerful of the two in explaining the variance of mhimpact. Thus, according to this sample of data, it is important for teachers to have access to technology for making an impact on the classroom, but is it much more important that they perceive that they have technology skills. Having skills explains almost 5 times more unique variance in mhimpact than having access to technology.

TERMS AND CONCEPTS

Collinearity Statistics Measures of the extent to which multilinearity exists among a set of predictor variables in a regression study.

Hierarchical Regression An interpretive technique in multiple regression studies in which the sequential impact of predictor variables on the outcome variable is measured.

Multicollinearity A measure of the overlap (or intercorrelation) among predictors in a multiple regression study. This overlap can obfuscate the overall relationship of the set of predictors to the outcome variable.

Variable Transformation The process of changing data by various procedures to accommodate outliers and problems of linearity (among other problems).

REAL-WORLD LAB XII: MULTIPLE LINEAR REGRESSION

In this lab, we will use the TAGLIT data as we did in the chapter example. However, we will use the factors extracted from elementary teachers rather than middle/high-school teachers. Can elementary teachers' technology impact scores be predicted by their technology skills and technology access scores?

Lab XII Questions

1. Are assumptions met for MLR?
2. What are the findings? Use Excel® and SPSS® to identify effect size, omnibus test, individual predictor tests, and the unstandardized regression equation.
3. What is the unique explanation of variance in eimpact from the predictor eaccess?
4. What conclusions can you draw about the research question from the findings?

REAL-WORLD LAB XII: MLR SOLUTIONS

1. Are the Assumptions Met for MLR?

- Variables are interval level. As in Chapter 15, I am treating all the factor index scores as interval data.
- Variables are normally distributed. Figure 16.14 shows that both eskills and eimpact are normally distributed. Eaccess appears to be negatively skewed and leptokurtic, however.

The standard skewness value (-9.28) is well beyond our suggested guideline of 2 to 3. However, recall that with large sample sizes (beyond 200 or so)

Descriptive Statistics

	N	Mean	Standard deviation	Skewness		Kurtosis	
	Statistic	Statistic	Statistic	Statistic	Standard Error	Statistic	Standard Error
Elementary Teachers technology skills factor	944	2.8407	0.25562	−0.189	0.080	0.406	0.159
Elementary Teachers technology impact factor	944	2.6240	0.26704	−0.083	0.080	−0.113	0.159
Elementary Teachers access to technology factor	944	2.8550	0.42637	−0.742	0.080	1.120	0.159
Valid N (listwise)	944						

FIGURE 16.14 The descriptive data for eaccess.

we need to rely on visual evidence as well as the numerical evidence because the standard errors with larger datasets will be very small and result in a higher skewness quotient. Figure 16.15 shows the histogram for ehaccess. As you can see, the histogram shows a very slight negative skew. It is reasonable to conclude from the graph that if there are normal distribution violations of this predictor, they are slight. I will discuss at length other criteria below for using this variable in the MLR analysis.

- Variances are equal; linear relationship. We will use a scattergram to show visual evidence for these two assumptions as we did in Chapter 15 for mhskills and mhimpact. I will use the SPSS® procedure for curvilinear relationships ("Curve Estimation") to show these results.

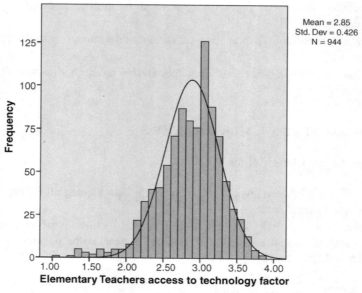

FIGURE 16.15 The ehaccess factor histogram.

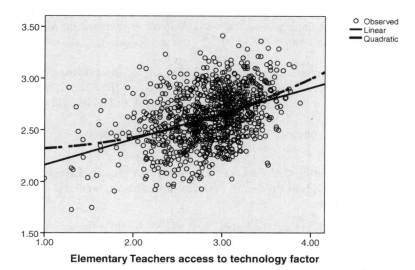

FIGURE 16.16 The SPSS® curve estimation scattergram with eaccess and eimpact.

As you can see in Figure 16.16, the two best-fit lines (linear and quadratic) are similar, but they diverge at the extremes where the values (dots) are sparse. The influence of the extreme values (at both ends) is problematic for deciding that this relationship is not linear. Authorities like Cohen et al. (2003) urge caution under these circumstances.

Figure 16.17 shows the SPSS® summary table showing the numerical curve fit analysis. Although the quadratic model is significant, there is a very minimal increase in the R^2 value over the linear model (0.004). Given the influence of the extreme scores, which I discussed above, I am not compelled by this summary to conclude a nonlinear relationship between eaccess and eimpact.

One additional piece of evidence for using eaccess as it exists is that, even if "transformed," there is very little impact on the overall R^2 of the model (from 0.158 to 0.163). I transformed eaccess using the accepted method of

Model Summary and Parameter Estimates

Dependent Variable is Elementary Teachers technology impact factor.

Equation	Model Summary					Parameter Estimates		
	R Square	F	df1	df2	Significance	Constant	b1	b2
Linear	0.158	177.125	1	942	0.000	1.913	0.249	
Quadratic	0.162	91.255	2	941	0.000	2.332	−0.071	0.059

The independent variable is Elementary Teachers access to technology factor.

FIGURE 16.17 The SPSS® curve fit summary for eaccess and eimpact.

"reflecting" and taking the square root of the raw scores to decrease curvilinearity. While this is an acceptable method, it makes interpretation more difficult (i.e., it changes the scale of the variables and it reverses the direction of the relationship), and the very small R^2 increase does not warrant using a transformed predictor.

I am leaving this predictor in the analysis because of the evidence I discussed above. As a researcher, you will be called upon to make difficult choices with your data. I left eaccess in this lab so that you could grapple with "real-world problems."

- Cases are independent of one another. This assumption is met.

2. What Are the Findings? Use Excel® and SPSS® to Identify Effect Size, Omnibus Test, Individual Predictor Tests, and the Unstandardized Regression Equation

In what follows, I will present the Excel® and SPSS® findings for the study including comments so that you can understand the output files as I did with the chapter example. I will begin with the Excel® findings and point out the additional findings from the SPSS® output.

Figure 16.18 presents the general findings including the R^2 and the s_{est}. According to these findings, eskills and eaccess together explain about 27% of the variance in eimpact.

Figure 16.19 shows the ANOVA summary table that we can use for the omnibus test of the model. As you can see, the F test indicates a significant

Regression Statistics	
Multiple R	0.52
R square	0.27
Adjusted R square	0.27
Standard error	0.23
Observations	944

FIGURE 16.18 The overall MLR output from Excel®.

ANOVA					
	df	SS	MS	F	Significance F
Regression	2	18.48	9.24	178.25	0.00
Residual	941	48.77	0.05		
Total	943	67.25			

FIGURE 16.19 The Excel® ANOVA summary table for the elementary teachers study.

	Coefficients	Standard Error	t Stat	P-value	Lower 95%	Upper 95%
Intercept	1.10	0.08	13.21	0.00	0.94	1.26
eskills	0.42	0.03	12.29	0.00	0.35	0.48
eaccess	0.12	0.02	5.86	0.00	0.08	0.16

FIGURE 16.20 The Excel® individual predictor coefficients for the elementary teachers study.

model ($F = 178.25$, $p < 0.000$) indicating that the F ratio is far into the exclusion area. There is only an extremely small probability that this is a chance finding. You can reproduce R^2 (0.27) by calculating the proportion of total variance explained by the regression (18.48/67.25) as we discussed in Chapter 15.

Figure 16.20 shows the individual predictor tests. From this output, you can examine the t tests for the two predictors and identify the (unstandardized) regression equation. The t tests indicate that *both predictors are significant predictors* of eimpact when the other predictor is held constant. The $X_{eskills}$ coefficient ($t = 12.29$) and $X_{eaccess}$ coefficient ($t = 5.86$) are both significant at $p < 0.05$.

The following is the MLR equation as it appears in "parameter language," but each of the elements are the same as in the bivariate equation:

$$Y_{pred} = \beta_0 + \beta_1 X_1 + \beta_2 X_2$$

β_0 indicates the intercept, or value of the coefficient a. Formally, it is the value of Y when the predictors $= 0$.

$\beta_1 X_1$ indicates the slope (β_1) of the first predictor (X_1).

$\beta_2 X_2$ indicates the slope (β_2) of the second predictor (X_2).

Replacing these values with the actual data from the results in Figure 16.20 results in the (unstandardized) regression equation for this study.

$$Y_{pred} = 1.10 + 0.42 X_{eskills} + 0.12 X_{eaccess}$$

Researchers can use this equation to predict values of eimpact when they know the values of eskills and eaccess. You would use the same process we discussed in Chapter 15 to calculate actual predictions. You could then construct confidence intervals around the predictions using the (multiple) standard error of estimate shown in Figure 16.18.

Note also in Figure 16.20 that the confidence intervals are shown for the regression coefficients. The confidence limits for neither predictor include the likelihood of a 0 slope, which is additional evidence of significant predictors.

Model Summary[c]

Model	R	R Square	Adjusted R Square	Standard error of the Estimate	Change Statistics				
					R Square Change	F Change	df1	df2	Significance F Change
1	0.498[a]	0.248	0.248	0.23165	0.248	311.182	1	942	0.000
2	0.524[b]	0.275	0.273	0.22766	0.026	34.313	1	941	0.000

[a]Predictors: (Constant), Elementary Teachers technology skills factor.
[b]Predictors: (Constant), Elementary Teachers technology skills factor, Elementary Teachers access to technology factor.
[c]Dependent Variable: Elementary Teachers technology impact factor.

FIGURE 16.21 The SPSS® MLR model summary for the elementary teacher study.

The SPSS® Findings

In what follows, I will point out some of the unique features of SPSS® that help to explain some of the nuances of the model as I did in the chapter example. Figure 16.21 shows the model summary information from SPSS® resulting from our specification.

Model 1 results show that eskills by itself explains approximately 25% (0.248) of the variance in eimpact, and that the change in the F test from this predictor is statistically significant. Model 2 results show that eaccess is also a signicant predictor (as determined by the "Significance F Change" results) and uniquely explains about 2.6% of the variance in eimpact.

The SPSS® ANOVA summary table shown in Figure 16.22 shows both Model 1 and Model 2 results to be statistically significant ($F = 311.182$, and $F = 178.25$, respectively). You can recreate the R^2 values using the SS values shown in the table in Figure 16.22.

ANOVA[c]

Model		Sum of Squares	df	Mean Square	F	Significance
1	Regression	16.698	1	16.698	311.182	0.000[a]
	Residual	50.549	942	0.054		
	Total	67.247	943			
2	Regression	18.477	2	9.238	178.250	0.000[b]
	Residual	48.770	941	0.052		
	Total	67.247	943			

[a]Predictors: (Constant), Elementary Teachers technology skills factor.
[b]Predictors: (Constant), Elementary Teachers technology skills factor, Elementary Teachers access to technology factor.
[c]Dependent Variable: Elementary Teachers technology impact factor.

FIGURE 16.22 The SPSS® ANOVA summary table for the elementary teachers study.

REAL-WORLD LAB XII: MLR SOLUTIONS

Coefficients[a]

Model		Unstandardized Coefficients		Standardized Coefficients	t	Significance	Correlations			Collinearity Statistics	
		B	Standard Error	Beta			Zero-order	Partial	Part	Tolerance	VIF
1	(Constant)	1.145	0.084		13.605	0.000					
	Elementary Teachers technology skills factor	0.521	0.030	0.498	17.640	0.000	0.498	0.498	0.498	1.000	1.000
2	(Constant)	1.098	0.083		13.215	0.000					
	Elementary Teachers technology skills factor	0.417	0.034	0.399	12.294	0.000	0.498	0.372	0.341	0.730	1.370
	Elementary Teachers access to technology factor	0.119	0.020	0.190	5.858	0.000	0.398	0.188	0.163	0.730	1.370

[a]Dependent Variable: Elementary Teachers technology impact factor

FIGURE 16.23 The SPSS® Coefficients output for the elementary teacher study.

The next SPSS® output concerns the tests for the individual predictors. Figure 16.23 shows the "Coefficients" output, which specifies the slope and intercept values for each of the models. Note again that the SPSS® results for Model 2 are the same as those from Excel® (see Figure 16.20.)

3. What Is the "Unique" Explanation of Variance in eimpact from the Predictor eaccess?

As you can see in Figure 16.23, the unique variance in eimpact explained by eaccess is .026 (or 2.6%) which is calculated by squaring the part correlation ($0.163^2 = 0.026$). This same value is shown in Figure 16.21 under Model 2 R-Square Change.

4. What Conclusions Can You Draw About the Research Question from the Findings?

The findings indicate that the two predictors, eskills and eaccess, are significant predictors of eimpact. Both predictors explain about 27% of the variance (i.e., $R^2 = 0.275$) in eimpact, a finding considered large according to the guidelines we discussed in Chapter 15. Each predictor is a significant predictor (as determined by the t values), but eskills is the more powerful of the two in explaining the variance of eimpact. According to the sample data, it is important for elementary teachers to have access to technology for making an impact on the classroom, but is it much more important that they perceive that they have technology skills. Having technology skills explains almost 5 times more unique variance in eimpact than having access to technology.

17

CHI SQUARE AND CONTINGENCY TABLE ANALYSIS

Up to now, we have explored statistical procedures that use continuous data that are interval level. These are the procedures you will encounter the most often in evaluation reports, newspaper articles and scholarly articles. However, we cannot conclude our study of statistics without discussing one of the most versatile and useful procedures that can be used with any level of data.

Chi Square is a statistical procedure that primarily uses nominal, or categorical, data. It works by examining frequency counts, or simply the number (frequency) of people or observations that fit into different categories. An example might be the number of people who prefer certain movies out of a range of movie possibilities, or the number of different types of automobiles in a parking lot.

The Chi Square procedure is used in two different ways. It is used in a goodness-of-fit test and in hypotheses tests of independence. We will discuss both of these uses in this chapter. Even though Chi Square uses nominal data, there is a Chi Square distribution of values with which we can determine exclusion values for our hypotheses tests. We will therefore treat the Chi Square procedures in the same fashion we have learned for hypothesis testing.

CONTINGENCY TABLES

Chi Square data are presented in spreadsheet form in rows and columns. The researcher can easily see how the data are arrayed across the categories of the variables. The tables of data containing the frequencies are called *contingency*

Understanding Educational Statistics Using Microsoft Excel® and SPSS®. By Martin Lee Abbott.
© 2011 John Wiley & Sons, Inc. Published 2011 by John Wiley & Sons, Inc.

tables because the data in the row cells are *contingent upon or are connected to* the data in the column cells. Statisticians and researchers often refer to the analysis of contingency tables as *cross-tabulation*, or simply *crosstabs*. If you look ahead in this chapter to Table 17.3, you can see an example of a contingency table. In that figure, 160 children are classified in terms of their sex and movie choice.

Having the data displayed in rows and columns according to the categories of the variables making up the contingency table is helpful to the researcher. Simple visual inspection may help to detect patterns not ordinarily apparent when the data are not placed in tables. However, the question that we have asked with other procedures in this book is, *How different do the data (in row and column cells) have to be before we could conclude that the data patterns are statistically significantly different*?

The answer to the question above is the reason researchers use Chi Square. The Chi Square procedure statistically analyzes the differences among the data in contingency tables to determine whether the patterns of difference are different enough to be considered statistically significant.

When researchers wish to present the results of their analyses, or to simply list the data in the tables, they use percentages instead of frequencies. This is because the frequencies in cells are often different across rows and columns. Therefore, percentages are a way to present the frequency data on a level playing field. Raw frequency differences are transformed to a common expression across the entire contingency table. We will explore these facets of contingency table analyses in this chapter and discuss some traditional rules for how to present the data in the tables.

One convention some researchers use when presenting data in contingency tables is to present the independent variable categories in columns and the dependent variable categories in rows.[1] In this way, the column data percentages are created to total 100% and the researcher can *compare values of the independent variable categories within rows of the dependent variable categories*. This enables a common way of interpreting the data from visual inspection and from Chi Square analyses. I will show how this works in this chapter by using some examples.

THE CHI SQUARE PROCEDURE AND RESEARCH DESIGN

The Chi Square procedure is very important because it is so versatile. It can be used with any kind of data. Often, researchers are limited as to what data they can gather and may only be able to use simple frequency counts. In some cases, researchers can use Chi Square with existing reports of some variable of interest.

---

[1] Other researchers organize contingency tables in the opposite fashion, with the independent variable categories in rows. As you will see in the analyses to follow, it does not matter which convention you use as long as you remember how to create the interpretation.

Post Facto Designs

Here are some examples of how researchers might use Chi Square in *post facto designs*:

- Determining whether frequency counts of crimes reported by neighborhood in a newspaper article represent a statistically significant finding.
- Comparing attitudes, choices, or behaviors among groups of school children.
- Windshield surveys of schools, businesses, or neighborhoods by classifying observed frequencies of human traffic patterns.
- Comparing the popularity of cafeteria food by classifying amount of waste.

You might recognize some of these as studies in which the researcher uses 'secondary data' (data that already exist) or 'unobtrusive' measures (data that are gathered in natural settings without the researcher 'intruding' into the research context).

Experimental Designs

Chi Square can also be used with *experimental* data. I will discuss one such study in which I used Chi Square to determine if there were pre–post differences among a group of students in an ethnic literature classroom (Trzyna and Abbott, 1991). This experimental design called for a special Chi Square design that uses repeated measures.

CHI SQUARE DESIGNS

I mentioned earlier that there are two primary uses of Chi Square to determine statistical significance. We will look at both of these in this chapter, but I want to make a distinction between the types at the outset.

GOODNESS OF FIT

Goodness-of-fit tests compare actual data to expected data distributions. This use of Chi Square involves one variable with several categories. The researcher wishes to determine whether the data that are seen (*observed frequencies*) are statistically different from the expectations of how the data *should* behave (*expected frequencies*). For example, if there are four movie choices available for a group of children, the expected distribution of choices is that there will be the same number of children who choose each movie. We compare the actual movie choices of children to see how closely these "real" choices compare to what we "expect."

Expected Frequencies—Equal Probability

One of the complexities of this kind of study is how to determine what is expected. Often, expectation is simply a matter of *equal probability*, as was evident in the movie choice example above. In the absence of any other information, the researcher would simply expect equal distributions of choices in each category. Thus, if there were 160 children and 4 movie choices, the expected number of choices for each movie would be 40. This is figured as follows:

$$f_e = \frac{N}{k}$$

We use f_e to represent expected frequency. So, in this formula, expected frequency is a matter of dividing the total number of subjects (N) by the number of categories (k) in the variable of interest. Therefore:

$$f_e = \frac{N}{k},$$
$$f_e = \frac{160}{4},$$
$$f_e = 40$$

If we have 160 children, and four movie choices, we would expect equal numbers of the children (40) to choose each of the movies.

Expected Frequencies—*A Priori* Assumptions

Another way to determine expected frequency is to use prior knowledge or theoretical assumptions about what *should* happen. Thus, a researcher may have knowledge (from other studies or on the basis of past observation) that one of the movie choices is much more popular. In this case, the researcher can determine the proportion or probability of each movie choice and then see how many children in the study *actually* choose each of the movies. It might look like this hypothetical finding:

Movie A	0.35 probability, resulting in an expected frequency (f_e) of 56 (0.35 * 160)
Movie B	0.25 probability, $f_e = 40$
Movie C	0.25 probability, $f_e = 40$
Movie D	0.15 probability, $f_e = 24$

THE CHI SQUARE TEST OF INDEPENDENCE

The second primary use of Chi Square is in research studies in which there are two or more variables involved. For example, the researcher may wish to determine whether the movie choices were equal, as in the above description, but also may question whether the gender of the child will have an impact on their movie choice. Thus, there are now two variables, movie choices and gender, which create a contingency table of data. The question is whether the values of the categories of one variable are in any way linked to the values of the other variable's categories. Do the cell frequencies of one variable influence the cell frequencies of the other?

The general expectation of chance in which one variable is not linked to another will result in what researchers call *independence*. That is, the categories of one variable are in no way connected or linked in a pattern to the categories of the other. An example would be that both boys and girls choose each of the four movies equally.

If the categories of data do show a pattern whereby the frequencies of some cells are much greater than the frequencies of other cells, we might speak of the variables being *dependent* on one another. Thus, if girls most often choose movie A and boys opt for movie C much more frequently, we might say that the choice of movie is *dependent* on the child's gender. A different way of thinking about this is asking whether knowing the values of one variable helps you to know anything about values of the other.

Researchers use Chi Square to determine whether there are statistically significant differences among the cell frequencies. In this way, they can detect patterns that might indicate relatedness between the study variables. This is what we call the test of independence. If we reject the null hypothesis (complete independence or no relationship among the cells), then we can conclude that the data are not independent, but instead that there is a dependent relationship between the study variables.

A FICTITIOUS EXAMPLE—GOODNESS OF FIT

Perhaps the easiest way to explore the nuances of Chi Square is to use an example. Because I introduced movie choices above, we can use this to show how to proceed with a goodness-of-fit test using Chi Square. This is also referred to as a *one-way Chi Square*. The "one" represents the fact that there is only one row of data (i.e., one variable with several categories). Thus, this example would be a 1×4 Chi Square, or a study with one row and four columns.

Let us suppose that a researcher conducts a brief study with 160 eighth-grade children. The research question is whether the children show a difference in their choice of movies. Here are some (fictitious) choices:

Movie A   Drooling Ghouls
Movie B   Blarney's Funny Adventure
Movie C   Happy, Lucky, and the Beached Whale
Movie D   Skateboard Bingo

Using what we expect to happen by chance, we can determine that the expected frequency is 40 for each movie, as we showed above.

$$f_e = \frac{160}{4},$$
$$f_e = 40$$

TABLE 17.1 The Fictitious Movie Choice Data Analysis

	Movie A	Movie B	Movie C	Movie D
f_o	56	30	34	40
f_e	40	40	40	40
$f_o - f_e$	16	−10	−6	0
$(f_o - f_e)^2$	256	100	36	0
$\dfrac{(f_o - f_e)^2}{f_e}$	6.4	2.5	0.9	0

$$\chi^2 = \sum \frac{(f_o - f_e)^2}{f_e} = (6.4 + 2.5 + .9 + 0) = 9.80$$

The question is, How do the *actual* choices (observed frequency, f_o) compare to the *expected* choices (expected frequencies, f_e)? For this, the researcher uses the general Chi Square (represented by χ^2) formula that compares f_o to f_e as follows:

$$\chi^2 = \sum \frac{(f_o - f_e)^2}{f_e}$$

As you can see, the formula sums up the squared differences between observed and expected frequencies which are divided by the expected frequencies. While this formula sums all the differences up, think of the process as the summing up of the statistical calculations *in each cell*. Table 17.1 shows how we might arrange the data so that we can make the required calculations.

As you can see from Table 17.1, we calculated $\chi^2 = 9.80$. As I described above, this is a general measure of how the differences between expected and observed frequencies are arrayed in the data table. The researcher must now decide if the 9.8 is a large enough Chi Square value to conclude that the movie choices are statistically different. To do this, we need to conduct a hypothesis test procedure as we did with the other statistical procedures.

1. *Null Hypothesis:* $f_o = f_e$. (There is no difference between what we expect to happen and what we observe to happen.)
2. *Alternative Hypothesis:* $f_o \neq f_e$. (There is a difference between what we expect to see and what we actually see.)
3. *Critical Value of Exclusion:* For this value, we use the Chi Square Table of Values. As we have seen with other statistical procedures, we need to establish a value of exclusion on the Chi Square distribution to compare with our actual, calculated value. If our calculated value exceeds this exclusion value, we would conclude that the results of our study are too large to be considered a chance finding.

    We must use degrees of freedom with the goodness-of-fit test to identify the appropriate comparison value from the Chi Square table of values. For the one-way Chi Square, the $df = k - 1$, where k is the number of categories. Thus, $df = 3$. For this study, the 0.05 critical value of exclusion is 7.82. We represent this as $\chi^2_{.05,3} = 7.82$.

A FICTITIOUS EXAMPLE—GOODNESS OF FIT

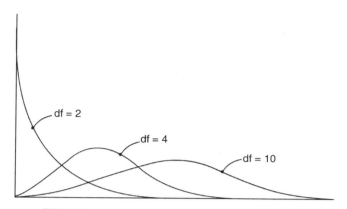

FIGURE 17.1 The Chi Square series of distributions.

Like the *T* distribution, the Chi Square distribution is actually several distributions that vary with the size of the sample. Thus, the degrees of freedom identify a separate Chi Square distribution that is used as a comparison distribution to establish the critical values of exclusion. *Chi Square is a directional test because of the shape of the distribution(s).* The exclusion value is located on the right side of the distribution. Figure 17.1 shows sketches of the general shapes of the Chi Square distribution at selected df values. As you can see, the shape of the distribution is quite different, depending on the degrees of freedom.

4. *Calculated Chi Square:* This is the Chi Square value we calculated above, $\chi^2 = 9.80$.
5. *Decision:* Because our calculated value (9.80) exceeds the critical value (7.82), we conclude that there is a statistically significant difference among the categories of movie choices at the 0.05 level, $p < 0.05$.
6. *Interpretation:* The students in the study group chose movies in statistically different numbers. The question for the researcher is how to report the findings of this study in the words of the study question. For this, I refer back to my earlier statement that researchers need to use percentages to report findings because categories may have unequal raw number values. Table 17.2 shows the fictitious data with appropriate percentages. Based on these percentages, we would conclude that most children prefer Movie A.

TABLE 17.2 The Reporting Data for the Fictitious Study

	Movie A	Movie B	Movie C	Movie D	
f_o	56	30	34	40	= 160
%	35% $\left(\frac{56}{160}\right)$	18.75% $\left(\frac{30}{160}\right)$	21.25% $\left(\frac{34}{160}\right)$	25% $\left(\frac{40}{160}\right)$	= 100%

Frequencies Versus Proportions

There is a technical point to be made about Chi Square and the kind of data we use. Most researchers use frequencies, or raw counts and numbers in the cells (e.g., the number of children who choose a certain movie). In this case, we are actually comparing two different distributions, an actual distribution of occurrences, and an expected distribution of occurrences. Statisticians refer to this as a *nonparametric* test because we are not comparing a *sample* set of data to a (known or unknown) *population*. A case in which we compare a sample to a population would be known as a *parametric* test because we refer to population values.

We can use Chi Square as a parametric test when we view the distributions as representing populations. This is seen most clearly where we express both expected and observed values as *proportions*, which can express population values. In the movie choice example above, I showed how we can create frequencies from expected probabilities or proportions. Treating the test as comparing sample proportions to population proportions would be considered a parametric test.

There is a relationship between these uses of the data, however. Calculating a Chi Square using proportions is equivalent to Chi Square calculated from frequencies through the following:

$$\chi^2_{\text{frequencies}} = (\chi^2_{\text{proportions}})^2 \times N$$

In practice, we can treat the data as parametric and perform hypothesis tests as we have with other statistical procedures. Most critical comparison tables for Chi Square use raw score frequencies (as does the one in this book). It is important to understand the difference between these two approaches and to understand that the data can be expressed in different ways.

EFFECT SIZE—GOODNESS OF FIT

Effect size for Chi Square is very basic because we are using categorical data. Essentially, we use *measures of association* to determine the impact of the relationship among our variables. Association measures include the correlation measures we used to show how changes in one variable were linked to changes in another.

In goodness of fit tests, we only have one variable with at least two categories. While there are a number of effect size measures for Chi Square, especially for the test of independence that I will discuss next, I will introduce the most common measure called *contingency coefficient* here, symbolized by C. The formula for C is

$$C = \sqrt{\frac{\chi^2}{\chi^2 + N}}$$

The contingency coefficient is essentially a correlation measure for nominal data. The Chi Square test compares observed and expected frequencies in an attempt to understand if there is a statistically significant difference between the two. Contingency coefficient is a correlation measure of the overall association between the two study variables.

Although this is a measure of association, we cannot interpret it as we did with r^2 (i.e., the proportion of variance in one variable explained by the other). Due to the nature of the data, we must refer to a set of values as we have with other effect size measures. Because C does not reach the maximum value of 1.00, the effect size values for small, medium, and large vary with the conditions of the study (e.g., number of categories). Cohen's (1988) suggestions are generally 0.10 (small), 0.30 (medium), and 0.50 (large), depending on the factors affecting C.

Using the fictitious data above, we can calculate C.

$$C = \sqrt{\frac{\chi^2}{\chi^2 + N}},$$

$$C = \sqrt{\frac{9.80}{9.80 + 160}},$$

$$C = 0.24$$

Using the criteria above, we can suggest a small effect size for the fictitious study.

CHI SQUARE TEST OF INDEPENDENCE

The goodness-of-fit test above is a very basic assessment of the difference among categories of one variable. With the test of independence, we can add the categories of a second variable that recognizes additional complexity of a study question and thereby create a contingency analysis. With two variables, we can express the category data in rows and columns.

Recall my earlier description of the test of independence as concerning whether the values of the categories of one variable are in any way linked to the values of the other variable's categories. That is, do the cell frequencies of one variable influence the cell frequencies of the other?

We can use the same formula for Chi Square that we used with the goodness-of-fit test:

$$\chi^2 = \sum \frac{(f_o - f_e)^2}{f_e}$$

Two-Way Chi Square

Because we use contingency tables with rows and columns for the test of independence, both of which may have multiple categories, the test is often referred to as the *two-way Chi Square*. Thus, if we added gender categories to the fictitious study above, this would be a 2×4 Chi Square since it would consist of two gender rows and four movie choice columns.

Assumptions

In this test, we need to note a couple of assumptions. As we have noted with other statistical procedures, the power of a test increases to the extent the assumptions are met. For the Chi Square test of independence, we need to address the following:

1. The categories of both variables should be independent of one another. This sounds like a redundancy in a test of independence, but the idea here is that the cases of the two variables are not related or structurally linked. If we add gender to the fictitious study above, for example, we might be concerned if many of the boys and girls were siblings. Siblings talk to one another and may even share similar movie choices.
2. The Chi Square test of independence works best if there is no expected cell frequency less than five. This small value tends to distort the value of the calculation. We will discuss the Yates Correction Factor as a possible remedy in a 2×2 table, but one solution is to collapse adjacent cells to avoid the problem. Of course, with sample sizes low enough to cause this problem, you might consider restructuring the study or getting additional cases.

A FICTITIOUS EXAMPLE—TEST OF INDEPENDENCE

As I mentioned above, the test of independence can recognize additional complexity. Let us assume the researcher was interested in gender effects as well as movie choices among a group of school children. This would yield the contingency table of frequencies shown in Table 17.3.

Creating Expected Frequencies

Please note that the data in Table 17.3 are *observed frequencies*. In order for us to conduct a Chi Square analysis, we need to know the expected frequencies for each

TABLE 17.3 The Fictitious Test of Independence Observed Frequency Data

	Movie A	Movie B	Movie C	Movie D	Totals (M_r)
Girls	20	18	22	35	95
Boys	36	12	12	5	65
Totals (M_c)	56	30	34	40	(Grand Total) GT = 160

A FICTITIOUS EXAMPLE—TEST OF INDEPENDENCE

of the cells. In the absence of *a priori* expected frequencies, we can calculate them from the following formula:

$$f_e = \frac{(M_r)(M_c)}{GT}$$

The language of this formula is quickly understood if you see that the M stands for marginal total. Thus:

M_r is the marginal total for rows.
M_c is the marginal total for columns.
GT is the grand total.

Here is an example using the observed frequency cell for girls who choose Movie A:

$$f_e = \frac{(M_r)(M_c)}{GT},$$

$$f_e = \frac{(95)(56)}{160},$$

$$f_e = 33.25$$

Therefore, the expected frequency for this cell (girls who choose Movie A) = 33.25, which, as you can see, is different from the observed frequency of 20.

The reason we need to use this method is that there are different numbers of boys ($N=65$) and girls ($N=95$) in the study. Also, the movie choices are uneven (i.e., 56, 30, 34, 40). *Because of the differences of category sizes in both variables, the expected frequencies cannot be equal in all cells.* If we had the same number of boys and girls, and the number of movie choices were all the same, we could simply divide the total (160) by the number of cells (8) to get the same expected frequency (20) for all cells, as we did in the goodness-of-fit test. However, the unequal sizes means that the weighting of each cell size is different.

We use the formula above to help us create appropriate cell sizes because we are using the marginal totals. *By multiplying the marginal totals associated with a given cell and dividing by the grand total, we can get a more precise (weighted) expected frequency for each cell.*

Table 17.4 shows the expected frequencies calculated for each cell in parentheses. If you calculate these expected frequencies, they should total the marginal totals just the same as the observed frequencies. All the expected frequencies together should equal 160.

If you calculate the expected frequencies as I did for the table, you should note that you actually would only have to calculate three of the row cells to derive all the rest of the expected frequencies. Look at Table 17.4 to see how this works. I highlighted the cell values to show that if you create these expected frequencies only, all the other expected frequencies in the other cells can be obtained by simply subtracting the sum of these calculated expected frequencies

TABLE 17.4 The Expected Frequencies for the Study

	Movie A	Movie B	Movie C	Movie D	Totals (M_r)
Girls	20 (33.25)	18 (17.81)	22 (20.19)	35 (23.75)	95
Boys	36 (22.75)	12 (12.19)	12 (13.81)	5 (16.25)	65
Totals (M_c)	56	30	34	40	(Grand Total) GT = 160

from the marginal total (M_r) of 95. Here is how we could calculate the expected frequency for the last cell in the first row (girls who choose Movie D):

$$33.25 + 17.81 + 20.19 = 71.25$$

Then,

$$95(M_r) - 71.25 = 23.75$$

You can use the same logic with the column expected frequencies, but here it is even simpler because there are only two cells per column. Thus, we calculate the expected frequency for boys who choose Movie A:

$$56(M_c) - 33.25 \text{(the already calculated expected frequency)} = 22.75$$

Degrees of Freedom for the Test of Independence

These calculations express the meaning of degrees of freedom. Only one of the girl (row) cells cannot change its value when the other three row cells are known. Thus all the row cells (4) are free to vary except for one. The same logic applies to column cells (movie choices). Because there are only two cell values making up each column, one is free to change its value, but the other one is not because it must equal a certain number to equal the marginal total for that column.

The degrees of freedom are therefore calculated as follows:

$$df = (\text{rows} - 1)(k - 1)$$

In our study,

$$df = (2 - 1)(4 - 1) = 3$$

Table 17.5 shows the calculated Chi Square based on the formula we used above in the goodness-of-fit test.

$$\chi^2 = \sum \frac{(f_o - f_e)^2}{f_e}$$

As you can see, the calculated χ^2 for this study was 26.52. Our assumptions were met because we did not have an expected frequency less than 5 and our study

A FICTITIOUS EXAMPLE—TEST OF INDEPENDENCE

TABLE 17.5 The Calculated Chi Square for the Test of Independence

	Girls				Boys			
	Movie A	Movie B	Movie C	Movie D	Movie A	Movie B	Movie C	Movie D
f_o	20	18	22	35	36	12	12	5
f_e	33.25	17.81	20.19	23.75	22.75	12.19	13.81	16.25
$f_o - f_e$	−13.25	0.19	1.81	11.25	13.25	−0.19	−1.81	−11.25
$(f_o - f_e)^2$	175.56	0.04	3.29	126.56	175.56	0.04	3.29	126.56
$\dfrac{(f_o - f_e)^2}{f_e}$	5.28	0.00	0.16	5.33	7.72	0.00	0.24	7.79
$\chi^2 = 26.52$	$(5.28 + 0.00 + 0.16 + 5.33 + 7.72 + 0.00 + 0.24 + 7.79 = 26.52)$							

description of subjects did not specify that the subjects were connected to each other (i.e., dependent). We can therefore proceed to the hypothesis test to determine whether the categories of the two variables show a *pattern of connection* or whether we can conclude that the variable categories are independent (do not demonstrate a connection of relationship).

Using the same hypothesis testing steps as with the goodness-of-fit test, we can determine whether the calculated Chi Square of 26.52 falls into the exclusion region of our comparison distribution.

1. *Null Hypothesis:* $f_o = f_e$. (There is no difference between what we expect to happen and what we observe to happen.)
2. *Alternative Hypothesis:* $f_o \neq f_e$. (There is a difference between what we expect to see and what we actually see.)
3. *Critical Value of Exclusion:* df = 3 (see above for calculation).

$$\chi^2_{.05,3} = 7.815$$

4. *Calculated Chi Square*

$$\chi^2 = 26.52$$

5. *Decision:* Because our calculated value (26.52) exceeds the critical value (7.815), we conclude that there is a statistically significant difference among the categories of girl's and boy's movie choices at the 0.05 level, $p < 0.05$.
6. *Interpretation:* The girls and boys in the study group chose movies in statistically different numbers. The specific interpretation of the direction of the findings is shown in Table 17.6. As you can see, I presented the table with percentages in the cells along with the frequency in the cell in parentheses.

I changed the nature of the table according to the protocol I noted above. That is, the independent variable should be placed in columns, and the dependent variable

TABLE 17.6 The Percentages of the Study Data for Interpretation of Findings

	Girls	Boys
Movie A	(20) 21.05%	(36) 55.38%
Movie B	(18) 18.95%	(12) 18.46%
Movie C	(22) 23.16%	(12) 18.46%
Movie D	(35) 36.84%	(5) 7.69%
Totals	(95) 100.00%	(65) 100.00%

should be placed in rows. *This way, you can create column percentages (equaling 100%) and interpret across the rows of the dependent variable categories.*

Using the table protocol allows you to frame the results in terms of independent and dependent variables that might be featured in the research design. Thus, we might interpret the findings as follows: Over twice the percentage of boys preferred Movie A (55.38% to 21.05%), whereas over four times the percentage of girls preferred Movie D (36.84% to 7.69%).

I presented the table in this fashion (with IV categories in columns and DV categories in rows) because it is more meaningful for interpretation. We will also use it this way in SPSS®.

SPECIAL 2 × 2 CHI SQUARE

There are special formulas for the 2 × 2 Chi Square tests of independence that allow you to calculate values directly from the frequencies in the contingency table rather than calculate f_e. Table 17.7 shows the layout of the 2 × 2 table with the cells labeled *a* through *d* for clarification. As you can see, the marginal totals simply add the cells across the rows and down the columns.

I present the alternate formula (without a calculated example) for four reasons.

1. The 2 × 2 table is fairly common in research, so you should be aware of alternate means for hand calculations.
2. There are special effect size calculations for this design that form the basis for effect size calculations that can be used with other types of tables.
3. The 2 × 2 table illustrates the use of a correction process for tables with low sample size.
4. This design is used with repeated measures studies, one of which I will discuss below.

TABLE 17.7 The 2 × 2 Chi Square Contingency Table

a	*b*	$a+b$
c	*d*	$c+d$
$a+c$	$b+d$	N

SPECIAL 2 × 2 CHI SQUARE

TABLE 17.8 The Example Movie Choice Data in a 2 × 2 Table

	Movie A	Movie B	Totals (M_r)
Girls	$a = 20$	$b = 18$	38
Boys	$c = 36$	$d = 12$	48
Totals (M_c)	56	30	(Grand Total) GT = 86

The Alternate 2 × 2 Formula

The formula for the 2 × 2 table, utilizing the cell identification letters (a through d, as shown in Table 17.7), is as follows:

$$\chi^2 = \frac{N(ad - bc)^2}{(a+b)(c+b)(a+c)(b+d)}$$

If you look closely at the formula, it generally expresses the ad to bc cell totals as a proportion of all the marginal totals. Thus, as the differences grow between the ad and bc axes, the Chi Square value increases.

To take an example of this special formula, consider Table 17.8, which is adapted from Table 17.3. The data in the table are simply the boys and girls choices between only the first two movie choices (Movie A and Movie B). Following the convention shown in Table 17.7, I labeled each cell as a, b, c, or d.

If you calculate Chi Square using the special formula, you will find that the result is the same had you used the general formula I introduced earlier in the chapter. The following analyses compare the two approaches showing the same calculated Chi Square result (with slight rounding differences).

$$\chi^2 = \frac{N(ad - bc)^2}{(a+b)(c+d)(a+c)(b+d)},$$

$$\chi^2 = \frac{86(240 - 648)^2}{(38)(48)(56)(30)},$$

$$\chi^2 = \frac{14,315,902}{3,064,320},$$

$$\chi^2 = 4.67$$

Now, using the general formula on the same data to calculate Chi Square, consider the data in Table 17.9. As you can see, the layout is similar to the general formula examples I introduced in earlier sections.

Effect Size in 2 × 2 Tables: Phi

In a 2 × 2 table under the circumstances I described above, the perfect relationship between the categories of two variables equals unity or a proportion of 1. These

TABLE 17.9 Using the General Chi Square Formula on the 2 × 2 Table

	Girls		Boys	
	Movie A	Movie B	Movie A	Movie B
Cell	a	b	c	d
f_o	20	18	36	12
f_e	24.74	13.26	31.26	16.74
$f_o - f_e$	−4.74	4.74	4.74	−4.74
$(f_o - f_e)^2$	22.468	22.468	22.468	22.468
$\dfrac{(f_o - f_c)^2}{f_e}$	0.908	1.694	0.719	1.342
$\chi^2 =$	(0.908 + 1.694 + 0.719 + 1.342) = 4.663			

special properties allow the researcher to express the relationships among the values in terms of a special effect size measure known as φ (phi). Phi varies between values of 0 and 1 and, when squared, can be expressed in terms of explained variance (i.e., the same as r^2). Thus, φ^2 is the proportion of variance in one variable explained by the other.

The formula for phi is very simple. As you can see, it is based on Chi Square and can be calculated simply by dividing χ^2 by N and taking the square root of the result:

$$\varphi = \sqrt{\frac{\chi^2}{N}}$$

The same criteria for judging C apply also to phi. Cohen's (1988) suggestions are generally 0.10 (small), 0.30 (medium), and 0.50 (large).

Correction for 2 × 2 Tables

Some researchers point out that it is important to correct 2 × 2 tables, especially those that have small sample sizes because they may not provide the best estimates. The Yates Correction for Continuity formula allows the researcher to adjust the calculations to make it more difficult to reject the null hypothesis and therefore provide a more conservative Chi Square result. Researchers are divided about whether to use the correction at all because it may be too conservative. If you use Chi Square with small samples, especially the special 2 × 2 table, be aware of this issue and investigate the use of the corrected formula. I present the formula here, but do not provide a calculated example. (The example above from Table 17.8 is inappropriate because the sample size is large.)

$$\chi^2 = \frac{N(|ad - bc| - \frac{N}{2})^2}{(a+b)(c+d)(a+c)(b+d)}$$

CRAMER'S V: EFFECT SIZE FOR THE CHI SQUARE TEST OF INDEPENDENCE

As you can see, this is simply the special Chi Square formula for a 2×2 table with the adjustment of subtracting $\frac{N}{2}$ from the $|ad - bc|$ difference before it is squared. (Note that the $\frac{N}{2}$ adjustment is subtracted from the absolute value of the $(ac - bc)$ difference because this value can be a negative value.) In practical terms, it adds one step to the overall set of steps in calculating Chi Square.

CRAMER'S V: EFFECT SIZE FOR THE CHI SQUARE TEST OF INDEPENDENCE

I return now to the question of effect size for the Chi Square Test of Independence. In earlier sections, I discussed C, the general effect size measure for Chi Square, especially appropriate for the goodness-of-fit test, and φ, the effect size measure especially appropriate for the 2×2 test of independence.

Tests of independence that are more complex than the 2×2 table (i.e., that have more than two rows and/or two categories) typically use another measure of effect size, Cramer's V. The φ value calculated from the 2×2 arrangement allows the researcher to express effect size as "variance explained," but we cannot do this with larger tables.

Cramer's V calculates effect size values that range between 0 and 1. Because it takes into account tables with different numbers of rows and column categories, the formula includes an adjustable feature.

$$\text{Cramer's } V = \sqrt{\frac{\chi^2}{N(\text{df}_{\text{smallest of } r \text{ or } c})}}$$

As you can see, the formula looks very much like the φ formula except it is modified by the number of rows and columns in the table. The following element of the formula captures the shape of the table and uses it to modify the φ formula:

$$\text{df}_{\text{smallest of } r \text{ or } c}$$

This value refers to the degrees of freedom calculation for the test of independence that I introduced earlier:

$$\text{df} = (r - 1)(k - 1)$$

Cramer's V uses the smaller of either $(r - 1)$ or $(k - 1)$ in the formula. Thus, in our fictitious example, we had a 2×4 table. In this case, we would use $(r - 1)$ because it represented the smallest number (i.e., $2 - 1$ as opposed to $4 - 1$). Cramer's V calculation for our fictitious test of independence would therefore be

$$\text{Cramer's } V = \sqrt{\frac{26.52}{160(2 - 1)}},$$

$$\text{Cramer's } V = \sqrt{\frac{26.52}{160(2 - 1)}},$$

$$\text{Cramer's } V = 0.407$$

The judgment for the magnitude of Cramer's V does not always use the guidelines we discussed for the other effect size measures (0.10 for small, 0.30 for medium, and 0.50 for large) that pertain to the other tests. This is because we must take into account the adjustment to the formula caused by the shape of the table. Once again, Cohen (1988) provides the adjusted set of guidelines. I refer you to Cohen's book for the complete set of figures.

In our example, because we had a 2×4 table, we could use the same 0.1, 0.3, and 0.5 guidelines for judging the effect size for Cramer's V as we did for C. Larger tables will have reduced magnitude effect size criteria that determine small, medium, and large effects. For example, if our table had been a 3×4, the effect size criteria for a 'large effect' would be 0.354. In this event, if we had calculated Chi Square to be the same 0.407, the effect size would have been determined to be large instead of medium. As is it however, we can conclude that our fictitious study using a 2×4 table showed a medium effect size.

REPEATED MEASURES CHI SQUARE

Recall that we made a distinction with other statistical procedures between those that used independent samples and those that used dependent samples. In Chapter 11, I identified dependent samples designs as those in which one group is somehow linked structurally to the other. Thus, in experimental designs, a group measured twice (e.g., pretest and posttest) would need to be treated differently than other group measures because the *same group of subjects would be measured twice*.

If the T test was used to detect a difference between two dependent samples in an experiment for example (i.e., pre–post, Time 1–Time 2, Matched samples), we would need to use a special T test that factored out the relatedness so the T test could see what the unique differences remained between the two samples. The same thing would be true for ANOVA procedures that measured the same group twice in a design.

You may recall that we referred to these kinds of measures in several ways:

- Repeated-measures tests
- Dependent-samples tests
- Paired-samples tests
- Within-subjects tests

I will talk about these dependent measures (especially dependent T and within subjects ANOVA) in a subsequent chapter because it is so important to research. Chi Square also has a special 'within subjects' design that I will mention here to give you an idea of what repeated measures are and how Chi Square can be used experimentally.

Recall that an experiment introduces a treatment (i.e., manipulation of the independent variable) and then takes a measure of the outcome (dependent) variable to

TABLE 17.10 The Dependent Samples Chi Square for the Grieving Study

		Post	
		Not Grieving	Grieving
Pre	Grieving	$a = 1$	$b = 9$
	Not Grieving	$c = 0$	$d = 17$

observe the effect of the treatment. Chi Square can be used with categorical data in these experimental designs. Look at Figure 11.2 to see the general design specification.

A number of years ago, a colleague and I published an article (Trzyna and Abbott, 1991) in which we reported a (dependent sample) Chi Square test with data we gathered from one of our courses. We were attempting to see what effect a college course on ethnic literature might have on grieving behavior. Using a series of protocols, we classified students as either grieving or not grieving at the beginning and end of the course (actually, we used data from several courses). Table 17.10 show the table of data we used for the Chi Square analysis.[2]

With this type of design, we measured the same students at different times (beginning and end of class). Therefore, we had to use a special dependent measures Chi Square test called the McNemar test. As you can see from the table, I labeled the cells a through d. There was one student (the upper left cell labeled a) grieving at the beginning of the class but not at the end. By contrast, there were 17 students not grieving at the beginning but grieving at the end of class (shown in cell d). The formula and example solution for this special test is

$$\chi^2_{McNemar} = \frac{(a-d)^2}{a+d}$$

$$\chi^2_{McNemar} = \frac{(1-17)^2}{1+17}$$

$$\chi^2_{McNemar} = 14.22$$

In our study, we therefore rejected the null hypothesis since 14.22 exceeded the exclusion value ($\chi^2_{0.5,1} = 3.84$). The treatment (class) had an effect on the students such that significantly more of the students were grieving at the end of class. The φ of 0.73 for this study indicated a large effect size. Phi Squared indicated that over 53% of the results of the grieving outcome were explained by categorizing the values of the intervention in this way ($\varphi^2 = 0.73^2 = 0.533$).

---

[2] This adapted table is used by permission of *College Literature*, West Chester University, 210 E. Rosedale Avenue, West Chester, PA 19382.

TABLE 17.11 Changing the Dependent Samples Chi Square Categories

		Post	
		Grieving	Not Grieving
Pre	Grieving	$a = 9$	$b = 1$
	Not Grieving	$c = 17$	$d = 0$

Repeated Measures Chi Square Table

As you can see, the focus in the cells is the change from cell a to cell d. That is, we want to determine the difference between *how many students had the condition before but not after and the number of students who did not have the condition at the beginning but did afterward*. If the null hypothesis is accurate, these two cells should be relatively equal. If the treatment had an effect, these two cells would show differences. (If the class results in grieving, then cell d would be much larger than cell a.) The formula essentially compares the differences between the two cells as a proportion of the total frequency of the two cells.

The McNemar formula will change depending on how you create the table. Just remember that the two critical cells are the ones I identified above. Here is an example of changing the table and how the formula would change. If you changed the position of the Post conditions as shown in Table 17.11, the formula would change to recognize the important two cells.

If you set the table up this way, the two critical cells are now b and c because you need to detect the difference between students who had the condition before but not after (cell b) compared to the number of students who did not have the condition at the beginning but did afterward (cell c). As you can see, we get the same calculated value (14.22), but you must be careful to set the table up to reflect these differences.

$$\chi^2_{McNemar} = \frac{(b-c)^2}{b+c},$$

$$\chi^2_{McNemar} = \frac{(1-17)^2}{1+17},$$

$$\chi^2_{McNemar} = 14.22$$

USING EXCEL® AND SPSS® WITH CHI SQUARE

Both Excel® and SPSS® will assist with Chi Square analyses, although SPSS® provides a more thorough summary of findings and is easier to use. I will demonstrate both programs using a sample of the TAGLIT data. In Chapter 14, we discussed correlation and the process for assessing the relationship between two interval level variables. We also discussed Spearman's rho as a way to calculate a correlation with ordinal data. If you recall, I used a TAGLIT variable ($ earmarked for technology) that was extremely skewed and therefore not suitable for a Pearson's *r* correlation.

USING EXCEL® AND SPSS® WITH CHI SQUARE

TABLE 17.12 Coding the Variables for the Chi Square Example

TAGLIT Variable	Coded TAGLIT Variable
School funding (in $) that schools chose to earmark for technology.	Earmark (0 if no funding was earmarked; 1 if *any* funding was earmarked for technology.)
Technology funding (in $) spent this year at the school for professional development.	PDexp (0 if no funding was spent on PD; 1 if *any* funding was spent on PD.)

I will return to this TAGLIT variable and similar variables that measure funding levels because they represent data that are too skewed for use in interval data procedures. This will show the versatility of Chi Square in managing any level of data that do not meet the assumptions of other (interval level) statistical procedures.

This example uses two TAGLIT variables that deal with funding levels. Table 17.12 shows the two variables and the coding of those variables for Chi Square analysis.

As you see in Table 17.12, I coded the two funding variables as 0 or 1, depending on whether any technology funding at all was earmarked or spent on professional development. The reason I recoded the variables is that both deal with funding and such variables are almost always skewed. Figures 17.2 and 17.3 show the histograms that confirm the extreme skew in both variables.

As I have mentioned in past chapters, there are ways of transforming variables that do not meet the assumptions for statistical procedures like these. The two

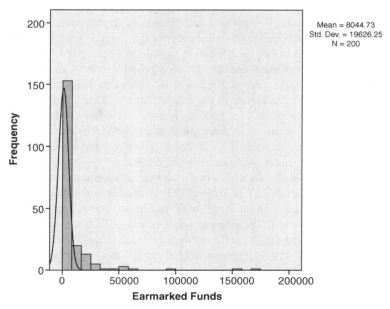

FIGURE 17.2 The "Earmarked Funds" variable with extreme positive skew.

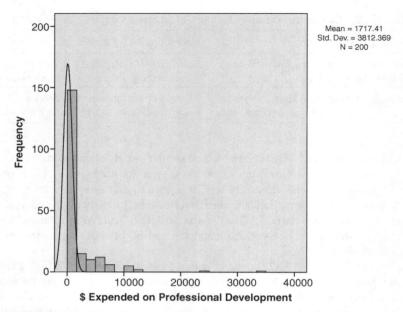

FIGURE 17.3 The "$ Expended on PD" variable with extreme positive skew.

variables I am using here *could* be transformed to meet the assumptions for the Pearson's r procedures, but there are many reasons why I might choose not to use transformations. Among these is that transforming variables sometimes creates difficulties for interpretation. For example, I might make a log transformation by using the logarithm of the values, or a related process that would force the data into a normal distribution, but it may be difficult to explain to a reader what this means!

Another reason for my using data as I have proposed it is that I am interested in a *question of difference rather than association*. By using Chi Square, I am interested to know how cases in one category of the independent variable compare on categories of the dependent variable. That is, how do the cases differ on the categories of comparison? Correlation measures association and therefore asks a different question. We end up using measures of association for effect size indications, but the primary Chi Square procedure is a procedure of difference.

I am recoding the study variables into categorical values for this example in order to show the versatility of Chi Square. This method is not without controversy, however, because someone might argue that creating categories does not use all the data sufficiently, and so on. These are not natural categories; rather I am creating them as categories based on (severely skewed) interval values. I defend this use of the data especially as a way to explore what differences might exist among the categories as created.

Table 17.13 shows the frequency data from the sample ($N = 200$) placed in the appropriate cells. Each cell includes a frequency of schools that indicate whether funding was earmarked or expended for PD. You can calculate this by hand

USING EXCEL® FOR CHI SQUARE ANALYSES

TABLE 17.13 The Chi Square Table of Values

	Earmark$—No	Earmark$—Yes	Totals (M_r)
PD Expenditure—No	53	61	114
PD Expenditure—Yes	16	70	86
Totals (M_c)	69	131	(Grand Total) GT = 200

according to the method I described earlier (Chi Square Test of Independence). Because this is a 2 × 2 table, you can use the special formula to calculate χ^2 if you do not want to calculate expected frequencies. The hand-calculated answer is $\chi^2 = 16.87, p < 0.05$ (because $\chi^2_{0.5,1} = 3.841$).

USING EXCEL® FOR CHI SQUARE ANALYSES

You can obtain the results from the Chi Square test of significance in Excel® *once you have created the contingency table from the database*. If you have a large database, this is cumbersome, but you can use the Count function to help create the table of values (as in Table 17.13).

Sort the Database

It helps to *sort the two variables* before you begin so that you can easily identify when both values are 0, when Earmark is 1 and PDexp is 0, and so forth. Using this method, you can reproduce the values in Table 17.13. Figure 17.4 shows a truncated Excel® database I used to create the frequency table. I only

Earmark	PDexp
0	0
0	0
0	0
0	1
0	1
0	1
1	0
1	0
1	0
1	1
1	1
1	1

FIGURE 17.4 The truncated database for the Chi Square example.

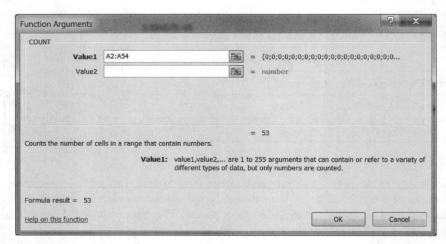

FIGURE 17.5 The Excel® Count function for summarizing data to use in Chi Square analyses.

listed 12 of the 200 cases to show how I coded the data. As you can see, if a school indicated they earmarked 0$ for technology, I coded the school "0" and likewise for PDexp. If the school earmarked ANY funds for technology, I coded the school "1" and likewise for PDexp.

The Excel® Count Function

Once you have coded and sorted the two variables, use the 'Formulas–Statistical–Count' selection from the menus, which will open the specification window shown in Figure 17.5. As you can see, I used this function to create the cell count for 'Earmark-0, PDexp-0.' This function simply reports the n size within the cells identified in the function. In the truncated database in Figure 17.4, this would represent the first three cases. As you can see in Figure 17.4, this represents 53 cases.

The Excel® CHITEST Function

When the contingency table cells are counted, you can use the Chi Test function to obtain results for the test of independence. Because this does not use the special (2×2) formula, you must calculate the expected frequencies in a separate table so Excel® can compare observed and expected frequencies.

Figure 17.6 shows the "CHITEST" function that you can obtain through the "Formulas–Statistical–CHITEST" menus. As you can see, I identified the cells for actual and expected frequencies from my spreadsheet. The CHITEST function immediately returns the results for the Chi Square test immediately below the cell

USING EXCEL® FOR CHI SQUARE ANALYSES

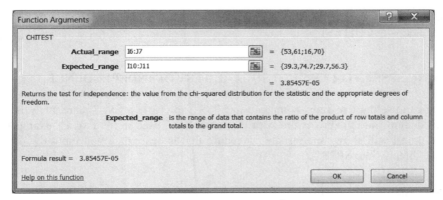

FIGURE 17.6 The CHITEST function in Excel® for Chi Square analysis.

range windows. In this case, Excel® returned the value 3.85457E-05, indicating that the Chi Square test of independence was statistically significant at the $p = 0.00004$ level. Thus, the calculated Chi Square value fell far into the exclusion values of the Chi Square distribution, indicating a likely non-chance finding.

The Excel® CHIDIST Function

Note that Excel® does not return the actual, calculated χ^2 (16.87), nor any indication of effect size. This makes it difficult to report effect size! You can use another Excel® function, CHIDIST, to return the probability of the finding according to the Chi Square distribution if you have already calculated χ^2. Use the Formulas–Statistical–CHIDIST menus to create the window shown in Figure 17.7.

As you can see, I simply placed the value of 16.87 in the X window and specified 1 for Deg_Freedom because our table is a 2 × 2. The function immediately returns the same probability value (or virtually equal since we didn't have the calculated

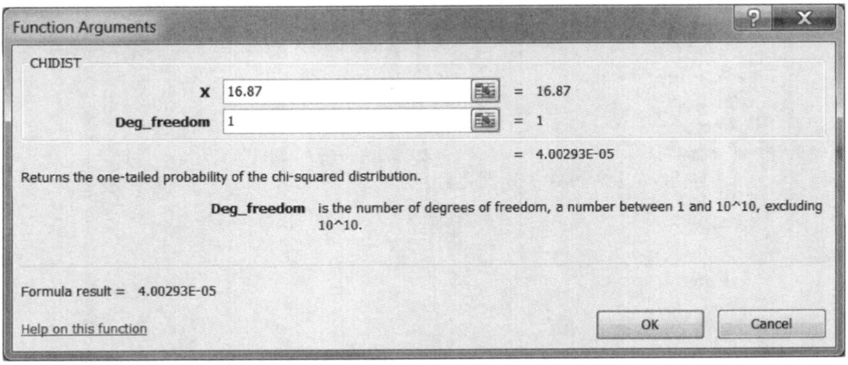

FIGURE 17.7 The Excel® CHIDIST function to identify the Chi Square probability.

value from Excel®, but rather used a rounded result from hand calculations) as we obtained with CHITEST.

USING SPSS® FOR THE CHI SQUARE TEST OF INDEPENDENCE

You can use the Crosstabs command in SPSS® to create a Chi Square analysis. I will demonstrate the procedure with the same data I used in the Excel® example, and then I will show an alternate procedure that is best when you input the data table directly into SPSS®.

The Crosstabs Procedure

SPSS® prepares a contingency table with the data you specify and provides a range of analyses. Starting with the database in which we code the variables as we did with Excel® above, we can choose the Crosstabs procedure through the main menu choices: Analyze–Descriptive–Crosstabs. Figure 17.8 shows the Crosstabs menu that results from this choice.

As you can see in Figure 17.8, I called for the column variable (Earmark) to be the independent variable and the row variable (PDexp) to be the dependent variable for this analysis according to the protocol I discussed above. This specification will result in a table similar to the one in Table 17.13.

The Statistics . . . button shown in Figure 17.8 allows the researcher to choose which statistical analyses are desired. Figure 17.9 shows this window in which I call

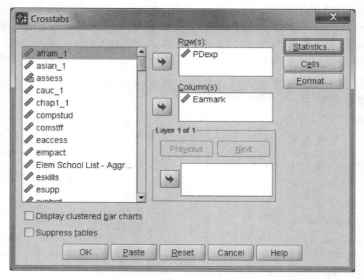

FIGURE 17.8 The SPSS® Crosstabs specification window.

USING SPSS® FOR THE CHI SQUARE TEST OF INDEPENDENCE

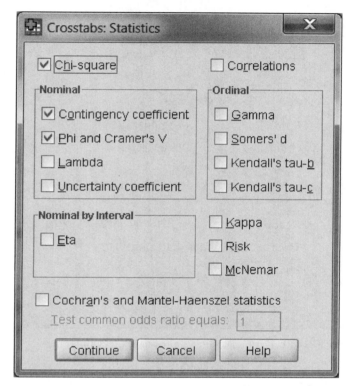

FIGURE 17.9 The Crosstabs: Statistics menu in SPSS®.

for the overall Chi Square statistics and a series of measures for nominal data: C, φ, and Cramer's V.

You also need to choose the Cells . . . button (just under the Statistics... button). In this menu, you can choose how the percentages are created, among other things. Figure 17.10 shows this menu choice in which I called for column percentages given the protocol I discussed above. I also requested expected frequencies so I could report them in a summary table.

This series of choices results in SPSS® output showing the Chi Square analyses needed for significance tests and effect sizes. I will show the relevant tables next. Figure 17.11 is the contingency table showing observed and expected frequencies and column percentages.

As you can see, Figure 17.11 looks like the contingency table shown in Table 17.12. Figure 17.11 provides the percentages we will need to interpret the results once we have tested the results for statistical significance. Figure 17.12 shows the Chi Square Test results from SPSS®.

As you can see in Figure 17.12, the Pearson Chi-Square analysis shows a statistically significant Chi Square result ($\chi^2 = 16.87$, $p < 0.0001$). This is the same result we saw earlier from the Excel® analyses. There are several other tests appropriate

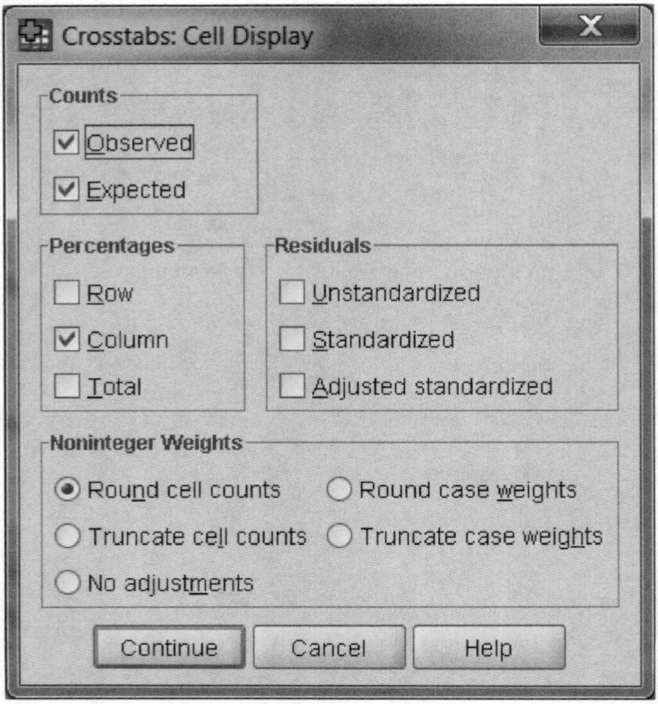

FIGURE 17.10 The SPSS® Crosstabs: Cell Display menu.

PDexp * Earmark Crosstabulation

			Earmark 0.00	Earmark 1.00	Total
PDexp	.00	Count	53	61	114
		Expected count	39.3	74.7	114.0
		% within Earmark	76.8%	46.6%	57.0%
	1.00	Count	16	70	86
		Expected count	29.7	56.3	86.0
		% within Earmark	23.2%	53.4%	43.0%
Total		Count	69	131	200
		Expected count	69.0	131.0	200.0
		% within Earmark	100.0%	100.0%	100.0%

FIGURE 17.11 The SPSS® crosstabs contingency table.

Chi-Square Tests

	Value	df	Asymptotic Significance (Two-Sided)	Exact Significance (Two-Sided)	Exact Significance (One-Sided)
Pearson Chi-Square	16.870[a]	1	0.000		
Continuity correction[b]	15.658	1	0.000		
Likelihood ratio	17.607	1	0.000		
Fisher's exact test				0.000	0.000
Linear-by-Linear association	16.785	1	0.000		
N of valid cases	200				

[a] 0 cells (0.0%) have expected count less than 5. The minimum expected count is 29.67.
[b] Computed only for a 2 × 2 table.

FIGURE 17.12 The SPSS® Chi square significance test output.

for Chi Square analyses that we will not cover in this book. I will only point out that the Yates Continuity Correction Factor is shown as Continuity Correction = 15.66, ($p < 0.000$). While I do not advocate the use of this result, you can note that the value, which is only created for the 2 × 2 table, is less than the Chi Square value because it is a more conservative estimate. In any case, this value is also statistically significant.

Effect sizes are shown in Figure 17.13. As you recall, I specified these in the Crosstabs procedure. Because this is a 2 × 2 table, we can use φ which indicates a value of 0.29. According to our guidelines, this is (nearly) a medium effect size according to the Table 17.9. If we square φ, we can interpret the value as the percent of variance in PDexp accounted for by Earmark. Thus, about 8.4% of the variance in PDexp is accounted for by Earmark ($\varphi^2 = 0.29^2 = 0.084$).

Analyzing the Contingency Table Data Directly

Another method of using SPSS® to calculate Chi Square is to create the summary table directly into the SPSS® spreadsheet rather than using the Crosstabs menus

Symmetric Measures

		Value	Approximate Significance
Nominal by Nominal	Phi	0.290	0.000
	Cramer's V	0.290	0.000
	Contingency Coefficient	0.279	0.000
N of Valid Cases		200	

FIGURE 17.13 The SPSS® effect size measures.

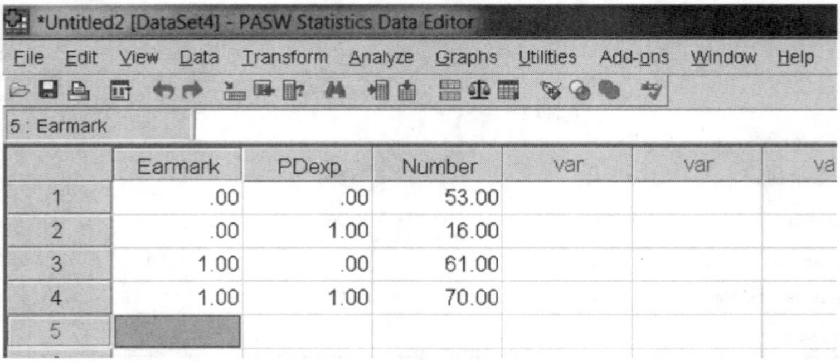

FIGURE 17.14 The SPSS® Chi Square data table.

with raw data as we did above. We will still end up using Crosstabs, but this alternate procedure allows you to simplify the data file prior to conducting the analyses by using a weight cases specification. Figure 17.14 shows the spreadsheet that we can create directly into SPSS®.

As you can see, the contingency table cells are identified by 1s and 0s and the observed frequencies are placed in a separate variable that I have called "Number." This arrangement simply identifies that, for example, 53 schools show no Earmark funds and no PDexp funds. The table shows that 61 schools had Earmark funds but no PDexp funds. The remaining two cells can be identified in this fashion.

Once in this format, you can choose the Data–Weight Cases menu, which will return the window shown in Figure 17.15. This calls for SPSS® to consider 53 lines of data with Earmark = 0 and PDexp = 0, and so forth. This is simply a shortcut method to avoid entering all 200 lines of data. As you can see, I called for the program to use the Number variable as a way to virtually recreate the raw score data file.

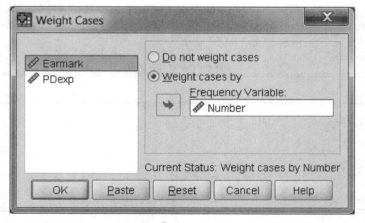

FIGURE 17.15 The SPSS® weight cases specification window.

Once you make this choice, you can use the Crosstabs menus to create the same Chi Square output that I described above. This method is much easier if you have a large sample of data to input.

Interpreting the Contingency Table

As I mentioned above, you need to interpret the analyses from the percentages in the contingency table once you have established a statistically significant finding. If you look at Figure 17.11, you will see that I followed the protocol of using the columns to create percentages that equal 100% so that we can interpret across rows of the dependent variable. In this case, we can point out that over twice the percentage of schools with Earmark funds designate PDexp funds compared to schools that do not have Earmark funds (53.4% to 23.2%). Alternatively, we can say that when schools do not show Earmark funds, they are much less likely to designate PDexp funds (76.8% to 46.6%).

The independent and dependent variables in this Chi Square analysis may be difficult to determine. As I have designated them, however, we can point out that when schools do not Earmark funds for technology, they typically do not spend technology funds for professional development. In this formulation, the PD expenditure appears to reflect the importance placed on technology signaled by earmarking funds for technology.

TERMS AND CONCEPTS

Chi Square Test of Independence Chi Square analyses involving frequencies from more than one study variable.

Contingency Coefficient An effect size measure for Chi Square analyses. This measure is essentially a correlation with nominal data and is used to show the strength of association among study variables.

Contingency Table Presentation of data in rows and columns to show how data in rows are contingent upon or connected to the data in column cells. Also called Crosstabs or Cross-tabulation analysis.

Cramer's V An effect size measure for the Chi Square Test of Independence with tables that exceed the 2×2 arrangement (i.e., that have more than two rows and/or categories).

Goodness of Fit Chi Square analyses that compare *actual* frequency data to *expected* data distributions.

McNemar Test A special Chi Square test used with repeated measures designs.

Phi Coefficient An effect size measure based upon Chi Square and typically used in studies with 2×2 tables. When Phi is squared, it expresses the proportion of variance in one variable explained by the other.

Yates Correction for Continuity A method for adjusting 2×2 Chi Square table data that have small sample sizes or small expected cell frequencies.

REAL-WORLD LAB XIII: CHI SQUARE

The Lab for Chi Square is from a small study I conducted a number of years ago in a subsidized housing development. The residents of this housing district were all low income, and there was a significant amount of crime prior to an intervention in which services were increased to residents and a new community policing model was established. This sample of residents ($N = 140$) answered a questionnaire that assessed, among other things, their fear of crime and whether they had been the victim of a crime. The survey was conducted after the intervention had taken place, so crime levels were decreasing.

The two variables in the analysis are:

- Victim—Whether respondents indicated they had been the victim of a crime.
- FEARHILO—Assessment of whether their fear of crime was high or low.

Conduct a Chi Square analysis on the data in Table 17.13 and summarize your findings.

REAL-WORLD LAB XIII: SOLUTIONS

Because this data table represents the special 2×2 Chi Square table, we can use the alternate formula for hand calculations. Once we do this, I will use Excel® and SPSS® to perform the test of independence.

Hand Calculations

The following are the hand calculations for the data in Table 17.13 using the alternate formula.

$$\chi^2 = \frac{N(ad - bc)^2}{(a+b)(c+d)(a+c)(b+d)},$$

$$\chi^2 = \frac{140(1872 - 660)^2}{(81)(59)(68)(72)},$$

$$\chi^2 = 8.79$$

TABLE 17.13 The Lab XIII Survey Data

Fear	Victim?	
	No	Yes
Low	48	33
High	20	39

REAL-WORLD LAB XIII: SOLUTIONS

We can now compare our actual χ^2 (8.79) to the tabled value $\chi^2_{0.05,1} = 3.841$. We can reject the null hypothesis ($p < 0.05$) and conclude that there is a statistically significant difference between the categories of Victim and FEARHILO.

The effect size is φ for this table, calculated as follows:

$$\varphi = \sqrt{\frac{\chi^2}{N}},$$

$$\varphi = \sqrt{\frac{8.79}{140}},$$

$$\varphi = 0.25$$

The effect size is small to medium and indicates that we can explain about 6% of the variance in one variable as a result of the other.

Using Excel® for Chi Square Analyses

As I demonstrated earlier in the chapter, you can use Excel® to perform the test of independence by using the CHIDIST function as shown in Figure 17.16. As you can see, this function returns the value of $p = 0.003$, indicating a very small probability of concluding a chance finding with these data.

Alternatively, you can use the CHITEST function by calculating the expected frequencies so that you can enter them from the spreadsheet into the CHITEST function menu. Table 17.14 shows the contingency table with the expected frequencies in separate cells. You need to create the table in this fashion so that you can easily use the CHITEST function. Figure 17.17 shows the CHITEST function with the observed and expected frequencies placed in the appropriate windows (Actual_range and Expected_range, respectively).

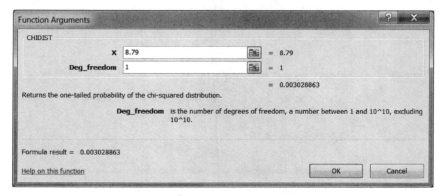

FIGURE 17.16 The Excel® CHIDIST results for the test of independence.

TABLE 17.14 The Contingency Table with Expected Frequencies

Fear	No	Yes
	Victim? (Observed)	
Low	48	33
High	20	39
	Victim? (Expected)	
Low	39.3	41.7
High	28.7	30.3

Figure 17.17 shows the results of the test of independence just below the observed and expected windows. As you can see, p is 0.003 (which is the same value as shown in Figure 17.16 with different rounding of values).

Using SPSS® for Chi Square Solutions

Figure 17.18 shows the SPSS® spreadsheet with the variable specification for the Chi Square analysis. Once you create the data in this fashion, you can weight the cases by the Number variable, which will allow you to use the crosstabs menu to complete the analysis.

Figure 17.19 shows the results of the Chi Square analysis. As you can see, $\chi^2 = 8.79$ and $p = 0.003$. Both of these findings confirm our hand calculations and Excel® analyses.

Figure 17.20 shows the effect size calculations. Recall that we can use Phi for this 2 × 2 table. The calculated $\varphi = .25$, which is the same value we calculated by hand above.

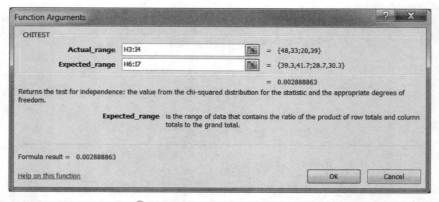

FIGURE 17.17 The Excel® CHITEST function with observed and expected frequencies identified.

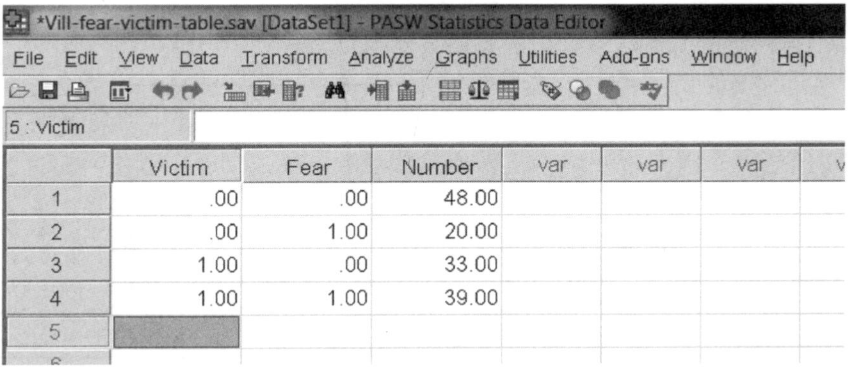

REAL-WORLD LAB XIII: SOLUTIONS

	Victim	Fear	Number	var	var	var	v
1	.00	.00	48.00				
2	.00	1.00	20.00				
3	1.00	.00	33.00				
4	1.00	1.00	39.00				
5							

FIGURE 17.18 The SPSS® variables used for the Chi Square analysis.

Chi-Square Tests

	Value	df	Asymptotic Significance (Two-Sided)	Exact Significance (Two-Sided)	Exact Significance (One-Sided)
Pearson Chi-Square	8.789[a]	1	0.003		
Continuity correction[b]	7.803	1	0.005		
Likelihood ratio	8.909	1	0.003		
Fisher's exact test				0.004	0.002
Linear-by-linear association	8.727	1	0.003		
N of valid cases	140				

[a] 0 cells (0.0%) have expected count less than 5. The minimum expected count is 28.66.

[b] Computed only for a 2 × 2 table.

FIGURE 17.19 The SPSS® Chi Square findings.

Symmetric Measures

		Value	Approximate Significance
Nominal by nominal	Phi	0.251	0.003
	Cramer's V	0.251	0.003
	Contingency Coefficient	0.243	0.003
N of valid cases		140	

FIGURE 17.20 The SPSS® Effect size findings.

FEARHILO * Victim Crosstabulation

			Victim		Total
			NO	Yes	
FEARHILO	Low	Count	48	33	81
			70.6%	45.8%	57.9%
	Hi	Count	20	39	59
		Percent within victim	29.4%	54.2%	42.1%
Total		Count	68	72	140
		Percent within victim	100.0%	100.0%	100.0%

FIGURE 17.21 The SPSS® contingency table output for the crosstabs analysis.

We can use the next part of the output for interpretation. Figure 17.21 shows the Crosstabs output for the contingency table according to the layout of our data. I called for column percentages in order to use the protocol of placing the predictor variable in columns and the outcome variable in rows.

As you can see from Figure 17.21, when the respondents have been victims of a crime, they are much more likely to have a high fear of crime. Over 54% of victims have a fear of crime compared to 29.4% of those not victimized. This difference in fear persists even when crime is on the decline as it was in our study.

18

REPEATED MEASURES PROCEDURES: T_{dep} AND ANOVA$_{ws}$

I have mentioned several times in this book that the researcher needs to consider a statistical procedure carefully before using it. The greatest power and confidence relies on using the correct statistical procedure with the available data according to the appropriate research design.

In this chapter, I am going to return to two statistical procedures that statisticians have adapted for use with "repeated measures" data. We discussed the independent T test (Chapter 11) and the general ANOVA procedure (Chapter 13), which were both designed for independent samples. While there are other such procedures, the dependent T test and the within-subjects ANOVA represent very important procedures that researchers can use in experimental and *post facto* situations in which the data are linked in some way.

Recall that I made a distinction between statistical procedures that used independent samples and those that used dependent samples. In Chapter 11, I identified dependent samples designs as those in which one group is somehow linked structurally to the other. Thus, in experimental designs, a group measured twice (e.g., pretest and post-test) would need to be treated differently than other group measures because the *same group of subjects would be measured twice*.

If the T test was used to detect a difference between two dependent samples in an experiment for example (i.e., pre–post, Time 1–Time 2, matched samples), we would need to use a special T test that factored out the relatedness so the T test could see what the unique differences remained between the two samples. The same thing would be true for ANOVA procedures that measured the same group twice in a design.

Understanding Educational Statistics Using Microsoft Excel® and SPSS®. By Martin Lee Abbott.
© 2011 John Wiley & Sons, Inc. Published 2011 by John Wiley & Sons, Inc.

As I noted in Chapter 17, in which I discussed the dependent samples Chi Square (McNemar Test), these kinds of measures have several names:

- Repeated-measures tests
- Dependent-samples tests
- Paired-samples tests
- Within-subjects tests

INDEPENDENT AND DEPENDENT SAMPLES IN RESEARCH DESIGNS

Independent and dependent sample data can be used in both experimental and *post facto* designs. Dependent samples are most commonly encountered in experimental or quasi-experimental designs in which a sample group is given a pretest and then administered a post-test after some intervention. It is used in some *post facto* research to measure a group on some attitude measure at different times.

Table 18.1 shows several possibilities for independent and dependent samples in research and *post facto* designs. This table highlights the dependent T test because there are only two groups. A similar table could be created for three or more groups that would emphasize the within-subjects ANOVA.

TABLE 18.1 Independent and Dependent Samples in Experimental and *Post Facto* Designs

	Experimental Designs	*Post Facto* Designs
Two Independent Samples (Between Subjects)	T_{ind} comparing post-test differences of two groups of *different subjects* after an intervention	T_{ind} comparing existing data on two groups of *different members*
	Example: Do the potential for violence (PFV) measures differ between two randomly chosen groups after one group watches the movie, 'Swimming in Gore'?	*Example: Do the PFV measures differ between a group of postal workers and a group of "save the puppies" advocates?*
Two Dependent Samples (Within Subjects)	T_{dep} comparing Pre–Post differences or Time 1–Time 2 differences of the *same group of subjects* after an intervention	T_{dep} test comparing *matched group* differences on an existing measure or outcome
	Example: Is the post-test PFV score higher than the pretest PFV scores of one group that watches "Swimming in Gore"?	*Example: Do the PFV scores differ between two classes (one in the urban core and one in the suburbs) of fourth-grade students matched on academic achievement?*

As I noted in Chapter 11, independent samples means that choosing subjects for one group has nothing to do with choosing subjects for the other group. Thus, if I randomly select Bob and assign him randomly to group 1 (the experimental group that watches "Swimming in Gore"), it has nothing to do with the fact that I choose Sally and assign her randomly to group 2 (the group that does not see 'Swimming in Gore'). The random selection of these group members is a very important assumption because it ensures the researcher that there are no built-in linkages between the groups' post-test scores. The power of randomization will result in the comparability of two groups chosen in this way.

Dependent samples consist of groups of subjects that have some structured linkage, like using the same people twice in a study. For example, I might use pretest PFV scores from both Bob and Sally and *compare these scores with their own post-test PFV scores*. Using dependent samples affects the ability of the randomness process to create comparable samples; in such cases, the researcher is assessing *individual* change (before-to-after measures) in the context of the experiment that is assessing *group* change.

Dependent samples also include *matched samples*, a situation in which we purposely choose people to be in two groups to be compared rather than choosing and assigning subjects randomly. For example, we might be concerned about gender for a given study and purposely assign equal numbers of men and women to two groups. In this case, the randomness criterion is violated in order to assure the researcher that the resulting groups have matching characteristics on a crucial study influence. This is typically the case where the researcher cannot control the conditions for random selection and assignment.

USING DIFFERENT T TESTS

When researchers use dependent samples, they need to use an independent T test that has been adjusted for the influence of the relatedness of the samples. This adjusted test is called a *dependent samples T test*. Other names for this test are: repeated measures T test, within subjects T test, and paired samples T test. Both Excel® and SPSS® refer to these as paired sample T tests.

There are two ways to calculate the dependent T test (T_{dep}). I will introduce both approaches in this chapter, but we will focus primarily on interpreting the Excel® and SPSS® procedures rather than the hand calculations.

THE DEPENDENT T-TEST CALCULATION: THE LONG FORMULA

The long formula for T_{dep} looks intimidating, but it makes conceptual sense. It is essentially the same as the formula for T_{ind} except that we must adjust the estimated standard error of difference s_D to account for the relatedness in the samples.

As you will recall from Chapter 11, these are the T_{ind} formulas:

$$t = \frac{(M_1 - M_2) - \mu_{M_1-M_2}}{s_D},$$

$$s_D = \sqrt{s_{m_1}^2 + s_{m_2}^2}$$

The formulas for T_{dep} are the same except for the estimated standard error of the difference using the adjusted formula below (s_{D_r}).

$$t_{dep} = \frac{(M_1 - M_2) - \mu_{M_1-M_2}}{s_{D_r}}$$

The T_{dep} formula thus becomes

$$t_{dep} = \frac{(M_1 - M_2)}{s_{D_r}}$$

Here is the formula for the adjusted standard error of difference:

$$s_{D_r} = \sqrt{s_{M_1}^2 + s_{M_2}^2 - [2r(s_{M_1})(s_{M_2})]}$$

As you can see, the only difference is that there is a second component included that is subtracted out of the s_D in the T_{ind} formula. I shaded this portion in the formula. Recall that I said that dependent samples are related (correlated) to one another and that the dependent T test must subtract out this relatedness.

If you look at the T_{dep} formula, the shaded portion that measures this relatedness is subtracted out of the formula. The shaded portion includes r, the Pearson's correlation coefficient. If the two sample groups are related (correlated), this r value will modify the measures that are used to calculate the T_{dep} ratio. However, if you have two independent samples, they will not be related, r will equal 0, and the shaded portion will therefore also equal 0. When this is the case, the s_{D_r} will be equivalent to s_D.

Example

I will use an example to show these calculations briefly. Recall the formulas we used to create the T_{ind} which we will use to create the T_{dep} values.

$$s_x = \sqrt{\frac{\sum x^2 - \frac{(\sum x)^2}{N}}{N-1}}, \quad s_D = \sqrt{s_{m_1}^2 + s_{m_2}^2}, \quad s_{m_1} = \frac{s_1}{\sqrt{n_1}}, \quad s_{m_2} = \frac{s_2}{\sqrt{n_2}}$$

THE DEPENDENT T-TEST CALCULATION: THE LONG FORMULA

The SS formulas can also be used since SS is part of the s_x calculation:

$$s_x = \sqrt{\frac{SS}{df}} \quad \text{or} \quad s_x = \sqrt{\frac{\sum x^2 - \frac{(\sum x)^2}{N}}{N-1}}$$

The data for this exercise are fictitious. The values represent the number of math errors committed by students before (Time 1) and after (Time 2) a constructivist math course in high school. Because the same students took the assessment at two different times, we will use the T_{dep} to see if the intervention (math course) was effective.

Table 18.2 shows the data in columns. Because of space limitations, I created two columns each for Time1 and Time2. Use the formulas above to calculate the Paired T ratio.

Table 18.3 shows the relevant calculations for this problem. Confirm the summary calculations and then use them to conduct a T_{dep} test.

TABLE 18.2 The Study Data

Time1	Time2	Time1	Time2
11	6.0	41	35.0
4	2.0	62	47.0
11	11.0	29	29.0
18	9.0	55	50.0
30	23.0	46	49.0
7	12.0	39	38.0
12	10.0	41	40.0
42	42.0	45	46.0
32	20.0	49	58.0
17	5.0	41	35.0
24	24.0	54	56.0
46	37.0	50	53.0
26	28.0	52	58.0
28	31.0	50	49.0
44	39.0	56	57.0
44	30.0	62	60.0
38	38.0	57	39.0
45	39.0	54	48.0
35	34.0	79	70.0
45	36.0	64	62.0
39	31.0	47	41.0
40	37.0	64	50.0
48	40.0		

TABLE 18.3 The Calculated Elements of the Study Data

	Time1	Time2
Means	40.51	36.76
SS	12,291.244	12,102.311
s_x	16.714	16.585
s_m	2.492	2.472
s_D	3.51	
r	0.936	
s_{D_r}	0.888	

Results

$$t_{dep} = \frac{(M_1 - M_2)}{s_{D_r}},$$

$$t_{dep} = \frac{(40.51 - 36.76)}{0.888},$$

$$t_{dep} = 4.223$$

The Paired T ratio for this study is 4.223. When we compare this to the tabled value of T to determine whether it is a significant ratio, we use df = $N-1$, where N is the *number of pairs of data*. Therefore $T_{dep.05,44} = 2.021$. Therefore, since our calculated $T_{dep} = 4.223$, we can reject the null hypothesis ($p < 0.05$). The difference between Time1 and Time2 is statistically different. The students committed fewer math errors in Time2 suggesting that the math course was successful.

Effect Size

We can use the same effect size formula for the T_{dep} as we did for the single-sample T test (in Chapter 10). The difference is that N is the number of pairs of data. Using 0.2, 0.5, and 0.8 as the criteria for small, medium, and large, as we did in Chapter 10, we can judge our calculated effect size for this finding ($d = 0.63$) as a medium effect.

$$d = \frac{T_{dep}}{\sqrt{N}},$$

$$d = \frac{4.223}{\sqrt{45}},$$

$$d = 0.63$$

THE DEPENDENT *T*-TEST CALCULATION: THE DIFFERENCE FORMULA

There is another way of hand calculating paired data that is shorter and follows a procedure you have already learned. As you recall from Chapter 10, we discussed the process for calculating the single sample *T* ratio. The formulas we used were

$$t = \frac{M - \mu}{S_m}, \quad S_m = \frac{S_x}{\sqrt{N}}$$

We can use this method to calculate a T ratio for paired data by subtracting values in one set of scores from paired values in the other set of scores to yield a single set of Difference scores. Once you have the Difference scores, you can use the formulas above to calculate the result. You can see how I included the Difference scores into the single sample *T* formula below to use it with our Difference change scores:

$$t = \frac{M - \mu}{S_m} \quad \text{becomes} \quad T_{\text{dep}} = \frac{M_{\text{diff}} - 0}{S_{m\text{-diff}}}$$

One matter of note in the altered formula that uses the Difference scores is the 0 in the place occupied by the population mean in the single sample *T* formula. This simply refers to the fact that the population of all differences between related group scores will be 0 because some differences will be negative and some will be positive. This would create a 0 sum. Therefore, the formula compares the *sample* mean Difference score to the *population of difference scores*, the latter of which is 0.

$$SS_{\text{diff}} = \sum D^2 - \frac{(\sum D)^2}{N},$$

$$S_{\text{diff}} = \sqrt{\frac{SS_{\text{diff}}}{N-1}},$$

$$S_{m\text{-diff}} = \frac{S_{\text{diff}}}{\sqrt{N}}$$

The bottom line is that you can use the same formula you used with the single sample *T* test with only slight variations. In this way, you do not have to calculate the Pearson's *r* value as you did in the long formula. By creating Difference scores, you subtract out the relatedness.

The two approaches (long formula and Difference formula) yield the same T_{dep} value. With larger datasets, we will use Excel® and SPSS® to make the calculations, so you do not need to be concerned at this point with the different formulas.

Here is how the Difference formulas and procedure would work using the data in Table 18.2. Table 18.4 shows the summary calculations which you can use to conduct the T_{dep} test. Remember that the df for this calculation is $N - 1$ (pairs).

TABLE 18.4 The Difference Procedure for Calculating T_{dep}

Time1	Time2
Mean	3.756
$\sum D$	169
$\sum D^2$	2193
SS_{diff}	1558.31
S_{diff}	5.95
$s_{m\text{-}diff}$	0.887

The T_{dep} Ratio from the Difference Method

When you use this method, use the modified formula below to calculate the T_{dep}.

$$T_{dep} = \frac{M_{diff} - 0}{s_{m\text{-}diff}},$$

$$T_{dep} = \frac{3.756 - 0}{0.887},$$

$$T_{dep} = 4.234$$

If you compare the T_{dep} value from the Difference method, you will find that it is equivalent to the T_{dep} calculated from the long method, with some rounding discrepancy.

T_{dep} AND POWER

As I have stated several times, *the maximum efficiency of statistical procedures is reached when the researcher uses the appropriate statistical measure with the study data*. To see how this works in the example study above, conduct T_{ind} with the data and compare the results to our T_{dep} result. Table 18.5 shows the comparison. As you can see, if we had not used the appropriate formula (T_{dep}), we would not have been able to reject the null hypothesis. However, because we did use T_{dep}, we rejected the null hypothesis and concluded that the math course was effective.

USING EXCEL® AND SPSS® TO CONDUCT THE T_{dep} ANALYSIS

Both Excel® and SPSS® have straightforward procedures for conducting the T_{dep} analysis. I will show the analysis of both procedures here starting with Excel®.

TABLE 18.5 The T_{ind} Comparison with T_{dep}

$T_{ind} = 1.07$	$T_{dep} = 4.234$
$T_{0.05,88} = 2.00$	$T_{0.05,44} = 2.021$
Decision: Do not reject H_0	Reject H_0, $p < 0.05$

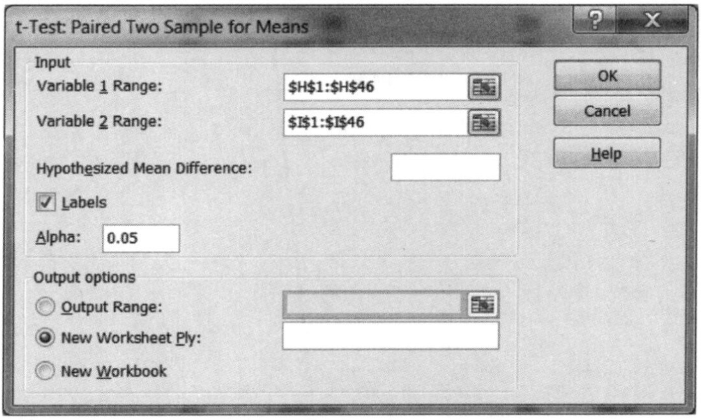

USING EXCEL® AND SPSS® TO CONDUCT THE T_{dep} ANALYSIS

FIGURE 18.1 The Excel® T_{dep} specification window.

T_{dep} with Excel®

The Excel® procedure is created through the Data–Data Analysis–"*t* Test: Paired Two Sample for Means" menu. When you make this choice, the screen in Figure 18.1 appears. As you can see, I specified the location of the data in the two Variable Range windows, and I indicated that the data included the variable label by checking the Labels box.

When I make this selection, Excel® returns the findings shown in Figure 18.2. As you can see, the T_{dep} value of 4.23 is shown (shaded cell), and the correlation (0.94) is shown. The values below the T_{dep} figure show the results of the hypothesis test(s) with their respective exclusion values. For example, the cell that includes $P(T \leq t)$ one-tail indicates that T_{dep} is significant at 0.00 (i.e., beyond $p < 0.05$) for the one-tailed *T* test which sets the exclusion value at 1.68. The last two cells confirm the significance level (0.00 or beyond $p < 0.05$) with a two-tailed test in which the exclusion value of 2.02.

t-Test: Paired Two Sample for Means

	Time1	Time2
Mean	40.51	36.76
Variance	279.35	275.05
Observations	45	45
Pearson correlation	0.94	
Hypothesized mean Difference	0	
df	44	
t Stat	4.23	
$P(T \leq t)$ one-tail	0.00	
t Critical one-tail	1.68	
$P(T \leq t)$ two-tail	0.00	
t Critical two-tail	2.02	

FIGURE 18.2 The T_{dep} findings from Excel® paired Two-Sample test.

REPEATED MEASURES PROCEDURES: T_{dep} AND ANOVA$_{ws}$

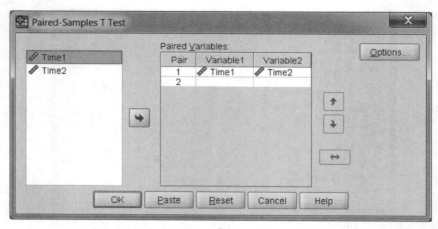

FIGURE 18.3 The SPSS® T_{dep} specification window.

T_{dep} with SPSS®

The T_{dep} analysis with SPSS® is equally straightforward. The T_{dep} analysis is accessed by using the Analyze–Compare Means–Paired-Samples T Test menu. This will create the window shown in Figure 18.3.

As you can see in Figure 18.3, I simply used the arrow button to indicate Time1 as "Variable1" and Time2 as "Variable2" for the analysis. This is all that is necessary. The default formulas and specifications will produce the analyses we need. When I choose "OK" SPSS® produces a series of output tables that I will use to complete the T_{dep} analysis. I show these output tables in Figures 18.4, 18.5 and 18.6.

Figure 18.4 shows the descriptive findings from the data. As you can see the s_m values (2.492 and 2.472) are produced and can be used with the long formula to check your hand calculations.

Figure 18.5 shows the correlation analysis between Time1 and Time2 which is necessary for the long formula calculation. In any case, it is important to note even if you choose to calculate the T_{dep} by hand with the Difference method.

Figure 18.6 shows the T_{dep} test findings. You can compare these to our hand calculations above. Note that the s_{D_r} value (0.888 in our hand calculations) is shown

Paired Samples Statistics

		Mean	N	Standard Deviation	Standard Error of the Mean
Pair 1	Time1	40.51	45	16.714	2.492
	Time2	36.76	45	16.585	2.472

FIGURE 18.4 The SPSS® descriptive output for T_{dep}.

Paired Samples Correlations

		N	Correlation	Significance
Pair 1	Time1 & Time2	45	0.936	0.000

FIGURE 18.5 The SPSS® correlation output for T_{dep}.

Paired Samples Test

		Paired Differences							
					95% Confidence Interval of the Difference				
		Mean	Standard Deviation	Standard Error of the Mean	Lower	Upper	t	df	Significance (Two-Tailed)
Pair 1	Time1–Time2	3.756	5.951	0.887	1.968	5.543	4.233	44	0.000

FIGURE 18.6 The SPSS® T_{dep}-test summary.

as the standard error of the mean. The T_{dep} value (4.233) is shown with the calculated significance ($p = 0.000$). The mean shown in the first column of data is the difference in means ($40.51 - 36.76 = 3.75$) from Time1 and Time2 that is used in the numerator of the T_{dep} formula. The "95% Confidence Interval of the Difference" values (1.968–5.543) indicates the confidence brackets that will contain the true population difference (i.e., Time1–Time2 difference) 95% of the time. Note that the df is 44, indicating 'pairs -1.'

WITHIN-SUBJECTS ANOVA (ANOVA$_{WS}$)

We now consider another extension of ANOVA. This time, we will discuss Within-Subjects ANOVA, which focuses on the repeated measures applied to the subjects of one group. The One-Way ANOVA (Chapter 12) and the Factorial ANOVA (Chapter 13) treatments of ANOVA focused on "between-subjects" applications. That is, do sample groups differ on some outcome measure? As you recall, we explored the differences in learning resulting from different noise conditions (One-Way ANOVA) and then extended this to include the differences between boys and girls (Factorial ANOVA) by including a second factor.

Experimental Designs

Table 18.6 shows the classic experimental design in which two groups (randomly selected in the "true" experiment and not in the "quasi" experiment) are compared on post-test scores after some intervention. The experimental group receives the intervention, and the control or comparison group does not receive the intervention.

TABLE 18.6 The Experimental Design

Pretest Scores	Experimental Group Treatment	Post-Test Scores
Pretest Scores	Comparison Group(s)	Post-Test Scores

If the post-test scores differ between the groups, the implication is that the treatment is responsible.

An example might be our factorial ANOVA (Chapter 13) in which we analyzed noise conditions (factor 1) and gender (factor 2) as influences on learning. To do this, we assumed that the experiment included randomization of subjects so that we did not have to include the pretests. The assumption is that if subjects are randomly chosen, the groups' pretest measures will be equal and therefore will not need to be included in the analyses. We simply compared the post-test (learning) scores for men and women in the different noise conditions.

There is also value in knowing how the individual subjects of each group change across the time of the experiment. If the experimental treatment is effective, it should cause changes in individuals' post-test scores compared to their pretest scores.

It is often difficult to achieve full randomization, so we cannot assume the pretest measures of different treatment groups (experimental and comparison groups) will be equal. Therefore, we need to take into account the differences that happen from pretest to post-test within both (all) treatment groups. *The experimental design that does not include full randomization therefore may include both within subjects and between subjects measures.* This is known as a *mixed design.* Figure 18.7 shows how this works.

In this chapter, I will discuss the within subjects element of this design because it is often the case that research simply focuses on how one group of subjects change over time. To extend our example from the T_{dep} section above, this might mean conducting math *assessments three times for each subject* in the experiment. Table 18.6 shows how this design looks.

Because we now have three measures for each student, we extend our T_{dep} test to a within-subjects ANOVA. This procedure will detect differences for each student among the three math assessments. Because the three math assessments are based on the same students, we must use a statistical procedure that factors out the

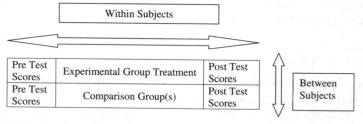

FIGURE 18.7 The mixed design that includes within-subjects and between-groups element.

USING SPSS® FOR WITHIN-SUBJECTS DATA

TABLE 18.7   Data for Within-Subjects Study with Three Categories

	Time1	Time2	Time3
Subject 1	11	6	3
Subject 2	4	2	3
Subject 3	11	11	4
	—	—	—
	Mean Time1	Mean Time2	Mean Time3

sameness so that we can see what the resulting differences are among the three assessments. Will the constructivist math teaching method result in declining math errors across the three assessments?

Post Facto Designs

Within-subjects procedures can be used with *post facto* designs as well because we are not necessarily assuming experimental manipulation. We might simply be interested in whether a group of students' math assessments are consistent across a certain time period (e.g., semester) or if they are erratic for some reason. The key issue is how the same students (or matched group of students) compare to themselves across different data collection periods or conditions.

WITHIN-SUBJECTS EXAMPLE

Table 18.7 shows the data we will use for this example. As you can see, it is the same data we used for the T_{dep} example with an additional assessment period for each student (Time3). For the purposes of this example, let us consider the time periods as Time1—beginning of school year, Time2—end of school year, and Time3—end of summer. Will the teaching approach produce consistent learning (lower math errors) over the course of the year including summer?

USING SPSS® FOR WITHIN-SUBJECTS DATA

For this example, I will not present the hand calculations but focus on the use of the statistical program to produce the findings. Excel® does not have a straightforward method for calculating the one way $ANOVA_{ws}$. I will therefore focus on the SPSS® program, which has detailed procedures for this design. I will discuss the procedure using the data in Table 18.7.

The output for $ANOVA_{ws}$ is quite complex. Therefore, I will present only the basic output to show how to interpret the primary findings.

Sphericity

One of the assumptions of $ANOVA_{ws}$ is that the variances among the subjects' (in this case) three time periods be approximately equal. That is, the variance between

TABLE 18.8 The Within-Subjects Example Data

Time1	Time2	Time3	Time1	Time2	Time3
11	6	3	41	35	40
4	2	3	62	47	40
11	11	4	29	29	40
18	9	6	55	50	40
30	23	7	46	49	41
7	12	8	39	38	41
12	10	9	41	40	41
42	42	9	45	46	44
32	20	12	49	58	47
17	5	18	41	35	48
24	24	20	54	56	49
46	37	21	50	53	49
26	28	22	52	58	51
28	31	27	50	49	52
44	39	30	56	57	52
44	30	32	62	60	56
38	38	33	57	39	57
45	39	34	54	48	62
35	34	36	79	70	62
45	36	36	64	62	62
39	31	37	47	41	63
40	37	38	64	50	75
48	40	40			

Time1 and Time2 should be equal, the variance between Time2 and Time3 should be equal, and the variance between Time1 and Time3 should be equal. This is like the assumption we saw for one-way ANOVA that was assessed by the Levene's Test. With the ANOVA$_{ws}$ we will assess the variances to make sure they are equivalent by the Mauchly's Test of Sphericity and then proceed with our analyses, depending on whether we have met or violated the assumptions.

THE SPSS® PROCEDURE

We create the specification for the ANOVA$_{ws}$ through the Analyze–General Linear Model–Repeated Measures menu. This creates the window shown in Figure 18.8. As you can see, I specified at the top that the within-subjects factor I will use is Time and that it has three measurements (specified in the "Number of Levels" window).

When I select "Define," the window shown in Figure 18.9 will appear. This summarizes my specification in the "Within-Subjects Variables" window. As you can see, there are several choices I can make to further specify the model on the right side of this window.

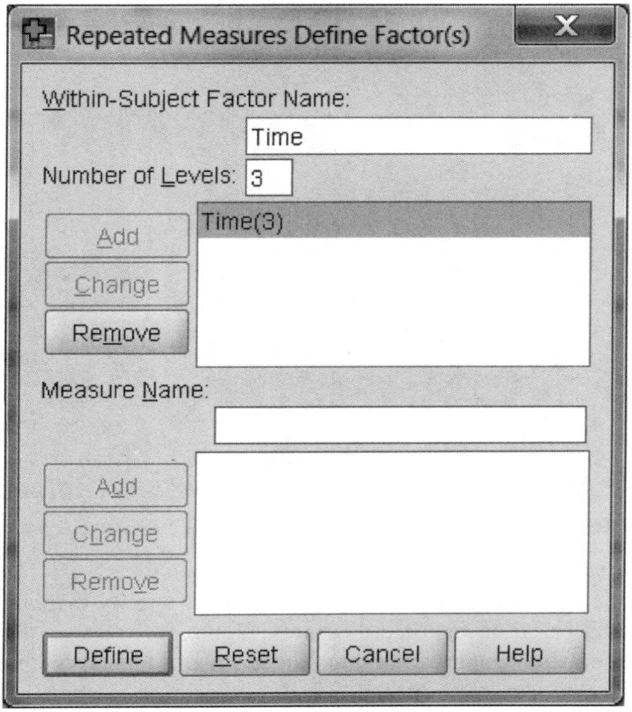

FIGURE 18.8 The SPSS® specification window for the ANOVA$_{ws}$ procedure.

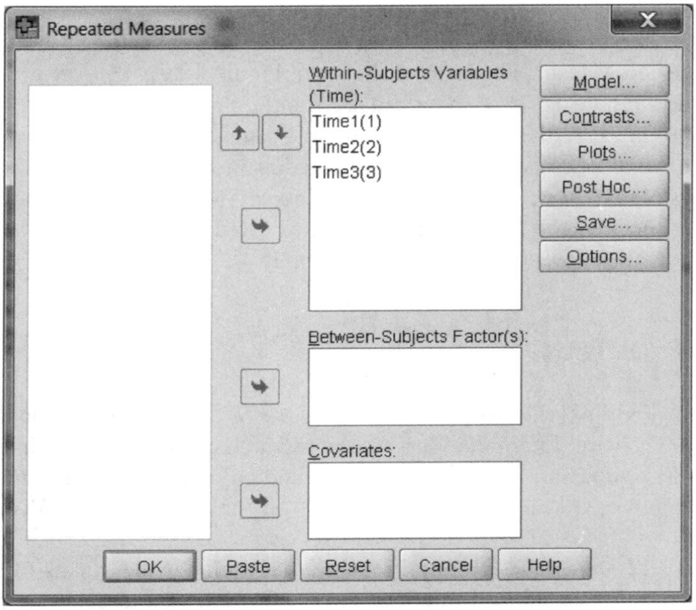

FIGURE 18.9 The "SPSS® Repeated Measures" window.

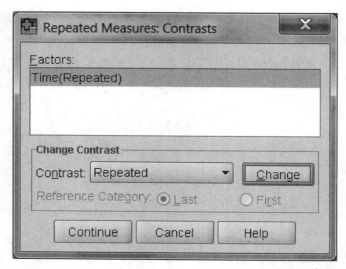

FIGURE 18.10 The "Contrasts" window to specify repeated contrasts.

Choosing the "Contrasts" button allows the user to specify that our comparison of conditions are based on repeated measures. Figure 18.10 shows this "Repeated Measures: Contrasts" window. As you can see, I clicked on "Time" and then chose the "Change Contrast" button, which allowed me to choose "Repeated," and then I clicked on "Change" to make sure the change registered in the window. This is shown in Figure 18.10, where the "Factors" window shows "Time(Repeated)."

When I make these selections, I next can choose the "Options" button in the main "Repeated Measures" window (shown in Figure 18.9). This creates the window shown in Figure 18.11. As you can see, I have chosen to "Display Means for" Time by moving it to the right-side window. I also checked the "Compare main effects" box and specified the "Bonferroni" adjustment from the "Confidence interval adjustment" button just below the "Display Means for:" window. Under 'Display' I checked the "Descriptive statistics" and "Estimates of effect size" boxes to show the appropriate results in the output.

THE SPSS® OUTPUT

In this book, I will provide only a basic look at the SPSS® $ANOVA_{ws}$ analyses and how to create general interpretations. As the research design increases in complexity, however, so does the design specification and output. You will need to consult additional sources if you choose to elaborate on the one-way $ANOVA_{ws}$ design shown here.

Figure 18.12 shows the Descriptive table that includes the means of the three time conditions for the 45 subjects. You can see that the SD measures are generally

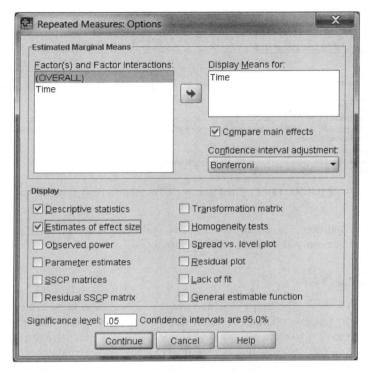

FIGURE 18.11 The "Options" menu for the ANOVA$_{ws}$ procedure.

equal, but the SD measure for Time3 is slightly larger. When we test the equivalence in variances (through the Mauchly's Test of Sphericity) we will assess whether this difference will be problematic.

The output for Mauchley's Test of Sphericity is shown in Figure 18.13. The critical value to look for is the "Significance" value shown in the middle of the table. In this case, the results show that there are significant differences among the variance measures of the three Time conditions. Thus, we have a violation of the

Descriptive Statistics

	Mean	Standard Deviation	N
Time1	40.51	16.714	45
Time2	36.76	16.585	45
Time3	35.49	18.817	45

FIGURE 18.12 The "Descriptive Statistics" output.

Mauchly's Test of Sphericity[b]

Measure:MEASURE_1

Within Subjects Effect	Mauchly's W	Approximate Chi Square	df	Significance	Epsilon[a]		
					Greenhouse–Geisser	Huynh–Feldt	Lower-bound
Time	0.694	15.697	2	0.000	0.766	0.788	0.500

Tests the null hypothesis that the error covariance matrix of the orthonormalized transformed dependent variables is proportional to an identity matrix.

[a]Design: Intercept; Within-Subjects Design: Time.
[b]May be used to adjust the degrees of freedom for the averaged tests of significance. Corrected tests are displayed in the Tests of Within-Subjects Effects table.

FIGURE 18.13 The Mauchley's Test of Sphericity results.

sphericity condition. These can be very sensitive tests, so we will take note of the violation in our analysis of the remaining output.

The Omnibus Test

Figure 18.14 shows the output for the within-subjects effects of Time. The first group of findings under Time show four different rows of output. The first "Sphericity Assumed" is like the one-way ANOVA F test in which there are equal variances. As you can see, this test is significant ($F = 7.830$, $p < 0.01$). If we had no sphericity violations, we would use this value to make our conclusions.

However, since we did observe a violation of sphericity, we need to use an adjusted F-test value. As you can see from Figure 18.14, SPSS® provides three such adjusted tests. The "Greenhouse–Geisser" test is shown in the second row and is a commonly used, conservative, test. As you can see, this F value is the same, but the significance level is slightly different due to the adjustment for sphericity. The result is the same, however. Time is a significant repeated measure condition ($F = 7.830$, $p = 0.002$).

Tests of Within-Subjects Effects

Measure:MEASURE_1

Source		Type III Sum of Squares	df	Mean Square	F	Significance	Partial Eta Squared
Time	Sphericity assumed	613.970	2	306.985	7.830	0.001	0.151
	Greenhouse–Geisser	613.970	1.532	400.870	7.830	0.002	0.151
	Huynh–Feldt	613.970	1.576	389.624	7.830	0.002	0.151
	Lower-bound	613.970	1.000	613.970	7.830	0.008	0.151
Error(Time)	Sphericity assumed	3450.030	88	39.205			
	Greenhouse–Geisser	3450.030	67.390	51.195			
	Huynh–Feldt	3450.030	69.335	49.759			
	Lower-bound	3450.030	44.000	78.410			

FIGURE 18.14 The within-subjects effects output for time.

THE SPSS® OUTPUT

Multivariate Tests[b]

Effect		Value	F	Hypothesis df	Error df	Significance	Partial Eta Squared
Time	Pillai's Trace	0.359	12.020[b]	2.000	43.000	0.000	0.359
	Wilks' Lambda	0.641	12.020[b]	2.000	43.000	0.000	0.359
	Hotelling's Trace	0.559	12.020[b]	2.000	43.000	0.000	0.359
	Roy's Largest Root	0.559	12.020[b]	2.000	43.000	0.000	0.359

[a]Design: Intercept; Within-Subjects Design: Time.
[b]Exact statistic

FIGURE 18.15 The effect size output for ANOVA$_{ws}$.

Effect Size

Figure 18.15 shows the "Multivariate Tests" results. These reveal the *effect size* findings for this study and are not impacted by sphericity. We have not discussed these so far, but many researchers use Wilks' Lambda as the effect size measure for ANOVA$_{ws}$ in this situation. Wilks' Lambda (Λ) is based on measuring unexplained variance, so the smaller the value, the stronger the effect.

As you can see from Figure 18.15, $\Lambda = 0.641$ and is significant ($p < 0.0001$). You will also note that in this test, Λ is complementary to Partial Eta Squared, which is not based on error, but on the size of the regression (i.e., Partial $\eta^2 = 1 - \Lambda$). As you can see, Partial $\eta^2 = 0.359$ and Wilks' Lambda $= 0.641$, suggesting that the impact of the time conditions on math errors is considered large (using the criteria we suggested in Chapter 13).

Post Hoc Analyses

Just as we did with the ANOVA tests in Chapter 12, we must perform post hoc analyses when the omnibus test is significant. Figure 18.16 shows the "Pairwise Comparisons" output that indicates which time conditions differ. As you can see, all the comparisons are significantly different except the difference between Time2 and Time3.

You can see that the differences among the means of the time conditions are large except for Time2–Time3 (1.267). The plot in Figure 18.17 provides visual comparisons in the plotting of the means. As you can see, there is a large difference (fewer math errors) between Time1 and Time2, but the difference in errors from Time2 to Time3 is much less.

The Interpretation

Taking all the output into account, we can say that the students do improve their math achievement across the time conditions. Each time period shows less math errors. The growth in understanding of math occurs mainly during the school year (Time1 to Time2). Although some growth continues into the summer, the effects are not significant.

Pairwise Comparisons

Measure:MEASURE_1

(I) Time	(J) Time	Mean Difference (I–J)	Standard Error	Significance[b]	95% Confidence Interval for Difference[b]	
					Lower Bound	Upper Bound
1	2	3.756[a]	0.887	0.000	1.547	5.964
	3	5.022[a]	1.453	0.004	1.405	8.639
2	1	−3.756[a]	0.887	0.000	−5.964	−1.547
	3	1.267	1.526	1.000	−2.532	5.065
3	1	−5.022[a]	1.453	0.004	−8.639	−1.405
	2	−1.267	1.526	1.000	−5.065	2.532

Based on estimated marginal means.
[a]The mean difference is significant at the 0.05 level.
[b]Adjustment for multiple comparisons: Bonferroni.

FIGURE 18.16 The *Post Hoc* output for the time study.

NONPARAMETRIC STATISTICS

In past chapters, I have presented the nonparametric counterparts of some parametric tests. Thus, for the Independent-Samples T test, I discussed the Mann–Whitney U test, and I presented a section on the Kruskal–Wallis Test when I discussed the One-Way ANOVA procedure.

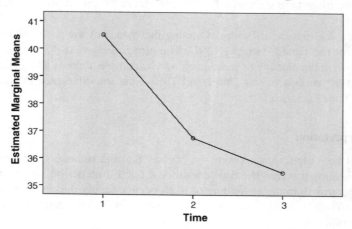

FIGURE 18.17 The comparison plot for the time conditions.

Both of the repeated measures tests in this chapter also have a nonparametric counterpart. With *two* related (dependent) samples of ordinal data the researcher would use the Wilcoxen test. If there are *more than two* dependent samples of less than interval data, the researcher would use the Friedman test. Both of these are available in SPSS® and can be accessed with the same "analyze" menu I demonstrated with the Mann–Whitney U and Kruskal–Wallis tests.

TERMS AND CONCEPTS

Dependent Samples Groups in which the members are structurally related, such as using the same group of subjects twice in an experiment (pre–post), or using "matched groups." Also known as paired samples, repeated measures, and within-subjects measures.

Friedman Test A nonparametric test used with more than two dependent samples (repeated measures) with ordinal data.

Greenhouse–Geisser Test This is one of the tests used by SPSS® to adjust values affected by sphericity.

Sphericity The assumption with repeated measures tests (like $ANOVA_{ws}$) that the variance of group differences are equal.

Wilcoxen Test A nonparametric test used with two dependent samples (repeated measures) with ordinal data.

Wilks' Lambda An effect size measure based on the amount of unexplained variance. Small results are considered stronger than large results.

REFERENCES

Abbott, M. L. *The Program Evaluation Prism: Using Statistical Methods to Discover Patterns.* Hoboken, NJ: John Wiley & Sons, 2010.

Abbott, M. L., Baker, D. B, and Pavese, E. P. *On the Road to Second-Order Change: What Washington State Educators Say about Collaboration, Instructional Enhancement, and Fundamental Change.* Washington School Research Center, Seattle, WA. Research Report #11. December 2008.

Abbott, M. L., Baker, D., Smith, K., and Trzyna, T. *Winning the Math Wars.* Seattle, WA: University of Washington Press, 2010.

Babbie, E. *The Practice of Social Research*, 12 ed. Wadsworth, 2010.

Bickel R. *Multilevel Analysis for Applied Research.* New York: Guilford Press, 2007.

Bohrnstedt, G. W., and Knoke, D. *Statistics for Social Data Analysis.* Itasca, IL: F. E. Peacock, 1982.

Campbell, D. T., and Stanley, J. C. *Experimental and Quasi-Experimental Designs for Research.* Chicago: Rand McNally, 1963.

Cohen, J., Cohen, P., West, S. G., and Aiken, L.S. *Applied Multiple Regression/Correlation Analysis for the Behavioral Sciences*, 3rd ed. Mahwah, NJ: Lawrence Erlbaum Associates, 2003.

Cohen, J. *Statistical Power Analysis for the Behavioral Science*, 2nd ed. Hillsdale, NJ: Lawrence Erlbaum Associates, 1988.

Field, A. *Discovering Statistics Using SPSS®*, 2nd ed. Thousand Oaks, CA: Sage, 2005.

Green, S. B., and Salkind, N. J. *Using SPSS® for Windows and Macintosh*, 5th ed. Pearson: Upper Saddle River, NJ, 2008.

Understanding Educational Statistics Using Microsoft Excel® and SPSS®. By Martin Lee Abbott.
© 2011 John Wiley & Sons, Inc. Published 2011 by John Wiley & Sons, Inc.

Hackman, R. J., and Oldham, G. R. *Work Redesign*. Reading, MA: Addison-Wesley, 1980.

Raudenbush, S. W., and Bryk, A. S. *Hierarchical Linear Models*, 2nd ed. Thousand Oaks: Sage, 2002.

Siegel, S., and Castellan, N. J. *Nonparametric Statistics for the Behavioral Sciences*, 2nd ed. New York: McGraw-Hill, 1988.

SPSS® for Windows, PASW Statistics 18, Release 18.02 (April 2, 2010) and Release 17.0.2.

Trzyna, T., and Abbott, M. Grieving in the Ethnic Literature Classroom, *College Literature*, Issue 18.3, October, 1991, pp. 1–14.

APPENDIX

STATISTICAL TABLES

Table A: Z-Score Table
Table B: Exclusion Values for the *T* Distribution
Table C: Critical (Exclusion) Values for the Distribution of *F*
Table D: Tukey's Range Test
Table E: Critical (Exclusion) Values for Pearson's Correlation Coefficient, *r*
Table F: Critical Values of the Chi Square Distribution

TABLE A Z-Score Table (Values Shown Are Percentages)

z	0	0.01	0.02	0.03	0.04	0.05	0.06	0.07	0.08	0.09
0.0	0.000	0.399	0.798	1.197	1.595	1.994	2.392	2.790	3.188	3.586
0.1	3.983	4.380	4.776	5.172	5.567	5.962	6.356	6.749	7.142	7.535
0.2	7.926	8.317	8.706	9.095	9.483	9.871	10.257	10.642	11.026	11.409
0.3	11.791	12.172	12.552	12.930	13.307	13.683	14.058	14.431	14.803	15.173
0.4	15.542	15.910	16.276	16.640	17.003	17.364	17.724	18.082	18.439	18.793
0.5	19.146	19.497	19.847	20.194	20.540	20.884	21.226	21.566	21.904	22.240
0.6	22.575	22.907	23.237	23.565	23.891	24.215	24.537	24.857	25.175	25.490
0.7	25.804	26.115	26.424	26.730	27.035	27.337	27.637	27.935	28.230	28.524
0.8	28.814	29.103	29.389	29.673	29.955	30.234	30.511	30.785	31.057	31.327
0.9	31.594	31.859	32.121	32.381	32.639	32.894	33.147	33.398	33.646	33.891
1.0	34.134	34.375	34.614	34.850	35.083	35.314	35.543	35.769	35.993	36.214
1.1	36.433	36.650	36.864	37.076	37.286	37.493	37.698	37.900	38.100	38.298
1.2	38.493	38.686	38.877	39.065	39.251	39.435	39.617	39.796	39.973	40.147
1.3	40.320	40.490	40.658	40.824	40.988	41.149	41.309	41.466	41.621	41.774
1.4	41.924	42.073	42.220	42.364	42.507	42.647	42.786	42.922	43.056	43.189
1.5	43.319	43.448	43.574	43.699	43.822	43.943	44.062	44.179	44.295	44.408
1.6	44.520	44.630	44.738	44.845	44.950	45.053	45.154	45.254	45.352	45.449
1.7	45.543	45.637	45.728	45.818	45.907	45.994	46.080	46.164	46.246	46.327
1.8	46.407	46.485	46.562	46.638	46.712	46.784	46.856	46.926	46.995	47.062
1.9	47.128	47.193	47.257	47.320	47.381	47.441	47.500	47.558	47.615	47.670
2.0	47.725	47.778	47.831	47.882	47.932	47.982	48.030	48.077	48.124	48.169
2.1	48.214	48.257	48.300	48.341	48.382	48.422	48.461	48.500	48.537	48.574
2.2	48.610	48.645	48.679	48.713	48.745	48.778	48.809	48.840	48.870	48.899
2.3	48.928	48.956	48.983	49.010	49.036	49.061	49.086	49.111	49.134	49.158
2.4	49.180	49.202	49.224	49.245	49.266	49.286	49.305	49.324	49.343	49.361
2.5	49.379	49.396	49.413	49.430	49.446	49.461	49.477	49.492	49.506	49.520
2.6	49.534	49.547	49.560	49.573	49.585	49.598	49.609	49.621	49.632	49.643
2.7	49.653	49.664	49.674	49.683	49.693	49.702	49.711	49.720	49.728	49.736
2.8	49.744	49.752	49.760	49.767	49.774	49.781	49.788	49.795	49.801	49.807
2.9	49.813	49.819	49.825	49.831	49.836	49.841	49.846	49.851	49.856	49.861
3.0	49.865	49.869	49.874	49.878	49.882	49.886	49.889	49.893	49.897	49.900
	50.000	46.017	45.224	44.828	44.433	44.038	43.644	43.251	42.858	42.465
	46.414	42.074	41.294	40.905	40.517	40.129	39.743	39.358	38.974	38.591
	38.209	37.828	37.448	37.070	36.693	36.317	35.942	35.569	35.197	34.827
	34.458	34.090	33.724	33.360	32.997	32.636	32.276	31.918	31.561	31.207
	30.854	30.503	30.153	29.806	29.460	29.116	28.774	28.434	28.096	27.760
	27.425	27.093	26.763	26.435	26.109	25.785	25.463	25.143	24.825	24.510
	24.196	23.885	23.576	23.270	22.965	22.663	22.363	22.065	21.770	21.476
	21.186	20.897	20.611	20.327	20.045	19.766	19.489	19.215	18.943	18.673
	18.406	18.141	17.879	17.619	17.361	17.106	16.853	16.602	16.354	16.109
	15.866	15.625	15.386	15.151	14.917	14.686	14.457	14.231	14.007	13.786
	13.567	13.350	13.136	12.924	12.714	12.507	12.302	12.100	11.900	11.702
	11.507	11.314	11.123	10.935	10.749	10.565	10.383	10.204	10.027	9.853
	9.680	9.510	9.342	9.176	9.012	8.851	8.692	8.534	8.379	8.226
	8.076	7.927	7.780	7.636	7.493	7.353	7.215	7.078	6.944	6.811
	6.681	6.552	6.426	6.301	6.178	6.057	5.938	5.821	5.705	5.592
	5.480	5.370	5.262	5.155	5.050	4.947	4.846	4.746	4.648	4.551
	4.457	4.363	4.272	4.182	4.093	4.006	3.920	3.836	3.754	3.673
	3.593	3.515	3.438	3.363	3.288	3.216	3.144	3.074	3.005	2.938
	2.872	2.807	2.743	2.680	2.619	2.559	2.500	2.442	2.385	2.330
	2.275	2.222	2.169	2.118	2.068	2.018	1.970	1.923	1.876	1.831
	1.786	1.743	1.700	1.659	1.618	1.578	1.539	1.500	1.463	1.426
	1.390	1.355	1.321	1.287	1.255	1.222	1.191	1.160	1.130	1.101
	1.072	1.044	1.017	0.990	0.964	0.939	0.914	0.889	0.866	0.842
	0.820	0.798	0.776	0.755	0.734	0.714	0.695	0.676	0.657	0.639
	0.621	0.604	0.587	0.570	0.554	0.539	0.523	0.508	0.494	0.480
	0.466	0.453	0.440	0.427	0.415	0.402	0.391	0.379	0.368	0.357
	0.347	0.336	0.326	0.317	0.307	0.298	0.289	0.280	0.272	0.264
	0.256	0.248	0.240	0.233	0.226	0.219	0.212	0.205	0.199	0.193
	0.187	0.181	0.175	0.169	0.164	0.159	0.154	0.149	0.144	0.139
	0.135	0.131	0.126	0.122	0.118	0.114	0.111	0.107	0.104	0.100

Source: Pearson, Karl, F.R.S. *Tables for Statisticians and Biometricians,* Part I, 3rd ed., University College, London: Biometric Laboratory, Cambridge University Press, 1930. Used with permission.

APPENDIX: STATISTICAL TABLES

TABLE B Exclusion Values for the T Distribution

One-Tailed	0.4	0.25	0.1	0.05	0.025	0.01	0.005	0.0025
Two-Tailed	0.8	0.5	0.2	0.1	0.05	0.02	0.01	0.005
Degrees of Freedom								
1	0.325	1.000	3.078	6.314	12.706	31.821	63.657	127.320
2	0.289	0.816	1.886	2.920	4.303	6.965	9.925	14.089
3	0.277	0.765	1.638	2.353	3.182	4.541	5.841	7.453
4	0.271	0.741	1.533	2.132	2.776	3.747	4.604	5.598
5	0.267	0.727	1.476	2.015	2.571	3.365	4.032	4.773
6	0.265	0.718	1.440	1.943	2.447	3.143	3.707	4.317
7	0.263	0.711	1.415	1.895	2.365	2.998	3.499	4.029
8	0.262	0.706	1.397	1.860	2.306	2.896	3.355	3.833
9	0.261	0.703	1.383	1.833	2.262	2.821	3.250	3.690
10	0.260	0.700	1.372	1.812	2.228	2.764	3.169	3.581
11	0.260	0.697	1.363	1.796	2.201	2.718	3.106	3.497
12	0.259	0.695	1.356	1.782	2.179	2.681	3.055	3.428
13	0.259	0.694	1.350	1.771	2.160	2.650	3.012	3.372
14	0.258	0.692	1.345	1.761	2.145	2.624	2.977	3.326
15	0.258	0.691	1.341	1.753	2.131	2.602	2.947	3.286
16	0.258	0.690	1.337	1.746	2.120	2.583	2.921	3.252
17	0.257	0.689	1.333	1.740	2.110	2.567	2.898	3.222
18	0.257	0.688	1.330	1.734	2.101	2.552	2.878	3.197
19	0.257	0.688	1.328	1.729	2.093	2.539	2.861	3.174
20	0.257	0.687	1.325	1.725	2.086	2.528	2.845	3.153
21	0.257	0.686	1.323	1.721	2.080	2.518	2.831	3.135
22	0.256	0.686	1.321	1.717	2.074	2.508	2.819	3.119
23	0.256	0.685	1.319	1.714	2.069	2.500	2.807	3.104
24	0.256	0.685	1.318	1.711	2.064	2.492	2.797	3.091
25	0.256	0.684	1.316	1.708	2.060	2.485	2.787	3.078
26	0.256	0.684	1.315	1.706	2.056	2.479	2.779	3.067
27	0.256	0.684	1.314	1.703	2.052	2.473	2.771	3.057
28	0.256	0.683	1.313	1.701	2.048	2.467	2.763	3.047
29	0.256	0.683	1.311	1.699	2.045	2.462	2.756	3.038
30	0.256	0.683	1.310	1.697	2.042	2.457	2.750	3.030
40	0.255	0.681	1.303	1.684	2.021	2.423	2.704	2.971
60	0.254	0.679	1.296	1.671	2.000	2.390	2.660	2.915
120	0.254	0.677	1.289	1.658	1.980	2.358	2.617	2.860
∞	0.253	0.674	1.282	1.645	1.960	2.326	2.576	2.807

Source: Pearson, E. S., and Hartley, H. O. *Biometrika Tables for Statisticians*, Volume I., 2nd ed. Cambridge, UK: University Press, 1962. Used with permission.

TABLE C Critical (Exclusion) Values for the Distribution of F

	Exclusion Level	df in the Numerator (Between)											
		1	2	3	4	5	6	7	8	9	10	11	12
1	0.05	161	200	216	225	230	234	237	239	241	242	243	244
	0.01	4052	4999	5403	5625	5764	5859	5928	5981	6022	6056	6082	6106
2	0.05	18.51	19.00	19.16	19.25	19.30	19.33	19.36	19.37	19.38	19.39	19.40	19.41
	0.01	98.49	99.00	99.17	99.25	99.30	99.33	99.36	99.37	99.39	99.40	99.41	99.42
3	0.05	10.13	9.55	9.28	9.12	9.01	8.94	8.88	8.84	8.81	8.78	8.76	8.74
	0.01	34.12	30.82	29.46	28.71	28.24	27.91	27.67	27.49	27.34	27.23	27.13	27.05
4	0.05	7.71	6.94	6.59	6.39	6.26	6.16	6.09	6.04	6.00	5.96	5.93	5.91
	0.01	21.20	18.00	16.69	15.98	15.52	15.21	14.98	14.80	14.66	14.54	14.45	14.37
5	0.05	6.61	5.79	5.41	5.19	5.05	4.95	4.88	4.82	4.78	4.74	4.70	4.68
	0.01	16.26	13.27	12.06	11.39	10.97	10.67	10.45	10.29	10.15	10.05	9.96	9.89
6	0.05	5.99	5.14	4.76	4.53	4.39	4.28	4.21	4.15	4.10	4.06	4.03	4.00
	0.01	13.74	10.92	9.78	9.15	8.75	8.47	8.26	8.10	7.98	7.87	7.79	7.72
7	0.05	5.59	4.74	4.35	4.12	3.97	3.87	3.79	3.73	3.68	3.63	3.60	3.57
	0.01	12.25	9.55	8.45	7.85	7.46	7.19	7.00	6.84	6.71	6.62	6.54	6.47
8	0.05	5.32	4.46	4.07	3.84	3.69	3.58	3.50	3.44	3.39	3.34	3.31	3.28
	0.01	11.26	8.65	7.59	7.01	6.63	6.37	6.19	6.03	5.91	5.82	5.74	5.67
9	0.05	5.12	4.26	3.86	3.63	3.48	3.37	3.29	3.23	3.18	3.13	3.10	3.07
	0.01	10.56	8.02	6.99	6.42	6.06	5.80	5.62	5.47	5.35	5.26	5.18	5.11
10	0.05	4.96	4.10	3.71	3.48	3.33	3.22	3.14	3.07	3.02	2.97	2.94	2.91
	0.01	10.04	7.56	6.55	5.99	5.64	5.39	5.21	5.06	4.95	4.85	4.78	4.71
11	0.05	4.84	3.98	3.59	3.36	3.20	3.09	3.01	2.95	2.90	2.86	2.82	2.79
	0.01	9.65	7.20	6.22	5.67	5.32	5.07	4.88	4.74	4.63	4.54	4.46	4.40
12	0.05	4.75	3.88	3.49	3.26	3.11	3.00	2.92	2.85	2.80	2.76	2.72	2.69
	0.01	9.33	6.93	5.95	5.41	5.06	4.82	4.65	4.50	4.39	4.30	4.22	4.16
13	0.05	4.67	3.80	3.41	3.18	3.02	2.92	2.84	2.77	2.72	2.67	2.63	2.60
	0.01	9.07	6.70	5.74	5.20	4.86	4.62	4.44	4.30	4.19	4.10	4.02	3.96
14	0.05	4.60	3.74	3.34	3.11	2.96	2.85	2.77	2.70	2.65	2.60	2.56	2.53
	0.01	8.86	6.51	5.56	5.03	4.69	4.46	4.28	4.14	4.03	3.94	3.86	3.80
15	0.05	4.54	3.68	3.29	3.06	2.90	2.79	2.70	2.64	2.59	2.55	2.51	2.48
	0.01	8.68	6.36	5.42	4.89	4.56	4.32	4.14	4.00	3.89	3.80	3.73	3.67
16	0.05	4.49	3.63	3.24	3.01	2.85	2.74	2.66	2.59	2.54	2.49	2.45	2.42
	0.01	8.53	6.23	5.29	4.77	4.44	4.20	4.03	3.89	3.78	3.69	3.61	3.55
17	0.05	4.45	3.59	3.20	2.96	2.81	2.70	2.62	2.55	2.50	2.45	2.41	2.38
	0.01	8.40	6.11	5.18	4.67	4.34	4.10	3.93	3.79	3.68	3.59	3.52	3.45
18	0.05	4.41	3.55	3.16	2.93	2.77	3.66	2.58	2.51	2.46	2.41	2.37	2.34
	0.01	8.28	6.01	5.09	4.58	4.25	4.01	3.85	3.71	3.60	3.51	3.44	3.37

df in the Denominator

TABLE 14.C (*Continued*)

		df in the Numerator (Between)										
	Exclusion Level	1	2	3	4	5	6	7	8	9	10	11	12
19	0.05	4.38	3.52	3.13	2.90	2.74	2.63	2.55	2.48	2.43	2.38	2.34	2.31
	0.01	8.18	5.93	5.01	4.50	4.17	3.94	3.77	3.63	3.52	3.43	3.36	3.30
20	0.05	4.35	3.49	3.10	2.87	2.71	2.60	2.52	2.45	2.40	2.35	2.31	2.28
	0.01	8.10	5.85	4.94	4.43	4.10	3.87	3.71	3.56	3.45	3.37	3.30	3.23
21	0.05	4.32	3.47	3.07	2.84	2.68	2.57	2.49	2.42	2.37	2.32	2.28	2.25
	0.01	8.02	5.78	4.87	4.37	4.04	3.81	3.65	3.51	3.40	3.31	3.24	3.17
22	0.05	4.30	3.44	3.05	2.82	2.66	2.55	2.47	2.40	2.35	2.30	2.26	2.23
	0.01	7.94	5.72	4.82	4.31	3.99	3.76	3.59	3.45	3.35	3.26	3.18	3.12
23	0.05	4.28	3.42	3.03	2.80	2.64	2.53	2.45	2.38	2.32	2.28	2.24	2.20
	0.01	7.88	5.66	4.76	4.26	3.94	3.71	3.54	3.41	3.30	3.21	3.14	3.07
24	0.05	4.26	3.40	3.01	2.78	2.62	2.51	2.43	2.36	2.30	2.26	2.22	2.18
	0.01	7.82	5.61	4.72	4.22	3.90	3.67	3.50	3.36	3.25	3.17	3.09	3.03
25	0.05	4.24	3.38	2.99	2.76	2.60	2.49	2.41	2.34	2.28	2.24	2.20	2.16
	0.01	7.77	5.57	4.68	4.18	3.86	3.63	3.46	3.32	3.21	3.13	3.05	2.99
26	0.05	4.22	3.37	2.98	2.74	2.59	2.47	2.39	2.32	2.27	2.22	2.18	2.15
	0.01	7.72	5.53	4.64	4.14	3.82	3.59	3.42	3.29	3.17	3.09	3.02	2.96
27	0.05	4.21	3.35	2.96	2.73	2.57	2.46	2.37	2.30	2.25	2.20	2.16	2.13
	0.01	7.68	5.49	4.60	4.11	3.79	3.56	3.39	3.26	3.14	3.06	2.98	2.93
28	0.05	4.20	3.34	2.95	2.71	2.56	2.44	2.36	2.29	2.24	2.19	2.15	2.12
	0.01	7.64	5.45	4.57	4.07	3.76	3.53	3.36	3.23	3.11	3.03	2.95	2.90
29	0.05	4.18	3.33	2.93	2.70	2.54	2.43	2.35	2.28	2.22	2.18	2.14	2.10
	0.01	7.60	5.42	4.54	4.04	3.73	3.50	3.33	3.20	3.08	3.00	2.92	2.87
30	0.05	4.17	3.32	2.92	2.69	2.53	2.42	2.34	2.27	2.21	2.16	2.12	2.09
	0.01	7.56	5.39	4.51	4.02	3.70	3.47	3.30	3.17	3.06	2.98	2.90	2.84
32	0.05	4.15	3.30	2.90	2.67	2.51	2.40	2.32	2.25	2.19	2.14	2.10	2.07
	0.01	7.50	5.34	4.46	3.97	3.66	3.42	3.25	3.12	3.01	2.94	2.86	2.80
34	0.05	4.13	3.28	2.88	2.65	2.49	2.38	2.30	2.23	2.17	2.12	2.08	2.05
	0.01	7.44	5.29	4.42	3.93	3.61	3.38	3.21	3.08	2.97	2.89	2.82	2.76
36	0.05	4.11	3.26	2.86	2.63	2.48	2.36	2.28	2.21	2.15	2.10	2.06	2.03
	0.01	7.39	5.25	4.38	3.89	3.58	3.35	3.18	3.04	2.94	2.86	2.78	2.72
38	0.05	4.10	3.25	2.85	2.62	2.46	2.35	2.26	2.19	2.14	2.09	2.05	2.02
	0.01	7.35	5.21	4.34	3.86	3.54	3.32	3.15	3.02	2.91	2.82	2.75	2.69
40	0.05	4.08	3.23	2.84	2.61	2.45	2.34	2.25	2.18	2.12	2.07	2.04	2.00
	0.01	7.31	5.18	4.31	3.83	3.51	3.29	3.12	2.99	2.88	2.80	2.73	2.66
42	0.05	4.07	3.22	2.83	2.59	2.44	2.32	2.24	2.17	2.11	2.06	2.02	1.99
	0.01	7.27	5.15	4.29	3.80	3.49	3.26	3.10	2.96	2.86	2.77	2.70	2.64
44	0.05	4.06	3.21	2.82	2.58	2.43	2.31	2.23	2.16	2.10	2.05	2.01	1.98
	0.01	7.24	5.12	4.26	3.78	3.46	3.24	3.07	2.94	2.84	2.75	2.68	2.62
46	0.05	4.05	3.20	2.81	2.57	2.42	2.30	2.22	2.14	2.09	2.04	2.00	1.97
	0.01	7.21	5.10	4.24	3.76	3.44	3.22	3.05	2.92	2.82	2.73	2.66	2.60

df in the Denominator

(*continued*)

TABLE 14.C (*Continued*)

df in the Denominator	Exclusion Level	df in the Numerator (Between)											
		1	2	3	4	5	6	7	8	9	10	11	12
48	0.05	4.04	3.19	2.80	2.56	2.41	2.30	2.21	2.40	2.08	2.03	1.99	1.96
	0.01	7.19	5.08	4.22	3.74	3.42	3.20	3.04	2.90	2.80	2.71	2.64	2.58
50	0.05	4.03	3.18	2.79	2.56	2.40	2.29	2.20	2.13	2.07	2.02	1.98	1.95
	0.01	7.17	5.06	4.20	3.72	3.41	3.18	3.02	2.88	2.78	2.70	2.62	2.56
55	0.05	4.02	3.17	2.78	2.54	2.38	2.27	2.18	2.11	2.05	2.00	1.97	1.93
	0.01	7.12	5.01	4.16	3.68	3.37	3.15	2.98	2.85	2.75	2.66	2.59	2.53
60	0.05	4.00	3.15	2.76	2.52	2.37	2.25	2.17	2.10	2.04	1.99	1.95	1.92
	0.01	7.08	4.98	4.13	3.65	3.34	3.12	2.95	2.82	2.72	2.63	2.56	2.50
65	0.05	3.99	3.14	2.75	2.51	2.36	2.24	2.15	2.08	2.02	1.98	1.94	1.90
	0.01	7.04	4.95	4.10	3.62	3.31	3.09	2.93	2.79	2.70	2.61	2.54	2.47
70	0.05	3.98	3.13	2.74	2.50	2.35	2.23	2.14	2.07	2.01	1.97	1.93	1.89
	0.01	7.01	4.92	4.08	3.60	3.29	3.07	2.91	2.77	2.67	2.59	2.51	2.45
80	0.05	3.96	3.11	2.72	2.48	2.33	2.21	2.12	2.05	1.99	1.95	1.91	1.88
	0.01	6.96	4.88	4.04	3.56	3.25	3.04	2.87	2.74	2.64	2.55	2.48	2.41
100	0.05	3.94	3.09	2.70	2.46	2.30	2.19	2.10	2.03	1.97	1.92	1.88	1.85
	0.01	6.90	4.82	3.98	3.51	3.20	2.99	2.82	2.69	2.59	2.51	2.43	2.36
125	0.05	3.92	3.07	2.68	2.44	2.29	2.17	2.08	2.01	1.95	1.90	1.86	1.83
	0.01	6.84	4.78	3.94	3.47	3.17	2.95	2.79	2.65	2.56	2.47	2.40	2.33
150	0.05	3.91	3.06	2.67	2.43	2.27	2.16	2.07	2.00	1.94	1.89	1.85	1.82
	0.01	6.81	4.75	3.91	3.44	3.14	2.92	2.76	2.62	2.53	2.44	2.37	2.30
200	0.05	3.89	3.04	2.65	2.41	2.26	2.14	2.05	1.98	1.92	1.87	1.83	1.80
	0.01	6.76	4.71	3.88	3.41	3.11	2.90	2.73	2.60	2.50	2.41	2.34	2.28
400	0.05	3.86	3.02	2.62	2.39	2.23	2.12	2.03	1.96	1.90	1.85	1.81	1.78
	0.01	6.7	4.66	3.83	3.36	3.06	2.85	2.69	2.55	2.46	2.37	2.29	2.23
1000	0.05	3.85	3.00	2.61	2.38	2.22	2.10	2.02	1.95	1.89	1.84	1.80	1.76
	0.01	6.66	4.62	3.80	3.34	3.04	2.82	2.66	2.53	2.43	2.34	2.26	2.20
∞	0.05	3.84	2.99	2.60	2.37	2.21	2.09	2.01	1.94	1.88	1.83	1.79	1.75
	0.01	6.63	4.60	3.78	3.32	3.02	2.80	2.64	2.51	2.41	2.32	2.24	2.18

Source: Snedecor, G. W., and Cochran, W. G. *Statistical Methods*. Ames, IA: Iowa State University Press, 1980. Used with permission.

TABLE D Tukey's Range Test (Upper 5% Points)

k = Num. of Groups	2	3	4	5	6	7	8	9	10
$MS_w df$									
1	18.00	27.00	32.80	37.10	40.40	43.10	45.40	47.40	49.10
2	6.09	8.30	9.80	10.90	11.70	12.40	13.00	13.50	14.00
3	4.50	5.91	6.82	7.50	8.04	8.48	8.85	9.18	9.46
4	3.93	5.04	5.76	6.29	6.71	7.05	7.35	7.60	7.83
5	3.64	4.60	5.22	5.67	6.03	6.33	6.58	6.80	6.99
6	3.46	4.34	4.90	5.31	5.63	5.89	6.12	6.32	6.49
7	3.34	4.16	4.68	5.06	5.36	5.61	5.82	6.00	6.16
8	3.26	4.04	4.53	4.89	5.17	5.40	5.60	5.77	5.92
9	3.20	3.95	4.42	4.76	5.02	5.24	5.43	5.60	5.74
10	3.15	3.88	4.33	4.65	4.91	5.12	5.30	5.46	5.60
11	3.11	3.82	4.26	4.57	4.82	5.03	5.20	5.35	5.49
12	3.08	3.77	4.20	4.51	4.75	4.95	5.12	5.27	5.40
13	3.06	3.73	4.15	4.45	4.69	4.88	5.05	5.19	5.32
14	3.03	3.70	4.11	4.41	4.64	4.83	4.99	5.13	5.25
15	3.01	3.67	4.08	4.37	4.60	4.78	4.94	5.08	5.20
16	3.00	3.65	4.05	4.33	4.56	4.74	4.90	5.03	5.15
17	2.98	3.63	4.02	4.30	4.52	4.71	4.86	4.99	5.11
18	2.97	3.61	4.00	4.28	4.49	4.67	4.82	4.96	5.07
19	2.96	3.59	3.98	4.25	4.47	4.65	4.79	4.92	5.04
20	2.95	3.58	3.96	4.23	4.45	4.62	4.77	4.90	5.01
24	2.92	3.53	3.90	4.17	4.37	4.54	4.68	4.81	4.92
30	2.89	3.49	3.84	4.10	4.30	4.46	4.60	4.72	4.83
40	2.86	3.44	3.79	4.04	4.23	4.39	4.52	4.63	4.74
60	2.83	3.40	3.74	3.98	4.16	4.31	4.44	4.55	4.65
120	2.80	3.36	3.69	3.92	4.10	4.24	4.36	4.48	4.56
∞	2.77	3.31	3.63	3.86	4.03	4.17	4.29	4.39	4.47

Source: Pearson, E. S., and Hartley, H. O. *Biometrika Tables for Statisticians,* Volume I, 2nd ed. Cambridge, UK: University Press, 1962. Used with permission.

TABLE E Critical (Exclusion) Values for Pearson's Correlation Coefficient, r

| One-Tailed | 0.05 | 0.025 | 0.01 | 0.005 | 0.0025 | 0.0005 |
Two-Tailed	0.1	0.05	0.02	0.01	0.005	0.001
Degrees of Freedom						
1	0.988	0.9969	0.9995	0.999877	0.9999692	0.99999877
2	0.9000	0.950	0.9800	0.99000	0.99500	0.99900
3	0.805	0.878	0.9343	0.9587	0.9740	0.99114
4	0.729	0.811	0.882	0.9172	0.9417	0.9741
5	0.669	0.754	0.833	0.875	0.9056	0.9509
6	0.621	0.707	0.789	0.834	0.870	0.9249
7	0.582	0.666	0.750	0.798	0.836	0.898
8	0.549	0.632	0.715	0.765	0.805	0.872
9	0.521	0.602	0.685	0.735	0.776	0.847
10	0.497	0.576	0.658	0.708	0.750	0.823
11	0.476	0.553	0.634	0.684	0.726	0.801
12	0.457	0.532	0.612	0.661	0.703	0.780
13	0.441	0.514	0.592	0.641	0.683	0.760
14	0.426	0.497	0.574	0.623	0.664	0.742
15	0.412	0.482	0.558	0.606	0.647	0.725
16	0.400	0.468	0.543	0.590	0.631	0.708
17	0.389	0.456	0.529	0.575	0.616	0.693
18	0.378	0.444	0.516	0.561	0.602	0.679
19	0.369	0.433	0.503	0.549	0.589	0.665
20	0.360	0.423	0.492	0.537	0.576	0.652
25	0.323	0.381	0.445	0.487	0.524	0.597
30	0.296	0.349	0.409	0.449	0.484	0.554
35	0.275	0.325	0.381	0.418	0.452	0.519
40	0.257	0.304	0.358	0.393	0.425	0.490
45	0.243	0.288	0.338	0.372	0.403	0.465
50	0.231	0.273	0.322	0.354	0.384	0.443
60	0.211	0.250	0.295	0.325	0.352	0.408
70	0.195	0.232	0.274	0.302	0.327	0.380
80	0.183	0.217	0.257	0.283	0.307	0.357
90	0.173	0.205	0.242	0.267	0.290	0.338
100	0.164	0.195	0.230	0.254	0.276	0.321

Source: Pearson, E. S., and Hartley, H. O. *Biometrika Tables for Statisticians*, Volume I, 2nd ed. Cambridge, UK: University Press, 1962. Used with permission.

TABLE F   Critical Values of the Chi Square Distribution

Degrees of Freedom (k)	p Value				
	0.1	0.05	0.02	0.01	0.001
1	2.706	3.841	5.412	6.635	10.827
2	4.605	5.991	7.824	9.210	13.815
3	6.251	7.815	9.837	11.345	16.266
4	7.779	9.488	11.668	13.277	18.467
5	9.236	11.070	13.388	15.086	20.515
6	10.645	12.592	15.033	16.812	22.457
7	12.017	14.067	16.622	18.475	24.322
8	13.362	15.507	18.168	20.090	26.125
9	14.684	16.919	19.679	21.666	27.877
10	15.987	18.307	21.161	23.209	29.588
11	17.275	19.675	22.618	24.725	31.264
12	18.549	21.026	24.054	26.217	32.909
13	19.812	22.362	25.472	27.688	34.528
14	21.064	23.685	26.873	29.141	36.123
15	22.307	24.996	28.259	30.578	37.697
16	23.542	26.296	29.633	32.000	39.252
17	24.769	27.587	30.995	33.409	40.790
18	25.989	28.869	32.346	34.805	42.312
19	27.204	30.144	33.687	36.191	43.820
20	28.412	31.410	35.020	37.566	45.315

Source: Pearson, E. S., and Hartley, H. O. *Biometrika Tables for Statisticians,* Volume I, (2nd ed.) Cambridge, UK: University Press, 1962. Used with permission.

INDEX

Aggregate scores, 21
Alpha error. See Type I error.
Analysis of variance (ANOVA), 257–336
  calculation of, 262–69
  components of variance, 260–61
  F-test, 262
  factorial (2XANOVA), 307–336
  interaction effect, See Interaction
  main effects, 310–11
  one way, 257–305
  simple effects, See Simple effects
  two-way within-subjects, 308
  within-subjects, 307–8, 329, 499–508
ANCOVA, 309, 327
Assumptions of
  ANOVA, 276–77
  bivariate regression, 408–9
  chi square test of independence, 462
  correlation, 360–62
  independent-samples T-test, 229–36
  multiple linear regression, 434–36
Average deviation, 89–91

Beta, 407–8, 419
Beta error. See Type II error

Bimodal distribution, See Distribution, bimodal
Bivariate regression, See Regression

Case study, 157, 169
Central limit theorem, 157–58, 169
Central tendency, See Descriptive statistics, central tendency
Chi square, 453–83
  contingency tables, 453–54, 483
  effect size, See Effect size
  expected frequencies, 456, 462–64
  frequencies versus proportions, 460
  goodness of fit, 455–461, 483
  repeated measures, 470–72, 483
  special 2X2, 466–67
  test of independence, 456–57, 461–72, 483
Coefficient of determination, 354, 374
Cohen's d, 192, 228, See also Effect size
Confidence interval,
  defined, 197, 204
  for independent t test, 227–28
  for population mean, 197–99

Confidence interval (*Continued*)
  for regression, 396–97, 419
  value of, 199
Contingency coefficient, 460, 483
Contingency tables, See Chi square, contingency tables
Continuous variable, 240
Control group, 150, 169
Convenience sample, 169
Correlation, 337–69, See also Pearson's r
  versus causation, 357
  contingency coefficient, See Contingency coefficient
  Cramer's V, See Cramer's V
  eta square, See Eta square
  nature of, 338–39
  partial eta square, See Partial eta square
  phi, See Phi
  problems, 356–58
  Spearman's rho, See Nonparametric statistics
  z score method, 349–51
Cramer's V, 469–470, 483
Crosstabs. See Chi Square, contingency tables
Cumulative proportions, 105, 108, 115–119, See also Percentile
  deriving sample scores, 130–31
  transforming to z scores, 128–29
Curvilinear relationship, 358, 409–12, 433

Deciles, 83, 96
Degrees of freedom
  in ANOVA, 266
  in chi square, 464–66
  in correlation, 353
  defined, 185–86, 204
  in independent T test, 221, 224
  in single sample T test, 185–87
Dependent variables, 169
Descriptive statistics, 41–99
  central tendency, 51, 54–60, 63, 71–75
  contrasted to inferential statistics, 44
  graphical methods, 66–71
  research applications, 41–44
  scales of measurement, 44–50, 74
  standard deviation, 87–97
  variability, 81–99
Diagnostics, 292, 412

Dichotomized variable, 339, 375
Distribution, 61–62
  bimodal, 54, 71
  data, 61–62, 73
  frequency, 66, 73
  of means, 160
  nature of, 61
  normal, See Normal distribution
  sampling, 178, 156–61
Distribution-free tests, See Nonparametric statistics
Dunnett test, 273

Effect size, 4–5, 168–69, 193
  ANOVA, factorial, 318–19
  ANOVA, one way, 269–71
  Chi square goodness of fit, 460
  Chi square 2x2 tables, 467
  Chi square test of independence, 469
  correlation, 354–56
  defined, 191, 205
  multiple linear regression, 430
  regression, 391–92, 414
  T test, dependent, 494
  T test, independent samples, 228
  T test, single mean, 191–93
  within-subjects ANOVA, 507
  Z test, 168
Error
  regression, See standard error, of estimate
  sampling, 157
Estimate
  biased and unbiased, 181, 204
  interval, See Confidence interval
  point, See Point estimate
Eta Square (η^2), 270
EXCEL®, xix–xx, 1, 3, 5–21
  ANOVA, 285–87
  bivariate regression, 401–404
  central tendency, 56–58
  chi square, 472–78
  correlation, 366–67
  cumulative proportions, 115–120
  data analysis procedures, 20
  data management, 7–9
  dependent t-test, 496–98
  entering formulas directly, 17–19
  F-test two sample for variances, 230
  independent t test, 236–38

menus, 9–16
missing cases and zeros, 20
parameter estimation, 181–82
percentiles, 84–87
real-world data with, 20–22
scattergram, 345–47
single sample T test, 203–4
standard deviation, 92–94, 180
transforming scores, 132–34
use of statistical functions, 17
use of statistics, 17
z scores, 115–120
Exclusion values, 160–61, 185–88, 224–25
Expected frequency, see Chi square, expected frequencies
Experiment, See Research design, experiment
Extreme score, 358, 375

F distribution, 230–33, 269
F ratio, 267–68, 317–18
F test, 296
  analysis of variance, see Analysis of variance (ANOVA)
  two sample for variances, 230
Factor analysis, 205
Familywise error, 259, 296
Frequency distribution, 66
Frequency polygon, 121
Friedman test, 509

Grade equivalent score, 132, 144
Graphs, See Histograms
Greenhouse-Geisser test, 506–7, 509
Group designs
  between, 246
  within, 247

Heteroscedasticity, 358, 375
Histograms, 67–71, See also Scattergram
Homogeneity of variance
  assumption of, 277
Homoscedasticity, 358, 375
HSD, See *Post Hoc* analyses
Hypothesis, 149
  alternative, 167
  null, 167
Hypothesis test, 167–68

Inferential statistics, 147, 156
  contrasted to descriptive statistics, 44
  defined, 73, 143
  populations and samples, 162
Interaction, 309
  ANOVA, 310–11
  charting, 311
  disordinal, 311, 328
  effect, 328
  ordinal, 311, 328
Interquartile range, 83, 96
Interval data, 48–50, 73
Interval estimate, See Confidence interval

Kruskal-Wallis test, See Nonparametric statistics
Kurtosis, 65, 73

Levels of measurement, See Scales of measurement
Levene's test, 233–35, 280, 296
Line of best fit, See Regression, line
Linear relationship, 342, 375

Main effects, See Analysis of variance (ANOVA)
MANCOVA, 309, 328
Mann-Whitney U test, See Nonparametric statistics
MANOVA, 309, 328
Matched groups, 212
Mauchley's test of sphericity, 505, 506, See also Sphericity
McNemar test. See Chi square, repeated measures
Mean, 52, 73
Mean squares, 265
Measurement, see Scales of measurement
Median, 52–53, 73
Mixed designs, 246
Mode, 54, 73
Multicollinearity, 431, 443, 445
Multiple comparisons, 258, 272
Multiple correlation, 417–19
Multiple linear regression, 419, 429–52
  coefficients, 430–31, 442
  elements of, 429–31
  entry schemes, 432, 439
  hierarchical, 439–42, 445

Multiple linear regression (*Continued*)
   multilevel analysis, 433
   squared part correlation, 443–44
   stepwise, 439
Multivariate, 309, 328

Nominal data, 45, 73
Nonparametric statistics, 243
   dependent *t*-test. See *T*-test,
      dependent-samples
   Kruskal-Wallis test, 293, 296
   Mann-Whitney U test, 243–46
   Spearman's rho (rank order correlation),
      369–374
   within-subjects ANOVA, See Analysis
      of variance, within-subjects
Normal curve equivalent (NCE)
   scores, 83, 96, 131
Normal distribution, 62, 101–126
   nature of, 101–103
   normal curve, 101–102, 131
   raw score distributions, 114
   standard, 102
   and z score, See Z score, distribution
Null hypothesis, See Hypothesis, null

Observed frequency, 458
Omnibus results
   ANOVA, 317–18
   bivariate regression, 414
   within subjects ANOVA, 506–7
One-tailed test, See Two-tailed and
   one-tailed tests
Ordinal data, 46–47, 73
Outlier, See Extreme score

Parameter
   defined, 162, 170
   estimation, 178–82, 187
Parametric statistics, 243
Partial correlation, 418–20
Partial eta square (η^2), 318, 328
Pearson's *r*, 340–41
Percentile, 82–83, See also Cumulative
   proportions
   calculating, 108
   defined, 82, 96
   identification, 84–87
   scores, 83–84

Phi, 467, 483
Point estimate, 197, 205
Pooled variance, See Variance, pooled
Population
   mean, 162
   and samples, 162
   standard deviation, 92, 96, 162,
      178–80
Post facto research, See Research design,
   post facto
Post hoc analyses, 271–76, 296
   Tukey's HSD test, 273–76
   varieties of, 272–3
   within subjects ANOVA, 507–8
Power, 193, 196–97, 205, 496
Practical significance, See Effect size
Predictor variable, see Regression, predictor
   variables
Pretest sensitivity, 213
Probability, 134–44
   determinism versus probability, 135
   elements of, 136
   empirical probability, 136–37
   exact probability, 141–43
   inside and outside areas, 139–41
   normal curve, 136–37
   relationship to z score, 137
   sampling, 170
Program Evaluation Prism, The, 20,
   41–42, 309, 340, 358, 408, 412,
   419, 429, 443
Proportional reduction in error
   (PRE), 339, 375

Quartiles, 83, 96
Quasi-experimental design, See Research
   design

r^2, see Coefficient of determination
Random sample,
   defined, 171
   stratified random sampling, 171
Randomization, 170
Range, 82, 96
Rank order correlation, See Nonparametric
   statistics
Ranked data, 369–71
Ranks, tied, 371–73
Ratio data, 50, 73

INDEX

Real-world data, 3, 20–22
Regression,
  bivariate, 383–428
  coefficients, 394, 407–8
  explaining variance, 397–401
  interpretation of, 390–91
  line, 383–88, 419–20
  nature of, 384–85
  predictor variables, 170, 383–4, 397–401
  slope, 386, 389
  y intercept, 386, 420
  z score formula, 392–93
Rejecting null hypothesis, 168
Repeated measures, 489–509
  chi square, See Chi square, repeated measures
  dependent t-test, 491–99
  within-subjects ANOVA, 499–508
Research design, 3–4, 148
  experiment, 150, 170, 177, 210–11, 499–501
  hypothesis, 149
  nature of, 154
  post facto, 153, 170, 178, 214–15, 339–40, 455, 501
  quasi-experimental, 151–52, 170
  sampling, 155, See also Sample(s) and Sampling distribution
  theory, 149
  variables, See Variable
Restricted range, 357–58, 375

Sample(s),
  convenience, 155, 169
  defined, 170
  dependent, 211, 246, 509
  equal and unequal, 229
  error, 157, 171
  independent, 211, 246
  matched, 212, 246
  mean, 163–66
  population, 162
  random, 171
  scores, 130–31
  snowball sampling, 171
  standard deviation, 97
  values, 143–44

Sampling distribution, 157–61, 171, See also Distribution, sampling
  of differences, 217–18
Scales of measurement, 44, 74
  choosing the correct statistical procedure, 50
  interval data, 48
  nominal data, 45
  ordinal data, 46
  ratio data, 50
Scattergram, 342–48, 375
  in multiple linear regression, 431
  in regression, 399–400, 416
Scheffe test, 273
School-level achievement data, 20
Significance, See Statistical significance
Simple effects
  in ANOVA, 312, 325–27, 329
Skewness, 63–65, 74
Slope, See Regression, line
Social research, 148
Sphericity, 501–502, 505–6, 509
Spearman's rho, See Nonparametric statistics
SPSS®, xix–xx, 1, 5, 23–39
  ANOVA, 287–92
  analysis functions, 39
  bivariate regression, 404–8
  central tendency, 58–60
  chi square, 478–83
  correlation, 367–69
  curvilinear relationships, 409–12
  dependent t-test, 495–96
  "explore" feature, 233–34
  factorial ANOVA, 321–27
  general features, 24–26
  independent-samples t test, 236, 239–43
  management functions, 26–39
  multiple linear regression, 438–45
  parameter estimation, 181–82
  percentiles, 84, 86–87
  scattergram, 347–48
  single sample T test, 200–203
  standard deviation, 92–96, 180
  syntax, 323–25, 329
  transforming scores, 132–34
  within-subjects ANOVA, 501–508
  z score, 119–20
Spuriousness, 41–44, 74

Standard deviation, 87–97
  calculation of, 88–91
  defined, 87, 97
  in descriptive statistics, 87–99
  sample and population, 92
Standard error
  of difference, 218–20
  of estimate, 394–96, 420
  estimated, of difference, 218–20
  estimated, of the mean, 183
  of the mean, 162–63, 171
Standard normal distribution, 121
Standard normal score, See z score
Stanine, 131–32, 144
Statistical significance, 160–61, 168, 171, 202, 241, 289, 295
Statistics, 171
Sum of squares, 91–92, 262–65, 351–52

T distribution, 187–88
T score, 132, 144
T test
  dependent-samples, 212, 491–99
  independent-samples, 209–46
  single sample, 175–204
  versus Z test, 175–76
Theory, 149, 171
Ties in ranks, See Ranks, tied
Treatment group, 171
Tukey's HSD test, See *Post hoc* analyses
Two-tailed and one-tailed tests, 193–96, 205
Type I error (alpha), 189–90, 205
Type II error (beta), 190–91, 193, 205

Univariate, 309, 329

Variable, 152, 171
  dependent, 153
  independent, 152–53, 170
  manipulated independent, 170
  nonmanipulated independent, 170
  transformation, 436, 445
Variance, 87–99
  between, 261, 296
  calculation of, 88–92
  components of, 260–61
  defined, 87, 97
  in descriptive statistics, 81–99
  pooled, 218–20
  total, 261, 266, 296
  within, 260, 296

Washington School Research Center, 43
Wilcoxen test, 509
Wilks' lambda, 507, 509
Winning the Math Wars, 2,

Yates correction for continuity in chi square, 468–69, 483

z score, 103–125
  and bivariate regression, 392
  calculating, 105–8, 111–14
  and correlation, 349
  creating rules for locating, 108–110
  and cumulative proportions, 105, 115
  defined, 121
  deriving sample scores from cumulative percentages, 130–31
  distribution, 102–111, 127, 176
  inflection point, 103
  table of values, 104–5
  transforming cumulative proportions to z scores, 128–29
  transforming to a raw score, 128–29
Z test, 166–69, 171
  elements, 169
  hypothesis test, See Hypothesis test
  versus T test, 175–76